WITHDRAWN

LIVERPOOL JMU LIBRARY

3 1111 01354 1741

VOLUME FOUR HUNDRED AND NINETY-TWO

METHODS IN ENZYMOLOGY

Biothermodynamics, Part D

METHODS IN ENZYMOLOGY

Editors-in-Chief

JOHN N. ABELSON AND MELVIN I. SIMON

Division of Biology
California Institute of Technology
Pasadena, California

Founding Editors

SIDNEY P. COLOWICK AND NATHAN O. KAPLAN

VOLUME FOUR HUNDRED AND NINETY-TWO

METHODS IN ENZYMOLOGY

Biothermodynamics, Part D

EDITED BY

MICHAEL L. JOHNSON

University of Virginia Health Science Center
Department of Pharmacology
Charlottesville, Virginia, USA

JO M. HOLT AND GARY K. ACKERS

Emeritus, Department of Biochemistry and Molecular Biophysics
Washington University School of Medicine
St. Louis, Missouri, USA

ELSEVIER

AMSTERDAM • BOSTON • HEIDELBERG • LONDON
NEW YORK • OXFORD • PARIS • SAN DIEGO
SAN FRANCISCO • SINGAPORE • SYDNEY • TOKYO
Academic Press is an imprint of Elsevier

Academic Press is an imprint of Elsevier
525 B Street, Suite 1900, San Diego, CA 92101-4495, USA
30 Corporate Drive, Suite 400, Burlington, MA 01803, USA
32 Jamestown Road, London NW1 7BY, UK

First edition 2011

Copyright © 2011, Elsevier Inc. All Rights Reserved.

No part of this publication may be reproduced, stored in a retrieval system or transmitted in any form or by any means electronic, mechanical, photocopying, recording or otherwise without the prior written permission of the publisher

Permissions may be sought directly from Elsevier's Science & Technology Rights Department in Oxford, UK: phone (+44) (0) 1865 843830; fax (+44) (0) 1865 853333; email: permissions@elsevier.com. Alternatively you can submit your request online by visiting the Elsevier web site at http://elsevier.com/locate/permissions, and selecting *Obtaining permission to use Elsevier material*

Notice
No responsibility is assumed by the publisher for any injury and/or damage to persons or property as a matter of products liability, negligence or otherwise, or from any use or operation of any methods, products, instructions or ideas contained in the material herein. Because of rapid advances in the medical sciences, in particular, independent verification of diagnoses and drug dosages should be made

For information on all Academic Press publications
visit our website at elsevierdirect.com

ISBN: 978-0-12-386003-3
ISSN: 0076-6879

Printed and bound in United States of America
11 12 13 10 9 8 7 6 5 4 3 2 1

Working together to grow
libraries in developing countries

www.elsevier.com | www.bookaid.org | www.sabre.org

ELSEVIER BOOK AID International Sabre Foundation

Contents

CONTRIBUTORS

Matthew Auton
Department of Medicine, Cardiovascular Research, Baylor College of Medicine, Houston, Texas, USA

Jannette Carey
Chemistry Department, Princeton University, Princeton, New Jersey, USA

Caroline Carr
Department of Pharmaceutical Sciences, University of Nebraska Medical Center, Omaha, Nebraska, USA

Patricia L. Clark
Department of Chemistry and Biochemistry, University of Notre Dame, Notre Dame, Indiana, USA

Natalya N. Degtyareva
Department of Chemistry, Furman University, Greenville, South Carolina, USA

Brian Domack
Department of Pharmaceutical Sciences, University of Nebraska Medical Center, Omaha, Nebraska, USA

W. Austin Elam
T.C. Jenkins Department of Biophysics, Johns Hopkins University, Baltimore, Maryland, USA

Rüdiger Ettrich
Department of Structure and Function of Proteins, Institute of Systems Biology and Ecology, Academy of Sciences of the Czech Republic, and University of South Bohemia, Nove Hrady, Czech Republic

Karen G. Fleming
T.C. Jenkins Department of Biophysics, Johns Hopkins University, Baltimore, Maryland, USA

Hernan G. Garcia
Department of Physics, California Institute of Technology, Pasadena, California, USA

Vincent J. Hilser
T.C. Jenkins Department of Biophysics, and Department of Biology, Johns Hopkins University, Baltimore, Maryland, USA

Luis Marcelo F. Holthauzen
Sealy Center for Structural Biology and Molecular Biophysics, University of Texas Medical Branch, Galveston, Texas, USA

Fany Iseka
Department of Pharmaceutical Sciences, University of Nebraska Medical Center, Omaha, Nebraska, USA

Lihua Jin
Chemistry Department, DePaul University, Chicago, Illinois, USA

Irine Khutsishvili
Department of Pharmaceutical Sciences, University of Nebraska Medical Center, Omaha, Nebraska, USA

Jonathan S. Kingsbury
Therapeutic Protein Research, Genzyme Corporation, Framingham, Massachusetts, USA

Jane Kondev
Department of Physics, Brandeis University Waltham, Massachusetts, USA

Thomas M. Laue
Department of Biochemistry and Molecular Biology, University of New Hampshire, Durham, New Hampshire, USA

Hui-Ting Lee
Department of Pharmaceutical Sciences, University of Nebraska Medical Center, Omaha, Nebraska, USA

Jing Li
T.C. Jenkins Department of Biophysics, Johns Hopkins University, Baltimore, Maryland, USA

Luis A. Marky
Department of Pharmaceutical Sciences; Department of Biochemistry and Molecular Biology, and Eppley Institute for Research in Cancer, University of Nebraska Medical Center, Omaha, Nebraska, USA

Milan Melichercik
Department of Structure and Function of Proteins, Institute of Systems Biology and Ecology, Academy of Sciences of the Czech Republic, and University of South Bohemia, Nove Hrady, Czech Republic

Abhijit Mishra
Department of Bioengineering, University of California, Los Angeles, California, USA

C. Preston Moon
T.C. Jenkins Department of Biophysics, Johns Hopkins University, Baltimore, Maryland, USA

Chris M. Olsen
Department of Pharmaceutical Sciences, University of Nebraska Medical Center, Omaha, Nebraska, USA

Nigel Orme
Garland Science Publishing, New York, USA

André J. Ouellette
Department of Pathology and Laboratory Medicine, USC Norris Cancer Center, Keck School of Medicine of the University of Southern California, Los Angeles, California, USA

Jeffrey T. Petty
Department of Chemistry, Furman University, Greenville, South Carolina, USA

Rob Phillips
Department of Applied Physics, California Institute of Technology, Pasadena, California, USA

Jörg Rösgen
Department of Biochemistry and Molecular Biology, Penn State College of Medicine, Hershey, Pennsylvania, USA

Jonathan P. Renn[1]
Department of Chemistry and Biochemistry, University of Notre Dame, Notre Dame, Indiana, USA

Nathan W. Schmidt
Department of Bioengineering, University of California, Los Angeles, California; Department of Physics, and Department of Materials Science, University of Illinois, Urbana-Champaign, Illinois, USA

Travis P. Schrank
Department of Biochemistry and Molecular Biology, University of Texas Medical Branch, Galveston, Texas, USA

Hollie Siebler
Department of Pharmaceutical Sciences, University of Nebraska Medical Center, Omaha, Nebraska, USA

Mikhail Sinev
Department of Biochemistry and Molecular Biology, Penn State College of Medicine, Hershey, Pennsylvania, USA

[1] Current address: Department of Molecular Biosciences, Northwestern University, Evanston, Illinois, USA

Thomas Stockner
Center for Physiology and Pharmacology, Institute of Pharmacology, Medical University of Vienna, Waehringerstrasse, Vienna, Austria

Rebecca Strawn
Chemistry Department, Princeton University, Princeton, New Jersey, USA

Kenneth P. Tai
Department of Pathology and Laboratory Medicine, USC Norris Cancer Center, Keck School of Medicine of the University of Southern California, Los Angeles, California, USA

Julie A. Theriot
Department of Biochemistry, Stanford University School of Medicine, Stanford, California, USA

Lela Waters
Department of Pharmaceutical Sciences, University of Nebraska Medical Center, Omaha, Nebraska, USA

Gerard C. L. Wong
Department of Bioengineering, University of California, Los Angeles, California; Department of Physics, and Department of Materials Science, University of Illinois, Urbana-Champaign, Illinois, USA

Wei-Feng Xue
Astbury Centre for Structural Molecular Biology, University of Leeds, Leeds, United Kingdom

PREFACE

This volume is the continuation in a series of *Methods in Enzymology* volumes, which promotes thermodynamics as an important tool for the study of biological systems.

One of many examples of biological thermodynamics is the cooperative binding of oxygen by hemoglobin. Cooperativity is inherently a thermodynamic phenomenon. Hemoglobin is the quintessential example of a ligand-binding protein. Most biochemistry textbooks explain that the hemoglobin tetramer exists in two structural states, a low-affinity structure without oxygen bound and a high-affinity structure with oxygen bound. This is the classic two-state allosteric model as presented by Monod, Wyman, and Changeux (1965, *J. Mol. Biol.*, **12**, 88–118) and extended by Ackers and Johnson (1981, *J. Mol. Biol.* **147**, 559–582). Unfortunately, this model tells us nothing about the specific molecular interactions that are altered by the binding of oxygen which forces the hemoglobin to shift to the alternative structural state. Thermodynamics provides the conceptual and mathematical framework, that is, a "logic tool," for the investigation of the specific molecular interactions, and the concomitant energetics, such as those that are altered by the binding of oxygen which forces the hemoglobin to shift to the alternative structural and/or association states.

Unfortunately, a large fraction of scientists have the impression that thermodynamic approaches are archaic, and, at best, ancillary to the central issues of biochemistry. One reason for this misconception is that thermodynamics is commonly either poorly or not at all taught in departments of chemistry, biochemistry, etc. Another reason for this narrow and insular perception is that thermodynamics is frequently equated with a single experimental technique (i.e., calorimetry). Sadly, thermodynamics has seldom been fused with developments in molecular biology, structural analysis, or computational chemistry. However, all these perceptions are far from accurate.

Nevertheless, branches of the U. S. government have twice acknowledged Josiah Williard Gibbs for his contributions to thermodynamics and thus indirectly acknowledged the importance of thermodynamics. The first acknowledgment was the U.S. Navy with the USNS Josiah Williard Gibbs which was a ship of the line between 1958 and 1971. The second example was the U.S. Postal Service by including him as one of four great American scientists on a series of postage stamps that were issued in 2005. "The greatest thermodynamicist of them all" (John Fenn, 2002 Nobel Prize in Chemistry).

Michael L. Johnson, Jo Holt and Gary K. Ackers

METHODS IN ENZYMOLOGY

Volume 89. Carbohydrate Metabolism (Part D)
Edited by Willis A. Wood

Volume 90. Carbohydrate Metabolism (Part E)
Edited by Willis A. Wood

Volume 91. Enzyme Structure (Part I)
Edited by C. H. W. Hirs and Serge N. Timasheff

Volume 92. Immunochemical Techniques (Part E: Monoclonal Antibodies and General Immunoassay Methods)
Edited by John J. Langone and Helen Van Vunakis

Volume 93. Immunochemical Techniques (Part F: Conventional Antibodies, Fc Receptors, and Cytotoxicity)
Edited by John J. Langone and Helen Van Vunakis

Volume 94. Polyamines
Edited by Herbert Tabor and Celia White Tabor

Volume 95. Cumulative Subject Index Volumes 61–74, 76–80
Edited by Edward A. Dennis and Martha G. Dennis

Volume 96. Biomembranes [Part J: Membrane Biogenesis: Assembly and Targeting (General Methods; Eukaryotes)]
Edited by Sidney Fleischer and Becca Fleischer

Volume 97. Biomembranes [Part K: Membrane Biogenesis: Assembly and Targeting (Prokaryotes, Mitochondria, and Chloroplasts)]
Edited by Sidney Fleischer and Becca Fleischer

Volume 98. Biomembranes (Part L: Membrane Biogenesis: Processing and Recycling)
Edited by Sidney Fleischer and Becca Fleischer

Volume 99. Hormone Action (Part F: Protein Kinases)
Edited by Jackie D. Corbin and Joel G. Hardman

Volume 100. Recombinant DNA (Part B)
Edited by Ray Wu, Lawrence Grossman, and Kivie Moldave

Volume 101. Recombinant DNA (Part C)
Edited by Ray Wu, Lawrence Grossman, and Kivie Moldave

Volume 102. Hormone Action (Part G: Calmodulin and Calcium-Binding Proteins)
Edited by Anthony R. Means and Bert W. O'Malley

Volume 103. Hormone Action (Part H: Neuroendocrine Peptides)
Edited by P. Michael Conn

Volume 104. Enzyme Purification and Related Techniques (Part C)
Edited by William B. Jakoby

Volume 105. Oxygen Radicals in Biological Systems
Edited by Lester Packer

Volume 106. Posttranslational Modifications (Part A)
Edited by Finn Wold and Kivie Moldave

Volume 107. Posttranslational Modifications (Part B)
Edited by Finn Wold and Kivie Moldave

Volume 108. Immunochemical Techniques (Part G: Separation and Characterization of Lymphoid Cells)
Edited by Giovanni Di Sabato, John J. Langone, and Helen Van Vunakis

Volume 109. Hormone Action (Part I: Peptide Hormones)
Edited by Lutz Birnbaumer and Bert W. O'Malley

Volume 110. Steroids and Isoprenoids (Part A)
Edited by John H. Law and Hans C. Rilling

Volume 111. Steroids and Isoprenoids (Part B)
Edited by John H. Law and Hans C. Rilling

Volume 112. Drug and Enzyme Targeting (Part A)
Edited by Kenneth J. Widder and Ralph Green

Volume 113. Glutamate, Glutamine, Glutathione, and Related Compounds
Edited by Alton Meister

Volume 114. Diffraction Methods for Biological Macromolecules (Part A)
Edited by Harold W. Wyckoff, C. H. W. Hirs, and Serge N. Timasheff

Volume 115. Diffraction Methods for Biological Macromolecules (Part B)
Edited by Harold W. Wyckoff, C. H. W. Hirs, and Serge N. Timasheff

Volume 116. Immunochemical Techniques
(Part H: Effectors and Mediators of Lymphoid Cell Functions)
Edited by Giovanni Di Sabato, John J. Langone, and Helen Van Vunakis

Volume 117. Enzyme Structure (Part J)
Edited by C. H. W. Hirs and Serge N. Timasheff

Volume 118. Plant Molecular Biology
Edited by Arthur Weissbach and Herbert Weissbach

Volume 119. Interferons (Part C)
Edited by Sidney Pestka

Volume 120. Cumulative Subject Index Volumes 81–94, 96–101

Volume 121. Immunochemical Techniques (Part I: Hybridoma Technology and Monoclonal Antibodies)
Edited by John J. Langone and Helen Van Vunakis

Volume 122. Vitamins and Coenzymes (Part G)
Edited by Frank Chytil and Donald B. McCormick

VOLUME 157. Biomembranes (Part Q: ATP-Driven Pumps and Related Transport: Calcium, Proton, and Potassium Pumps)
Edited by SIDNEY FLEISCHER AND BECCA FLEISCHER

VOLUME 158. Metalloproteins (Part A)
Edited by JAMES F. RIORDAN AND BERT L. VALLEE

VOLUME 159. Initiation and Termination of Cyclic Nucleotide Action
Edited by JACKIE D. CORBIN AND ROGER A. JOHNSON

VOLUME 160. Biomass (Part A: Cellulose and Hemicellulose)
Edited by WILLIS A. WOOD AND SCOTT T. KELLOGG

VOLUME 161. Biomass (Part B: Lignin, Pectin, and Chitin)
Edited by WILLIS A. WOOD AND SCOTT T. KELLOGG

VOLUME 162. Immunochemical Techniques (Part L: Chemotaxis and Inflammation)
Edited by GIOVANNI DI SABATO

VOLUME 163. Immunochemical Techniques (Part M: Chemotaxis and Inflammation)
Edited by GIOVANNI DI SABATO

VOLUME 164. Ribosomes
Edited by HARRY F. NOLLER, JR., AND KIVIE MOLDAVE

VOLUME 165. Microbial Toxins: Tools for Enzymology
Edited by SIDNEY HARSHMAN

VOLUME 166. Branched-Chain Amino Acids
Edited by ROBERT HARRIS AND JOHN R. SOKATCH

VOLUME 167. Cyanobacteria
Edited by LESTER PACKER AND ALEXANDER N. GLAZER

VOLUME 168. Hormone Action (Part K: Neuroendocrine Peptides)
Edited by P. MICHAEL CONN

VOLUME 169. Platelets: Receptors, Adhesion, Secretion (Part A)
Edited by JACEK HAWIGER

VOLUME 170. Nucleosomes
Edited by PAUL M. WASSARMAN AND ROGER D. KORNBERG

VOLUME 171. Biomembranes (Part R: Transport Theory: Cells and Model Membranes)
Edited by SIDNEY FLEISCHER AND BECCA FLEISCHER

VOLUME 172. Biomembranes (Part S: Transport: Membrane Isolation and Characterization)
Edited by SIDNEY FLEISCHER AND BECCA FLEISCHER

Volume 190. Retinoids (Part B: Cell Differentiation and Clinical Applications)
Edited by Lester Packer

Volume 191. Biomembranes (Part V: Cellular and Subcellular Transport: Epithelial Cells)
Edited by Sidney Fleischer and Becca Fleischer

Volume 192. Biomembranes (Part W: Cellular and Subcellular Transport: Epithelial Cells)
Edited by Sidney Fleischer and Becca Fleischer

Volume 193. Mass Spectrometry
Edited by James A. McCloskey

Volume 194. Guide to Yeast Genetics and Molecular Biology
Edited by Christine Guthrie and Gerald R. Fink

Volume 195. Adenylyl Cyclase, G Proteins, and Guanylyl Cyclase
Edited by Roger A. Johnson and Jackie D. Corbin

Volume 196. Molecular Motors and the Cytoskeleton
Edited by Richard B. Vallee

Volume 197. Phospholipases
Edited by Edward A. Dennis

Volume 198. Peptide Growth Factors (Part C)
Edited by David Barnes, J. P. Mather, and Gordon H. Sato

Volume 199. Cumulative Subject Index Volumes 168–174, 176–194

Volume 200. Protein Phosphorylation (Part A: Protein Kinases: Assays, Purification, Antibodies, Functional Analysis, Cloning, and Expression)
Edited by Tony Hunter and Bartholomew M. Sefton

Volume 201. Protein Phosphorylation (Part B: Analysis of Protein Phosphorylation, Protein Kinase Inhibitors, and Protein Phosphatases)
Edited by Tony Hunter and Bartholomew M. Sefton

Volume 202. Molecular Design and Modeling: Concepts and Applications (Part A: Proteins, Peptides, and Enzymes)
Edited by John J. Langone

Volume 203. Molecular Design and Modeling: Concepts and Applications (Part B: Antibodies and Antigens, Nucleic Acids, Polysaccharides, and Drugs)
Edited by John J. Langone

Volume 204. Bacterial Genetic Systems
Edited by Jeffrey H. Miller

Volume 205. Metallobiochemistry (Part B: Metallothionein and Related Molecules)
Edited by James F. Riordan and Bert L. Vallee

Volume 206. Cytochrome P450
Edited by Michael R. Waterman and Eric F. Johnson

Volume 207. Ion Channels
Edited by Bernardo Rudy and Linda E. Iverson

Volume 208. Protein–DNA Interactions
Edited by Robert T. Sauer

Volume 209. Phospholipid Biosynthesis
Edited by Edward A. Dennis and Dennis E. Vance

Volume 210. Numerical Computer Methods
Edited by Ludwig Brand and Michael L. Johnson

Volume 211. DNA Structures (Part A: Synthesis and Physical Analysis of DNA)
Edited by David M. J. Lilley and James E. Dahlberg

Volume 212. DNA Structures (Part B: Chemical and Electrophoretic Analysis of DNA)
Edited by David M. J. Lilley and James E. Dahlberg

Volume 213. Carotenoids (Part A: Chemistry, Separation, Quantitation, and Antioxidation)
Edited by Lester Packer

Volume 214. Carotenoids (Part B: Metabolism, Genetics, and Biosynthesis)
Edited by Lester Packer

Volume 215. Platelets: Receptors, Adhesion, Secretion (Part B)
Edited by Jacek J. Hawiger

Volume 216. Recombinant DNA (Part G)
Edited by Ray Wu

Volume 217. Recombinant DNA (Part H)
Edited by Ray Wu

Volume 218. Recombinant DNA (Part I)
Edited by Ray Wu

Volume 219. Reconstitution of Intracellular Transport
Edited by James E. Rothman

Volume 220. Membrane Fusion Techniques (Part A)
Edited by Nejat Düzgüneş

Volume 221. Membrane Fusion Techniques (Part B)
Edited by Nejat Düzgüneş

Volume 222. Proteolytic Enzymes in Coagulation, Fibrinolysis, and Complement Activation (Part A: Mammalian Blood Coagulation Factors and Inhibitors)
Edited by Laszlo Lorand and Kenneth G. Mann

Volume 223. Proteolytic Enzymes in Coagulation, Fibrinolysis, and Complement Activation (Part B: Complement Activation, Fibrinolysis, and Nonmammalian Blood Coagulation Factors)
Edited by Laszlo Lorand and Kenneth G. Mann

Volume 224. Molecular Evolution: Producing the Biochemical Data
Edited by Elizabeth Anne Zimmer, Thomas J. White, Rebecca L. Cann, and Allan C. Wilson

Volume 225. Guide to Techniques in Mouse Development
Edited by Paul M. Wassarman and Melvin L. DePamphilis

Volume 226. Metallobiochemistry (Part C: Spectroscopic and Physical Methods for Probing Metal Ion Environments in Metalloenzymes and Metalloproteins)
Edited by James F. Riordan and Bert L. Vallee

Volume 227. Metallobiochemistry (Part D: Physical and Spectroscopic Methods for Probing Metal Ion Environments in Metalloproteins)
Edited by James F. Riordan and Bert L. Vallee

Volume 228. Aqueous Two-Phase Systems
Edited by Harry Walter and Göte Johansson

Volume 229. Cumulative Subject Index Volumes 195–198, 200–227

Volume 230. Guide to Techniques in Glycobiology
Edited by William J. Lennarz and Gerald W. Hart

Volume 231. Hemoglobins (Part B: Biochemical and Analytical Methods)
Edited by Johannes Everse, Kim D. Vandegriff, and Robert M. Winslow

Volume 232. Hemoglobins (Part C: Biophysical Methods)
Edited by Johannes Everse, Kim D. Vandegriff, and Robert M. Winslow

Volume 233. Oxygen Radicals in Biological Systems (Part C)
Edited by Lester Packer

Volume 234. Oxygen Radicals in Biological Systems (Part D)
Edited by Lester Packer

Volume 235. Bacterial Pathogenesis (Part A: Identification and Regulation of Virulence Factors)
Edited by Virginia L. Clark and Patrik M. Bavoil

Volume 236. Bacterial Pathogenesis (Part B: Integration of Pathogenic Bacteria with Host Cells)
Edited by Virginia L. Clark and Patrik M. Bavoil

Volume 237. Heterotrimeric G Proteins
Edited by Ravi Iyengar

Volume 238. Heterotrimeric G-Protein Effectors
Edited by Ravi Iyengar

Volume 239. Nuclear Magnetic Resonance (Part C)
Edited by Thomas L. James and Norman J. Oppenheimer

Volume 240. Numerical Computer Methods (Part B)
Edited by Michael L. Johnson and Ludwig Brand

Volume 241. Retroviral Proteases
Edited by Lawrence C. Kuo and Jules A. Shafer

Volume 242. Neoglycoconjugates (Part A)
Edited by Y. C. Lee and Reiko T. Lee

Volume 243. Inorganic Microbial Sulfur Metabolism
Edited by Harry D. Peck, Jr., and Jean LeGall

Volume 244. Proteolytic Enzymes: Serine and Cysteine Peptidases
Edited by Alan J. Barrett

Volume 245. Extracellular Matrix Components
Edited by E. Ruoslahti and E. Engvall

Volume 246. Biochemical Spectroscopy
Edited by Kenneth Sauer

Volume 247. Neoglycoconjugates (Part B: Biomedical Applications)
Edited by Y. C. Lee and Reiko T. Lee

Volume 248. Proteolytic Enzymes: Aspartic and Metallo Peptidases
Edited by Alan J. Barrett

Volume 249. Enzyme Kinetics and Mechanism (Part D: Developments in Enzyme Dynamics)
Edited by Daniel L. Purich

Volume 250. Lipid Modifications of Proteins
Edited by Patrick J. Casey and Janice E. Buss

Volume 251. Biothiols (Part A: Monothiols and Dithiols, Protein Thiols, and Thiyl Radicals)
Edited by Lester Packer

Volume 252. Biothiols (Part B: Glutathione and Thioredoxin; Thiols in Signal Transduction and Gene Regulation)
Edited by Lester Packer

Volume 253. Adhesion of Microbial Pathogens
Edited by Ron J. Doyle and Itzhak Ofek

Volume 254. Oncogene Techniques
Edited by Peter K. Vogt and Inder M. Verma

Volume 255. Small GTPases and Their Regulators (Part A: Ras Family)
Edited by W. E. Balch, Channing J. Der, and Alan Hall

Volume 256. Small GTPases and Their Regulators (Part B: Rho Family)
Edited by W. E. Balch, Channing J. Der, and Alan Hall

Volume 257. Small GTPases and Their Regulators (Part C: Proteins Involved in Transport)
Edited by W. E. Balch, Channing J. Der, and Alan Hall

Volume 258. Redox-Active Amino Acids in Biology
Edited by Judith P. Klinman

Volume 259. Energetics of Biological Macromolecules
Edited by Michael L. Johnson and Gary K. Ackers

Volume 260. Mitochondrial Biogenesis and Genetics (Part A)
Edited by Giuseppe M. Attardi and Anne Chomyn

Volume 261. Nuclear Magnetic Resonance and Nucleic Acids
Edited by Thomas L. James

Volume 262. DNA Replication
Edited by Judith L. Campbell

Volume 263. Plasma Lipoproteins (Part C: Quantitation)
Edited by William A. Bradley, Sandra H. Gianturco, and Jere P. Segrest

Volume 264. Mitochondrial Biogenesis and Genetics (Part B)
Edited by Giuseppe M. Attardi and Anne Chomyn

Volume 265. Cumulative Subject Index Volumes 228, 230–262

Volume 266. Computer Methods for Macromolecular Sequence Analysis
Edited by Russell F. Doolittle

Volume 267. Combinatorial Chemistry
Edited by John N. Abelson

Volume 268. Nitric Oxide (Part A: Sources and Detection of NO; NO Synthase)
Edited by Lester Packer

Volume 269. Nitric Oxide (Part B: Physiological and Pathological Processes)
Edited by Lester Packer

Volume 270. High Resolution Separation and Analysis of Biological Macromolecules (Part A: Fundamentals)
Edited by Barry L. Karger and William S. Hancock

Volume 271. High Resolution Separation and Analysis of Biological Macromolecules (Part B: Applications)
Edited by Barry L. Karger and William S. Hancock

Volume 272. Cytochrome P450 (Part B)
Edited by Eric F. Johnson and Michael R. Waterman

Volume 273. RNA Polymerase and Associated Factors (Part A)
Edited by Sankar Adhya

VOLUME 274. RNA Polymerase and Associated Factors (Part B)
Edited by SANKAR ADHYA

VOLUME 275. Viral Polymerases and Related Proteins
Edited by LAWRENCE C. KUO, DAVID B. OLSEN, AND STEVEN S. CARROLL

VOLUME 276. Macromolecular Crystallography (Part A)
Edited by CHARLES W. CARTER, JR., AND ROBERT M. SWEET

VOLUME 277. Macromolecular Crystallography (Part B)
Edited by CHARLES W. CARTER, JR., AND ROBERT M. SWEET

VOLUME 278. Fluorescence Spectroscopy
Edited by LUDWIG BRAND AND MICHAEL L. JOHNSON

VOLUME 279. Vitamins and Coenzymes (Part I)
Edited by DONALD B. MCCORMICK, JOHN W. SUTTIE, AND CONRAD WAGNER

VOLUME 280. Vitamins and Coenzymes (Part J)
Edited by DONALD B. MCCORMICK, JOHN W. SUTTIE, AND CONRAD WAGNER

VOLUME 281. Vitamins and Coenzymes (Part K)
Edited by DONALD B. MCCORMICK, JOHN W. SUTTIE, AND CONRAD WAGNER

VOLUME 282. Vitamins and Coenzymes (Part L)
Edited by DONALD B. MCCORMICK, JOHN W. SUTTIE, AND CONRAD WAGNER

VOLUME 283. Cell Cycle Control
Edited by WILLIAM G. DUNPHY

VOLUME 284. Lipases (Part A: Biotechnology)
Edited by BYRON RUBIN AND EDWARD A. DENNIS

VOLUME 285. Cumulative Subject Index Volumes 263, 264, 266–284, 286–289

VOLUME 286. Lipases (Part B: Enzyme Characterization and Utilization)
Edited by BYRON RUBIN AND EDWARD A. DENNIS

VOLUME 287. Chemokines
Edited by RICHARD HORUK

VOLUME 288. Chemokine Receptors
Edited by RICHARD HORUK

VOLUME 289. Solid Phase Peptide Synthesis
Edited by GREGG B. FIELDS

VOLUME 290. Molecular Chaperones
Edited by GEORGE H. LORIMER AND THOMAS BALDWIN

VOLUME 291. Caged Compounds
Edited by GERARD MARRIOTT

VOLUME 292. ABC Transporters: Biochemical, Cellular, and Molecular Aspects
Edited by SURESH V. AMBUDKAR AND MICHAEL M. GOTTESMAN

Volume 293. Ion Channels (Part B)
Edited by P. Michael Conn

Volume 294. Ion Channels (Part C)
Edited by P. Michael Conn

Volume 295. Energetics of Biological Macromolecules (Part B)
Edited by Gary K. Ackers and Michael L. Johnson

Volume 296. Neurotransmitter Transporters
Edited by Susan G. Amara

Volume 297. Photosynthesis: Molecular Biology of Energy Capture
Edited by Lee McIntosh

Volume 298. Molecular Motors and the Cytoskeleton (Part B)
Edited by Richard B. Vallee

Volume 299. Oxidants and Antioxidants (Part A)
Edited by Lester Packer

Volume 300. Oxidants and Antioxidants (Part B)
Edited by Lester Packer

Volume 301. Nitric Oxide: Biological and Antioxidant Activities (Part C)
Edited by Lester Packer

Volume 302. Green Fluorescent Protein
Edited by P. Michael Conn

Volume 303. cDNA Preparation and Display
Edited by Sherman M. Weissman

Volume 304. Chromatin
Edited by Paul M. Wassarman and Alan P. Wolffe

Volume 305. Bioluminescence and Chemiluminescence (Part C)
Edited by Thomas O. Baldwin and Miriam M. Ziegler

Volume 306. Expression of Recombinant Genes in Eukaryotic Systems
Edited by Joseph C. Glorioso and Martin C. Schmidt

Volume 307. Confocal Microscopy
Edited by P. Michael Conn

Volume 308. Enzyme Kinetics and Mechanism (Part E: Energetics of Enzyme Catalysis)
Edited by Daniel L. Purich and Vern L. Schramm

Volume 309. Amyloid, Prions, and Other Protein Aggregates
Edited by Ronald Wetzel

Volume 310. Biofilms
Edited by Ron J. Doyle

Volume 311. Sphingolipid Metabolism and Cell Signaling (Part A)
Edited by Alfred H. Merrill, Jr., and Yusuf A. Hannun

Volume 312. Sphingolipid Metabolism and Cell Signaling (Part B)
Edited by Alfred H. Merrill, Jr., and Yusuf A. Hannun

Volume 313. Antisense Technology
(Part A: General Methods, Methods of Delivery, and RNA Studies)
Edited by M. Ian Phillips

Volume 314. Antisense Technology (Part B: Applications)
Edited by M. Ian Phillips

Volume 315. Vertebrate Phototransduction and the Visual Cycle (Part A)
Edited by Krzysztof Palczewski

Volume 316. Vertebrate Phototransduction and the Visual Cycle (Part B)
Edited by Krzysztof Palczewski

Volume 317. RNA–Ligand Interactions (Part A: Structural Biology Methods)
Edited by Daniel W. Celander and John N. Abelson

Volume 318. RNA–Ligand Interactions (Part B: Molecular Biology Methods)
Edited by Daniel W. Celander and John N. Abelson

Volume 319. Singlet Oxygen, UV-A, and Ozone
Edited by Lester Packer and Helmut Sies

Volume 320. Cumulative Subject Index Volumes 290–319

Volume 321. Numerical Computer Methods (Part C)
Edited by Michael L. Johnson and Ludwig Brand

Volume 322. Apoptosis
Edited by John C. Reed

Volume 323. Energetics of Biological Macromolecules (Part C)
Edited by Michael L. Johnson and Gary K. Ackers

Volume 324. Branched-Chain Amino Acids (Part B)
Edited by Robert A. Harris and John R. Sokatch

Volume 325. Regulators and Effectors of Small GTPases
(Part D: Rho Family)
Edited by W. E. Balch, Channing J. Der, and Alan Hall

Volume 326. Applications of Chimeric Genes and Hybrid Proteins
(Part A: Gene Expression and Protein Purification)
Edited by Jeremy Thorner, Scott D. Emr, and John N. Abelson

Volume 327. Applications of Chimeric Genes and Hybrid Proteins
(Part B: Cell Biology and Physiology)
Edited by Jeremy Thorner, Scott D. Emr, and John N. Abelson

Volume 328. Applications of Chimeric Genes and Hybrid Proteins (Part C: Protein–Protein Interactions and Genomics)
Edited by Jeremy Thorner, Scott D. Emr, and John N. Abelson

Volume 329. Regulators and Effectors of Small GTPases (Part E: GTPases Involved in Vesicular Traffic)
Edited by W. E. Balch, Channing J. Der, and Alan Hall

Volume 330. Hyperthermophilic Enzymes (Part A)
Edited by Michael W. W. Adams and Robert M. Kelly

Volume 331. Hyperthermophilic Enzymes (Part B)
Edited by Michael W. W. Adams and Robert M. Kelly

Volume 332. Regulators and Effectors of Small GTPases (Part F: Ras Family I)
Edited by W. E. Balch, Channing J. Der, and Alan Hall

Volume 333. Regulators and Effectors of Small GTPases (Part G: Ras Family II)
Edited by W. E. Balch, Channing J. Der, and Alan Hall

Volume 334. Hyperthermophilic Enzymes (Part C)
Edited by Michael W. W. Adams and Robert M. Kelly

Volume 335. Flavonoids and Other Polyphenols
Edited by Lester Packer

Volume 336. Microbial Growth in Biofilms (Part A: Developmental and Molecular Biological Aspects)
Edited by Ron J. Doyle

Volume 337. Microbial Growth in Biofilms (Part B: Special Environments and Physicochemical Aspects)
Edited by Ron J. Doyle

Volume 338. Nuclear Magnetic Resonance of Biological Macromolecules (Part A)
Edited by Thomas L. James, Volker Dötsch, and Uli Schmitz

Volume 339. Nuclear Magnetic Resonance of Biological Macromolecules (Part B)
Edited by Thomas L. James, Volker Dötsch, and Uli Schmitz

Volume 340. Drug–Nucleic Acid Interactions
Edited by Jonathan B. Chaires and Michael J. Waring

Volume 341. Ribonucleases (Part A)
Edited by Allen W. Nicholson

Volume 342. Ribonucleases (Part B)
Edited by Allen W. Nicholson

Volume 343. G Protein Pathways (Part A: Receptors)
Edited by Ravi Iyengar and John D. Hildebrandt

Volume 344. G Protein Pathways (Part B: G Proteins and Their Regulators)
Edited by Ravi Iyengar and John D. Hildebrandt

Volume 381. Oxygen Sensing
Edited by Chandan K. Sen and Gregg L. Semenza

Volume 382. Quinones and Quinone Enzymes (Part B)
Edited by Helmut Sies and Lester Packer

Volume 383. Numerical Computer Methods (Part D)
Edited by Ludwig Brand and Michael L. Johnson

Volume 384. Numerical Computer Methods (Part E)
Edited by Ludwig Brand and Michael L. Johnson

Volume 385. Imaging in Biological Research (Part A)
Edited by P. Michael Conn

Volume 386. Imaging in Biological Research (Part B)
Edited by P. Michael Conn

Volume 387. Liposomes (Part D)
Edited by Nejat Düzgüneş

Volume 388. Protein Engineering
Edited by Dan E. Robertson and Joseph P. Noel

Volume 389. Regulators of G-Protein Signaling (Part A)
Edited by David P. Siderovski

Volume 390. Regulators of G-Protein Signaling (Part B)
Edited by David P. Siderovski

Volume 391. Liposomes (Part E)
Edited by Nejat Düzgüneş

Volume 392. RNA Interference
Edited by Engelke Rossi

Volume 393. Circadian Rhythms
Edited by Michael W. Young

Volume 394. Nuclear Magnetic Resonance of Biological Macromolecules (Part C)
Edited by Thomas L. James

Volume 395. Producing the Biochemical Data (Part B)
Edited by Elizabeth A. Zimmer and Eric H. Roalson

Volume 396. Nitric Oxide (Part E)
Edited by Lester Packer and Enrique Cadenas

Volume 397. Environmental Microbiology
Edited by Jared R. Leadbetter

Volume 398. Ubiquitin and Protein Degradation (Part A)
Edited by Raymond J. Deshaies

Volume 417. Functional Glycomics
Edited by Minoru Fukuda

Volume 418. Embryonic Stem Cells
Edited by Irina Klimanskaya and Robert Lanza

Volume 419. Adult Stem Cells
Edited by Irina Klimanskaya and Robert Lanza

Volume 420. Stem Cell Tools and Other Experimental Protocols
Edited by Irina Klimanskaya and Robert Lanza

Volume 421. Advanced Bacterial Genetics: Use of Transposons and Phage for Genomic Engineering
Edited by Kelly T. Hughes

Volume 422. Two-Component Signaling Systems, Part A
Edited by Melvin I. Simon, Brian R. Crane, and Alexandrine Crane

Volume 423. Two-Component Signaling Systems, Part B
Edited by Melvin I. Simon, Brian R. Crane, and Alexandrine Crane

Volume 424. RNA Editing
Edited by Jonatha M. Gott

Volume 425. RNA Modification
Edited by Jonatha M. Gott

Volume 426. Integrins
Edited by David Cheresh

Volume 427. MicroRNA Methods
Edited by John J. Rossi

Volume 428. Osmosensing and Osmosignaling
Edited by Helmut Sies and Dieter Haussinger

Volume 429. Translation Initiation: Extract Systems and Molecular Genetics
Edited by Jon Lorsch

Volume 430. Translation Initiation: Reconstituted Systems and Biophysical Methods
Edited by Jon Lorsch

Volume 431. Translation Initiation: Cell Biology, High-Throughput and Chemical-Based Approaches
Edited by Jon Lorsch

Volume 432. Lipidomics and Bioactive Lipids: Mass-Spectrometry–Based Lipid Analysis
Edited by H. Alex Brown

Volume 433. Lipidomics and Bioactive Lipids: Specialized Analytical Methods and Lipids in Disease
Edited by H. Alex Brown

Volume 434. Lipidomics and Bioactive Lipids: Lipids and Cell Signaling
Edited by H. Alex Brown

Volume 435. Oxygen Biology and Hypoxia
Edited by Helmut Sies and Bernhard Brüne

Volume 436. Globins and Other Nitric Oxide-Reactive Protiens (Part A)
Edited by Robert K. Poole

Volume 437. Globins and Other Nitric Oxide-Reactive Protiens (Part B)
Edited by Robert K. Poole

Volume 438. Small GTPases in Disease (Part A)
Edited by William E. Balch, Channing J. Der, and Alan Hall

Volume 439. Small GTPases in Disease (Part B)
Edited by William E. Balch, Channing J. Der, and Alan Hall

Volume 440. Nitric Oxide, Part F Oxidative and Nitrosative Stress in Redox Regulation of Cell Signaling
Edited by Enrique Cadenas and Lester Packer

Volume 441. Nitric Oxide, Part G Oxidative and Nitrosative Stress in Redox Regulation of Cell Signaling
Edited by Enrique Cadenas and Lester Packer

Volume 442. Programmed Cell Death, General Principles for Studying Cell Death (Part A)
Edited by Roya Khosravi-Far, Zahra Zakeri, Richard A. Lockshin, and Mauro Piacentini

Volume 443. Angiogenesis: *In Vitro* Systems
Edited by David A. Cheresh

Volume 444. Angiogenesis: *In Vivo* Systems (Part A)
Edited by David A. Cheresh

Volume 445. Angiogenesis: *In Vivo* Systems (Part B)
Edited by David A. Cheresh

Volume 446. Programmed Cell Death, The Biology and Therapeutic Implications of Cell Death (Part B)
Edited by Roya Khosravi-Far, Zahra Zakeri, Richard A. Lockshin, and Mauro Piacentini

Volume 447. RNA Turnover in Bacteria, Archaea and Organelles
Edited by Lynne E. Maquat and Cecilia M. Arraiano

Volume 464. Liposomes, Part F
Edited by Nejat Düzgüneş

Volume 465. Liposomes, Part G
Edited by Nejat Düzgüneş

Volume 466. Biothermodynamics, Part B
Edited by Michael L. Johnson, Gary K. Ackers, and Jo M. Holt

Volume 467. Computer Methods Part B
Edited by Michael L. Johnson and Ludwig Brand

Volume 468. Biophysical, Chemical, and Functional Probes of RNA Structure, Interactions and Folding: Part A
Edited by Daniel Herschlag

Volume 469. Biophysical, Chemical, and Functional Probes of RNA Structure, Interactions and Folding: Part B
Edited by Daniel Herschlag

Volume 470. Guide to Yeast Genetics: Functional Genomics, Proteomics, and Other Systems Analysis, 2nd Edition
Edited by Gerald Fink, Jonathan Weissman, and Christine Guthrie

Volume 471. Two-Component Signaling Systems, Part C
Edited by Melvin I. Simon, Brian R. Crane, and Alexandrine Crane

Volume 472. Single Molecule Tools, Part A: Fluorescence Based Approaches
Edited by Nils G. Walter

Volume 473. Thiol Redox Transitions in Cell Signaling, Part A Chemistry and Biochemistry of Low Molecular Weight and Protein Thiols
Edited by Enrique Cadenas and Lester Packer

Volume 474. Thiol Redox Transitions in Cell Signaling, Part B Cellular Localization and Signaling
Edited by Enrique Cadenas and Lester Packer

Volume 475. Single Molecule Tools, Part B: Super-Resolution, Particle Tracking, Multiparameter, and Force Based Methods
Edited by Nils G. Walter

Volume 476. Guide to Techniques in Mouse Development, Part A Mice, Embryos, and Cells, 2nd Edition
Edited by Paul M. Wassarman and Philippe M. Soriano

Volume 477. Guide to Techniques in Mouse Development, Part B Mouse Molecular Genetics, 2nd Edition
Edited by Paul M. Wassarman and Philippe M. Soriano

Volume 478. Glycomics
Edited by Minoru Fukuda

CHAPTER ONE

A Thermodynamic Approach for the Targeting of Nucleic Acid Structures Using Their Complementary Single Strands

Hui-Ting Lee,* Caroline Carr,* Hollie Siebler,* Lela Waters,* Irine Khutsishvili,* Fany Iseka,* Brian Domack,* Chris M. Olsen,* *and* Luis A. Marky*,†,‡

Contents

Abstract

The main focus of our investigations is to further our understanding of the physicochemical properties of nucleic acid structures. We report on a thermodynamic approach to study the reaction of a variety of intramolecular nucleic

* Department of Pharmaceutical Sciences, University of Nebraska Medical Center, Omaha, Nebraska, USA
† Department of Biochemistry and Molecular Biology, University of Nebraska Medical Center, Omaha, Nebraska, USA
‡ Eppley Institute for Research in Cancer, University of Nebraska Medical Center, Omaha, Nebraska, USA

Methods in Enzymology, Volume 492
ISSN 0076-6879, DOI: 10.1016/B978-0-12-381268-1.00013-6
© 2011 Elsevier Inc.
All rights reserved.

acid structures with their respective complementary strands. Specifically, we have used a combination of isothermal titration (ITC) and differential scanning calorimetry (DSC) and spectroscopy techniques to determine standard thermodynamic profiles for the reaction of a triplex, G-quadruplex, hairpin loops, pseudoknot, and three-arm junctions with their complementary strands. Reaction enthalpies are measured directly in ITC titrations, and compared with those obtained indirectly from Hess cycles using DSC unfolding data. All reactions investigated yielded favorable free energy contributions, indicating that each single strand is able to invade and disrupt the corresponding intramolecular DNA structure. These favorable free energy terms are enthalpy-driven, resulting from a favorable compensation of exothermic contributions due to the formation of additional base-pair stacks in the duplex product, and endothermic contributions, from the disruption of base stacking contributions of the reactant single strands.

The overall results provide a thermodynamic approach that can be used in the targeting of nucleic acids, especially the secondary structures formed by mRNA, with oligonucleotides for the control of gene expression.

1. Introduction

The formation of a variety of DNA secondary structures, such as hairpin loops, triplexes, G-quadruplexes, and i-motifs, is well documented (Gehring *et al.*, 1993; Rich, 1993). These noncanonical DNA secondary structures have been postulated to be involved in a variety of biological functions (Bock *et al.*, 1992; Crooke, 1999; Firulli *et al.*, 1994; Fox, 1990; Han and Hurley, 2000; Helene, 1991, 1994; Juliano *et al.*, 2001; Mills *et al.*, 2002; Rando *et al.*, 1995; Wang *et al.*, 1993) and may also be important causal factors in human diseases such as cancer and the aging of the cell (Beal and Dervan, 1991; Brown *et al.*, 1998; Huard and Autexier, 2002; Mills *et al.*, 2002; Simonsson *et al.*, 1998; Zahler *et al.*, 1991). The targeting of these structures with nucleic acid oligonucleotides may stop their biological function (Folini *et al.*, 2002; Koeppel *et al.*, 2001). In general, nucleic acid oligonucleotides, as drugs, present an exquisite selectivity and are able to discriminate targets that differ by a single nucleotide (Crooke, 1999). Therefore, oligonucleotides may be used for the control of gene expression (Helene, 1991, 1994). There are three main approaches for the use of oligonucleotides as modulators of gene expression: the antigene, antisense, and more recently, small interfering RNA (siRNA), all target specific nucleic acid structures (Mahato *et al.*, 2005). In the antigene strategy, a single strand binds to the major groove of a DNA duplex, forming a triple helix (Soyfer and Potaman, 1996) that inhibits transcription, by competing with the binding of proteins that activate the transcriptional machinery (Helene, 1994; Maher *et al.*, 1989). In the antisense strategy, a single strand binds to messenger RNA, forming a DNA/RNA hybrid duplex that inhibits translation, by sterically blocking the correct assembly of the translational machinery or by inducing an RNase H-mediated cleavage of the DNA/RNA hybrid duplex (Helene, 1991). The third and most recent approach in using relatively short oligonucleotides to control gene expression is RNA interference (RNAi). RNAi is a major regulatory pathway in eukaryotic cells that silences gene expression through siRNAs that bind specifically to target sequences (Ghildiyal and Zamore, 2009; Malone and Hannon, 2009; Moazed, 2009).

There are advantages and disadvantages in these strategies. In the direct targeting of a gene, the antigene strategy offers some advantages over the other strategies; there are only two copies of a particular gene, whereas there is a large continuous supply of the mRNA gene transcript. Therefore, blocking the transcription of the gene itself prevents repopulation of the mRNA pool, allowing a more efficient and lasting inhibition of gene expression (Fox, 2000; Vasquez *et al.*, 2001). The main disadvantage is that the oligonucleotide needs to cross the nuclear membrane and access its DNA target within the densely packed chromatin structure (Brown *et al.*, 1998), while in the targeting of mRNA, there is a need to know the potential secondary/tertiary structures that the target sequence is involved.

Common disadvantages of the use of oligonucleotides for targeting purposes are (a) oligonucleotides need to cross the hydrophobic cellular membranes; for instance, nucleic acid duplexes do not cross these membranes because of their hydrophilicity (Cantor and Schimmel, 1980), and (b) the fast degradative action of nucleases. These disadvantages can be circumvented by using single strands that are somewhat hydrophobic and by modification of its constituent phosphate or sugar groups. Another possibility is the use of stable intramolecular DNA structures containing end loops; the unpaired nucleobases of the loops offsets its hydrophobic–hydrophilic balance toward slightly more hydrophobic, allowing them to interact better with polycationic micelles and/or enabling them to cross-cellular membranes. These polycations can be used as delivery vectors, protecting the oligonucleotide from the action of nucleases.

From a thermodynamic point of view, successful control of gene expression depends on the effective binding of a DNA sequence to its target with tight affinity and specificity. This can be provided by using a long sequence of 15–20 bases in length (Crooke, 1999); strong specificity is conferred by hydrogen bonding in the formation of Watson–Crick and/or Hoogsteen base pairs, while high affinity is provided by the large negative free energy upon the formation of a duplex or triplex, thereby competing efficiently with the proteins involved in transcription or translation.

In this work, we have investigated reactions involving the interaction of a variety of intramolecular DNA structures (G-quadruplex, triplex, hairpin loops, hairpins containing bulges and internal loops, pseudoknots, and three-arm junctions) with their respective complementary strands, that is, we use DNA single strands to target intramolecular DNA complexes. We are actually mimicking the targeting of the potential secondary structures formed by mRNA using complementary strands. We seek answers to the following questions: Is the single strand able to invade the intramolecular complex, disrupting the complex and forming a stable duplex product? And consequently, what are the thermodynamic contributions favoring the formation of these product duplexes? To this end, we have used a combination of isothermal titration (ITC) and differential scanning calorimetric (DSC) techniques to determine reaction enthalpies and their associated thermodynamic profiles. The results show that all reactions yielded favorable free energy terms, resulting from exothermic enthalpies, which are due to the formation of additional base-pair stacks in the formation of duplex products involving the loop unpaired nucleobases of the intramolecular complexes.

2. Materials and Methods

2.1. Materials

All oligonucleotides were synthesized by the Core Synthetic Facility of the Eppley Research Institute at UNMC, HPLC purified, and desalted by column chromatography using G-10 Sephadex exclusion chromatography. The sequences of oligonucleotides used in this work and their designation are shown in Table 1.1. The concentrations of the oligomer solutions were determined at 260 nm and 80 °C using a Perkin-Elmer Lambda-10 spectrophotometer and the molar extinction coefficients shown in the last column of Table 1.1. These values were obtained by extrapolation of the tabulated values for dimers and monomeric bases (Cantor *et al.*, 1970) at 25–80 °C using procedures reported previously (Marky *et al.*, 1983, 2007). The extinction coefficients of the duplexes (not shown in this table) are simply calculated by averaging the molar extinction coefficients of its component complementary single strands. Inorganic salts from Sigma were reagent grade and used without further purification. Typical measurements were made in appropriate buffer solutions: 10 m*M* sodium phosphate with 200 m*M* NaCl at pH 6.2; 10 m*M* sodium phosphate with 200 m*M* NaCl at pH 7; or 10 m*M* Cs–HEPES with 100 m*M* KCl at pH 7.5; and adjusted to the desired salt concentration with NaCl or KCl, and pH with either NaOH, HCl, or CsOH, respectively. All oligonucleotide solutions were prepared by dissolving the dry and desalted ODNs in buffer, heating the solution to 90 °C for 5 min, and cooling to room temperature over 25 min.

Table 1.1 Sequences, designations, and molar extinction coefficients of oligonucleotides

Oligonucleotide sequence	Designation	ε^{260}
d($G_2T_2G_2TGTG_2T_2G_2$)	*G-quadruplex*	146
d($C_2A_2C_2ACAC_2A_2C_2$)	*G-quadCS*	146
d($A_7C_5T_7CT_3CT_7$)	*Triplex*	279
d($A_7GA_3GA_7GG$)	*TripCS*	285
d(GTAACGCAATGTTAC)	*Hairpin*	144
d(GTAACTTGCG)	*HairpCS*	97
d($GCGCT_3GTAACT_5GTTACGCGC$)	*Bhairpin*	231
d($GCGCT_3GTAACT_5GTTACT_3GCGC$)	*ILhairpin*	255
d(AAGTTACAAAGCGC)	*BILhairpCS*	154
d($CGCGCGT_4GAAATTCGCGCGT_4A_2T_3C$)	*Pseudoknot*	292
d(GA_3TTA_4)	*PseudoCS*	130
d($TGCGCT_5GCGCGTGCT_5GCACA$)	*Dumbbell*	245
d($GTCGA_5C$)	*DumbbCS*	112
d($GA_3T_2GCGCT_5GCGCGTGCT_5GCACA_2T_3C$)	*Hammer*	347
d(GA3T2GTGCA5G)	*HammCS*	192

2.2. Isothermal titration calorimetry

The heat for the reaction of a particular DNA secondary structure with its complementary strand was measured directly by ITC calorimetry using the Omega titration calorimeter from Microcal (Northampton, MA) or the iTC_{200} (GE-Healthcare). A 100-μL syringe was used to inject the titrant with the Omega calorimeter; typically, five to seven injections of 7–10 μL of titrant with at least 20-fold higher concentration than the oligonucleotide contained in the 1.4-mL reaction cell were used in this calorimeter and mixing was effected by stirring the syringe at 400 rpm. A 40-μL syringe was used to inject the titrant with the iTC_{200}, 13–16 injections of 2.35 μL injections of titrant with at least 10-fold higher concentration than the oligomer in the 0.2-mL reaction cell were used in this calorimeter, and mixing was effected by stirring this syringe at 1000 rpm. All complementary strands were placed in the reaction cell to decrease the heat contribution from base–base stacking interactions, if any. The reaction heat of each injection was measured by integration of the area of the injection curve, corrected for the dilution heat of the titrant, and normalized by the moles of titrant added to yield the reaction enthalpy, ΔH_{ITC} (Wiseman *et al.*, 1989). All titrations were designed to obtain mainly the ΔH_{ITC} for each targeting reaction, by averaging the heat of at least three injections, which correspond to the formation of the duplex products.

2.3. Differential scanning calorimetry

The total heat required for the unfolding of each oligonucleotide, reagent, or product, of these targeting reactions was measured with a VP-DSC from Microcal. Standard thermodynamic profiles and transition temperatures, T_Ms, were obtained from a DSC experiment using the following relationships (Marky and Breslauer, 1987): $\Delta H_{cal} = \int \Delta C_p(T)\, dT$; $\Delta S_{cal} = \int \Delta C_p(T) / T\, dT$, and the Gibbs equation, $\Delta G = \Delta H_{cal} - T\Delta S_{cal}$, where ΔC_p is the anomalous heat capacity of the ODN solution during the unfolding process, ΔH_{cal} is the unfolding enthalpy, and ΔS_{cal} is the entropy of unfolding; both the latter terms are temperature-dependent. Additional DSC experiments were obtained as a function of salt concentration to determine indirectly the associated heat capacity contributions, which were determined from the slopes of the lines of the ΔH_{cal} versus T_M plots (Shikiya and Marky, 2005; Soto *et al.*, 2006).

2.4. UV melting curves

Absorbance versus temperature profiles were measured at 260, 275, and/or 297 nm with a thermoelectrically controlled Aviv Spectrophotometer Model 14DS UV–Vis (Lakewood, NJ) or Perkin-Elmer Lambda-10 UV–Vis Spectrophotometer. The temperature was scanned at a heating rate of ~0.6 °C/min, and shape analysis of the melting curves yielded T_Ms (Marky and Breslauer, 1987). The transition molecularity for the unfolding of a particular complex was obtained by monitoring the T_M as a function of the strand concentration. Intramolecular complexes show a T_M-independence on strand concentration, while the T_M of intermolecular complexes does depend on strand concentration (Marky and Breslauer, 1987). Furthermore, these T_M-dependences were used to estimate the T_Ms of the concentration of duplex products formed in the ITC titrations, allowing to correct the associated unfolding heats and to include heat capacity contributions.

2.5. Overall experimental approach

A cartoon of the investigated targeting reactions is shown in Fig. 1.1A–F. All reactions have been designed to yield a favorable free energy contribution in enthalpy-driven reactions; since the $\Delta G°$ term basically depends on the associated T_Ms and reaction heats, we have concentrated on measuring these two parameters, as will be illustrated later. However, we also determined standard thermodynamic profiles for each targeting reaction. Initially, we use ITC titrations to measure directly the heat (ΔH_{ITC}) for each of the first six reactions investigated, by averaging the heats of at least three to four injections under unsaturated conditions. We then use DSC unfolding to determine the T_Ms and unfolding heats for the reactants and products of a

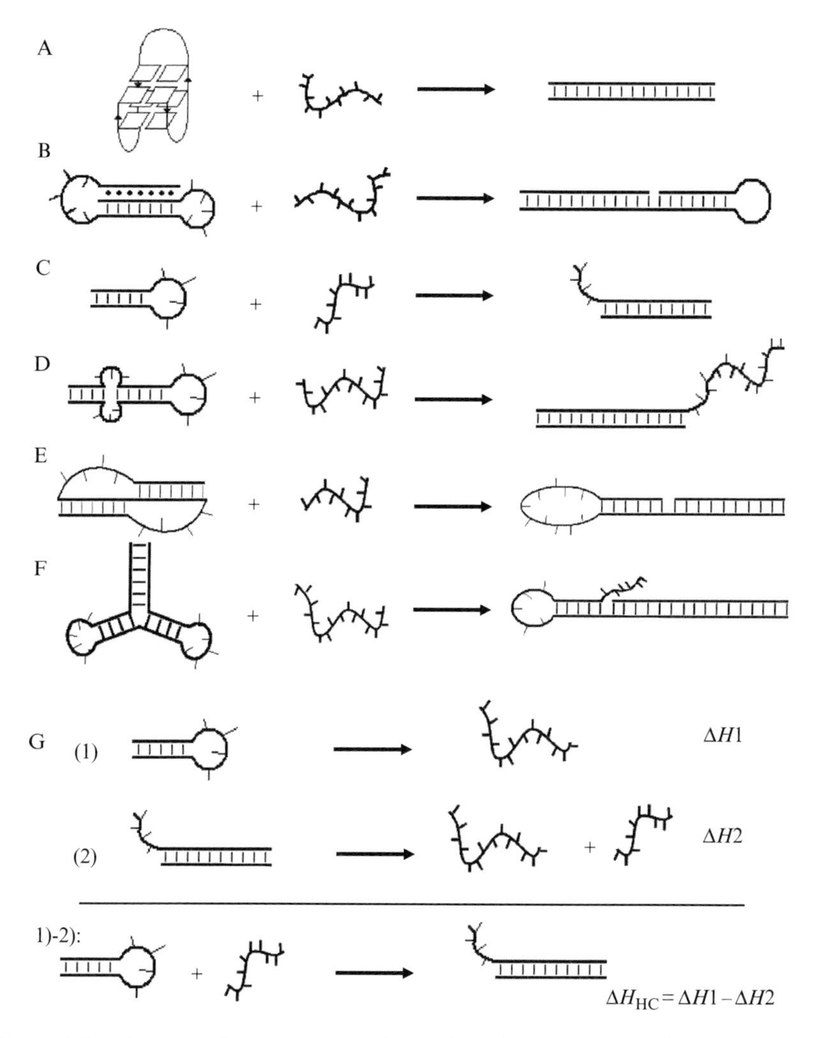

Figure 1.1 Cartoon of targeting reactions. Complementary strands were used to target the following DNA secondary structures: (A) G-quadruplex, (B) *Triplex*, (C) *Hairpin*, (D) *hairpin with an internal loop*, (E) Pseudoknot, and (F) *Three-way junction*. (G) Example of a Hess cycle, the coupling of the DSC unfolding reactions of reactants and products yields the ITC targeting reaction shown in (C).

given reaction. Furthermore, DSC unfolding experiments are carried out for each reactant and product as a function of salt to estimate indirectly their heat capacity effects. The unfolding thermodynamic profiles are then used to set up Hess cycles (Fig. 1.1G) to yield the enthalpy (ΔH_{HC}) and other thermodynamic profiles for each targeting reaction, allowing us to compare the results obtained from DSC experiments with the ITC data. UV melting curves are carried out for the oligonucleotide reactants and products as a function of strand concentration to determine their T_M-dependences on strand concentration. The resulting data of these latter experiments are then used to correct the ΔH_{HC}s for both the ODN concentration used in the ITC experiments and for heat capacity effects, by obtaining the T_Ms for the unfolding of the oligonucleotide product at the concentration used in ITC titrations.

UV melting curves are carried out for the oligonucleotide reactants and products as a function of strand concentration to determine their T_M-dependences on strand concentration. The resulting data are then used to correct the ΔH_{HC}s for both the ODN concentration used in ITC experiments and for heat capacity effects, by obtaining the T_Ms for unfolding of the oligonucleotide product at the concentration used in ITC titrations. To determine the free energy, $\Delta G°_{ITC}$, for each targeting reaction, we use the following relationship: $\Delta G°_{ITC} = \Delta G°_{HC}\ (\Delta H_{ITC}/\Delta H_{HC})$ (Marky *et al.*, 1996; Rentzeperis *et al.*, 1994), where $\Delta G°_{HC}$ is calculated from the DSC data in a similar way as with the ΔH_{HC} terms. Finally, the Gibbs equation is used to determine the $T\Delta S_{ITC}$ parameter, where T is the temperature of the ITC experiments.

3. Results and Discussion

3.1. Targeting an intramolecular G-quadruplex, G2

In these reactions, the complementary strand, *G-quadCS*, is used to target the thrombin aptamer, a G-quadruplex, to form a fully complementary duplex (*G-quadDup*), see Fig. 1.2A. The ITC titrations for this reaction were carried out at 14 °C, and are shown in Fig. 1.2C. The single strand *G-quadCS* was placed in the reaction cell to diminish single-stranded stacking contributions, if any; the time between injections was set at 6–10 min to allow for complete reaction. After correcting each injection for the titrant dilution heat (*G-quadruplex*), the average heat of injections two to four of each titration yielded exothermic heats of −55 μcal, which after normalization by the concentration of the limiting reagent yielded ΔH_{ITC}s of −84 ± 5 kcal/mol for the formation of the duplex product (*G-quadDup*). The favorable heat of this reaction is due to the net compensation of an exothermic heat from the formation of base-pair stacks and the endothermic heats of breaking both a G-quartet stack and base–base stacking of the loops of *G2*.

The DSC unfolding curves of the reactant and product for this reaction are shown in Fig. 1.2D, and standard thermodynamic profiles for the unfolding of each reactant and product at 14 °C are shown in Table 1.2. The DSC curves of the *G-quadruplex* reactant and *G-quadDup* product show

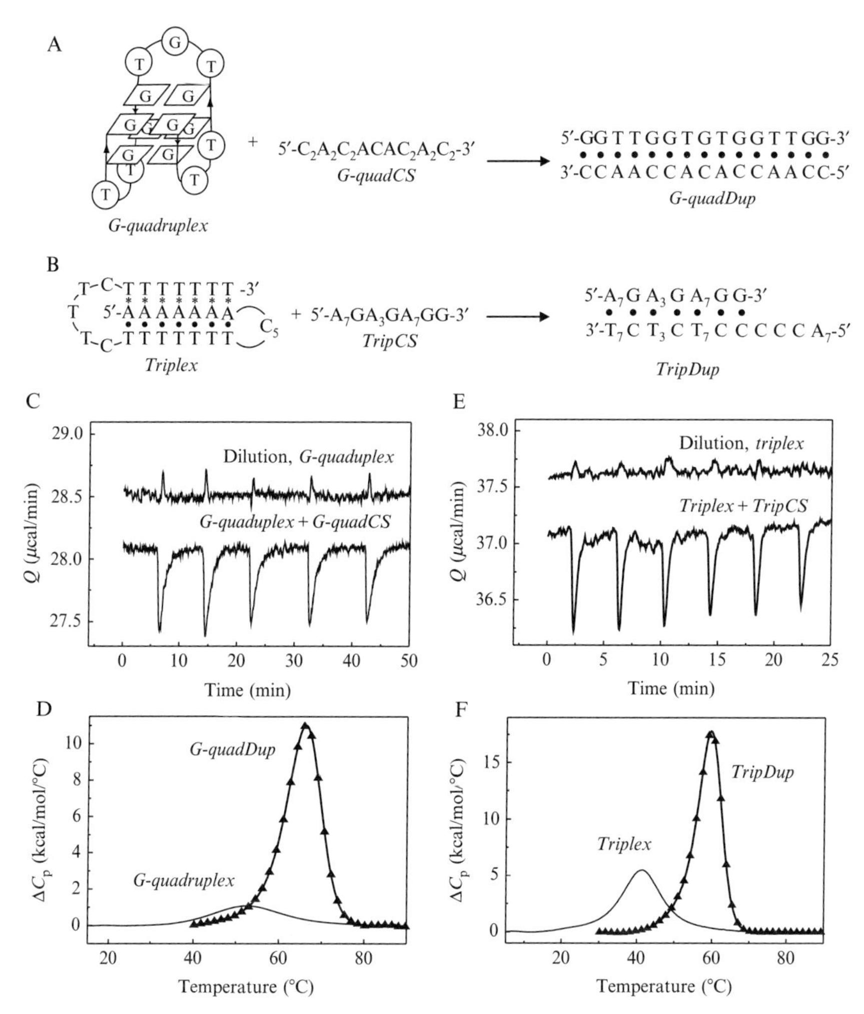

Figure 1.2 Targeting *G-quadruplex* and *Triplex*. (A, B) Cartoon of targeting reactions; (C, E) ITC titration curves, curves are offset for clarity; and (D, F) DSC curves of reactants and products. All experiments were carried out in 10 m*M* Cs–HEPES, 100 m*M* KCl at pH 7.5 (*G-quaduplex* targeting) or 10 m*M* NaPi, and 200 m*M* NaCl at pH 7.0 (*Triplex* targeting).

monophasic transitions with negligible heat capacity effects between the initial and final states. The *G-quadruplex* unfolded with a T_M of 52.6 °C and an unfolding enthalpy of 23.7 kcal/mol, consistent with previous reports (Kankia and Marky, 2001; Olsen *et al.*, 2006), indicating that this reactant is a stable intramolecular G-quadruplex that forms completely at 14 °C. The *G-quadDup* product yielded a T_M of 65.6 °C and an enthalpy of

Table 1.2 Thermodynamic unfolding profiles for targeting a G-quadruplex and Triplex[a]

Molecule	T_M (°C)	ΔH_{cal} (kcal/mol)	$\Delta G°_T$ (kcal/mol)	$T\Delta S_{cal}$ (kcal/mol)	ΔC_p (kcal/°C-mol)
Targeting of *G-quadruplex*					
G-quadruplex	52.6 ± 0.5	23.7 ± 1	2.8 ± 0.2	21 ± 1	+0.5 ± 0.2
G-quadDup	65.6 ± 0.5	118 ± 5	17.9 ± 1.1	100 ± 4	+2.0 ± 1.0
Targeting of *Triplex*					
Triplex	38.3 ± 0.5	83 ± 3	6.7 ± 0.5	76 ± 3	−0.6 ± 0.1
TripDup	58.9 ± 0.5	156 ± 6	21.6 ± 1.5	134 ± 5	+0.2 ± 0.2

[a] The $\Delta G°_T$ and $T\Delta S_{cal}$ terms correspond to temperatures of 14 °C (*G-quadruplex* and *G-quadDup*) and 13 °C (*Triplex* and *TripDup*). All experiments were carried out in 10 m*M* Cs–HEPES buffer at pH 7.5 and 100 m*M* KCl (*G-quadruplex* and *G-quadDup*) or 10 m*M* phosphate buffer at pH 7.0 and 200 m*M* NaCl (*Triplex* and *TripDup*).

118 kcal/mol, in good agreement with the enthalpy of 108 kcal/mol estimated from nearest-neighbor parameters (Breslauer *et al.*, 1986; SantaLucia *et al.*, 1996). Additional DSC experiments were carried out at different salt concentrations (data not shown) to determine indirectly the associated heat capacity effects for each molecule from the slopes of the lines of the ΔH_{cal} versus T_M plots. We obtained ΔC_p values of 0.46 ± 0.16 kcal/°C-mol (*G-quadruplex*) and 2.0 ± 1.0 kcal/°C-mol (*G-quadDup*; Lee *et al.*, 2008). The ΔH_{cal}s of *G-quadDup* in this range of salt concentrations are similar (within experimental error), yielding ΔC_ps with large errors, which are consistent with the negligible heat capacities effects observed in the DSC curves.

UV melting curves (data not shown) of *G-quadruplex* and *G-quadDup* at several strand concentrations showed that the *G-quadruplex* unfolds with similar T_Ms of its DSC curve, indicating the formation of an intramolecular complex (Kankia and Marky, 2001; Olsen *et al.*, 2006). *G-quadDup* unfolds with T_Ms smaller than the corresponding T_M of the DSC curves, carried out at higher strand concentrations, indicating that *G-quadDup* is forming a bimolecular duplex at low temperatures. We use this T_M-dependence to determine the actual heat contribution for the formation of *G-quadDup* at the strand concentration used in the ITC experiments at 14 °C.

We use the DSC unfolding enthalpies of the duplex product to create Hess cycles that correspond to the targeting reactions, that is, by subtracting the unfolding enthalpy of *G-quadDup* from that of *G-quadruplex* reactant. This Hess cycle corresponds to temperatures equal to the T_Ms, which are extrapolated to the temperature of the ITC targeting reaction, yielding an enthalpy value of −94 ± 5 kcal/mol for the formation of *G-quadDup*. After correcting this enthalpy for the heat capacity effect of *G-quadruplex*, we obtained a ΔH_{HC} value of −112 ± 8 kcal/mol (*G-quadDup*; Table 1.2), which is 18 kcal/mol more exothermic. Further inclusion of the duplex product heat capacity yielded ΔH_{HC} of -13 ± 60 kcal/mol (*G2-dupB*); due to the large experimental error in this ΔC_p, we do not use this latter ΔH_{HC} in our discussion. Therefore, the resulting ΔH_{HC} is more exothermic than the ΔH_{ITC}s obtained from ITC titrations (Table 1.2), by 28 ± 8 kcal/mol.

In this targeting reaction, there are several enthalpy contributions that need to be discussed in order to understand the differences in the exothermic heats obtained isothermally and from the DSC unfolding reactions. In the ITC reactions, there are endothermic contributions from disrupting the single G-quartet stack of the *G-quadruplex*, disruption of base–base stacking of the complementary strands (*G-quadCS*), and the release of electrostricted water molecules from the *G-quadruplex* (Gasan *et al.*, 1990). These endothermic contributions are completely overridden by exothermic contributions due to the formation of base-pair stacks in the duplex product (*G-quadDup*), the release of structural water from the complementary strand, and the immobilization of electrostricted water by the duplex products (Marky and Kupke, 2000; Zieba *et al.*, 1991). In the DSC Hess cycles, similar types of enthalpy contributions take place; however, the enthalpy contributions for the unfolding of the *G-quadruplex*, the formation of product duplex, the release of electrostricted water molecule by *G2*, and the uptake of electrostricted water molecules by the duplex product, have similar magnitudes compared to those of the ITC reactions. In contrast, the removal of structural water from the complementary strand and the extent of its base–base stacking contribute to a lesser extent because their formation in the DSC experiments takes place at higher temperatures, >60 °C. Both of these latter contributions decrease the overall enthalpy observed in the ITC targeting reactions and may account for the difference between ΔH_{ITC} and ΔH_{HC}.

3.2. Targeting an intramolecular triplex of the pyrimidine motif

In this targeting reaction, an intramolecular triplex (*Triplex*) is targeted with its partial complementary strand (*TripCS*) to form a duplex with a dangling end (*TripDup*), see Fig. 1.2B. The sequence, CTTTC, of the second loop of *Triplex* is designed to prevent more than two continuous guanines on *TripCS*, which may form aggregates or G-quadruplexes.

The ITC titration for the reaction of *Triplex* with *TripCS* was carried out at 13 °C, Fig. 1.2E. The complementary strand was placed in the reaction cell to reduce base–base stacking contributions. After correcting each injection for the dilution heat of the titrant (*Triplex*), the average heat of injections 2–5 of this titration was normalized by the concentration of the limiting reagent yielded a ΔH_{ITC}s of -39.9 ± 2 kcal/mol. This favorable heat is due to the net compensation of an exothermic heat for the formation of base-pair stacks and the endothermic heats of both breaking base-triplet stacks and base–base stacking of the constrained loops of *Triplex*.

The DSC unfolding curves of the reactant and product of this reaction are shown in Fig. 1.2F, and standard unfolding thermodynamic profiles of each reactant and product at 13 °C are shown in Table 1.2. The unfolding of the reactant *Triplex* and of the duplex product *TripDup* show clear monophasic transitions, while no transition is observed with the reactant single strand (data not shown). All DSC curves show negligible heat capacity effects between the initial and final states. *Triplex* unfolds with a T_M of 38.3 °C and an endothermic heat of 82.5 kcal/mol, in excellent agreement with previous reports of a similar triplex (Shikiya and Marky, 2005; Soto *et al.*, 2002, 2006). This intramolecular triplex forms 100% at the measuring temperature of the ITC experiments. *TripDup* yielded a T_M and enthalpy of 16.3 °C and 156 kcal/mol, respectively, in excellent agreement with the enthalpy of 156 kcal/mol, estimated from DNA nearest-neighbor parameters (Breslauer *et al.*, 1986; SantaLucia *et al.*, 1996). UV melting experiments were carried out as a function of strand concentration (data not shown); the T_Ms for *TripDup* are smaller than the calorimetric T_Ms that were carried out at higher strand concentrations. This shows that the T_M of this duplex product depends on strand concentration and indicates that it is forming a bimolecular duplex at low temperatures.

DSC scans were carried out in different salt concentrations to determine the associated heat capacity effects for the unfolding of reactant and product data not shown. We obtained ΔC_p values of -0.57 ± 0.10 kcal/°C-mol (*Triplex*; Shikiya and Marky, 2005; Soto *et al.*, 2006), and 0.22 ± 0.21 kcal/°C-mol (*TripDup*; Lee *et al.*, 2008). The ΔC_ps of *TripDup* is fairly small with large experimental errors, in agreement with the measured negligible heat capacity of the DSC curves.

Similar to the previous section, we use the DSC unfolding enthalpies of the triplex reactant and each duplex product to create Hess cycles that correspond to each targeting reaction. We subtract the unfolding enthalpy of the duplex (*TripDup*) from that of the reactant (*Triplex*). This Hess cycle corresponds to temperatures equal to the T_Ms, which are extrapolated to the temperature of the ITC targeting reaction, by correcting the T_M for the strand concentration used in the ITC titration. This exercise yielded ΔH_{HC} values of -73.5 ± 7 kcal/mol for the formation of *TripDup*. However, the additional correction for the heat capacity effects of *Triplex* yielded a ΔH_{HC} value of -60 ± 7 kcal/mol. If this ΔH_{HC} value is further corrected for the heat capacities of the product, we obtain a ΔH_{HC}s of -50 ± 12 kcal/mol, which has a large uncertainty and is not used in the discussion of the next section. Similar to the previous targeting reaction, the ΔH_{HC}s value is more exothermic than the one obtained in ITC (Table 1.3), by an average heat of -34 kcal/mol. The net exothermic enthalpies obtained in this targeting reaction correspond to a complete override of the endothermic heat

Table 1.3 Thermodynamic unfolding profiles for targeting stem-loop motifs[a]

Molecule	Transition	T_M (°C)	ΔH_{cal} (kcal/mol)	$\Delta G°_T$ (kcal/mol)	$T\Delta S_{cal}$ (kcal/mol)	ΔC_p (kcal/°C-mol)
Targeting of *Hairpin*						
Hairpin		66.6	38.4	5.8	32.6	0.31 ± 0.06
HairpDup	1st	48.6	36.1	3.8	32.3	
	2nd	66.6	38.4	5.8	32.6	
	Total		74.5	9.6	54.9	0.57 ± 0.05
Targeting of *Bhairpin*						
Bhairpin		57.1	75.1	8.4	66.7	−0.07 ± 0.08
BhairDup		57	92.8	10.8	76.4	−0.01 ± 0.03
ILhairpin targeting						
ILhairpin		58.5	86.1	9.7	82.0	0.08 ± 0.38
ILhairDup		58.8	111.6	13.0	98.6	−0.04 ± 0.02

[a] The $\Delta G°_T$ and $T\Delta S_{cal}$ terms correspond to temperatures of 15 °C (*Hairpin* and *HairpDup*) and 20 °C (all others). All experiments were carried out in 10 m*M* phosphate buffer at pH 7.0 and 100 m*M* NaCl.

contributions by the exothermic heat contributions. Endothermic contributions include the disruption of base-triplet stacks of *Triplex*, disruption of base–base stacking contributions of the complementary strand (*TripCS*), and release of electrostricted water molecules from *Triplex*; while exothermic contributions include the formation of base-pair stacks in the duplex product (*TripDup*), the release of structural water from *Triplex* and *TripCS*, and the immobilization of electrostricted water by the product duplex (Lee *et al.*, 2008; Marky and Kupke, 2000; Zieba *et al.*, 1991).

The comparison of these heat contributions, in the context of the ITC versus DSC targeting reactions, indicates similar magnitudes for the following enthalpy contributions: unfolding of *Triplex*, formation of product duplex, release of both electrostricted water and structural molecules from *Triplex*, and the uptake of electrostricted water molecules by *TripDup*. On the other hand, and similar to the previous targeting reaction, the removal of structural water from the complementary strand and the extent of its base–base stacking contributions, are contributing to a lesser extent in the DSC reactions. Both contributions decrease the exothermicity of the ITC targeting reactions and may account for its difference with the Hess cycle enthalpy.

3.3. Targeting stem-loop motifs

In the first reaction of Fig. 1.3, a complementary strand (*HairpinCS*) is used to target a hairpin loop (*Hairpin*) to form a duplex with a dangling end (*HairpDup*), Fig. 1.3A. The ITC titrations for these reactions were carried

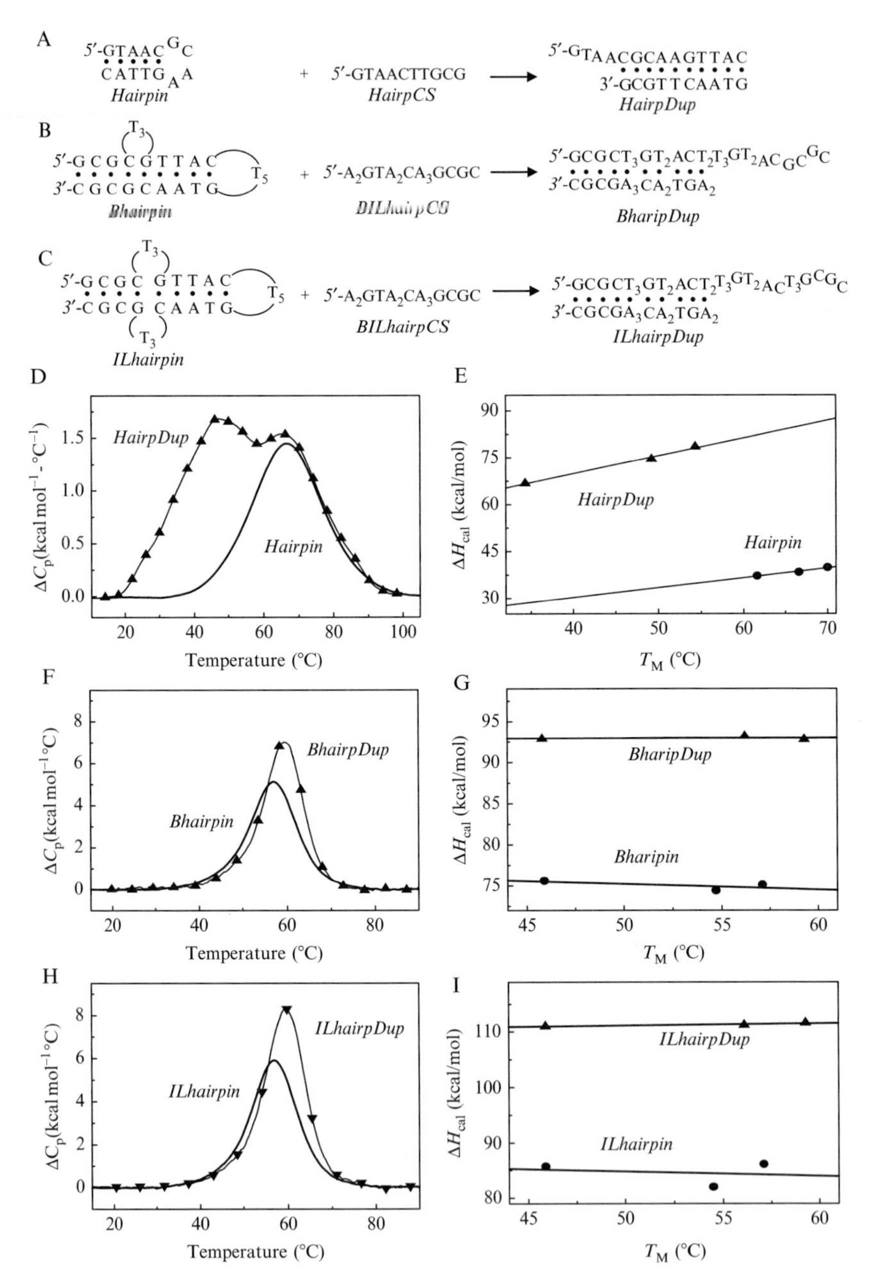

Figure 1.3 Targeting *Hairpin*, *Bhairpin*, and *ILhairpin*. (A–C) Cartoon of targeting reactions; (D, F, H) DSC curves of reactants and products; and (E, G, I) Determination of heat capacity effects from the ΔH_{cal}-dependence on T_M. All experiments were carried out in 10 mM NaPi, 100 mM NaCl at pH 7.0.

out at 15 °C (data not shown). The complementary strand (*HairpinCS*) with lower concentration was placed in the reaction cell and the time between injections was set at 4–12 min to allow for complete reaction. After correcting each injection for the dilution heat of the titrant (*Hairpin*), the average heat of the first two to four injections yielded an exothermic heat of −46.5 μcal, which after normalization by the concentration of the limiting reagent yielded ΔH_{ITC}s of −27.0 ± 1.1 kcal/mol for the formation of the duplex product (*HairpDup*). The favorable heat of this reaction is due to a net compensation of exothermic heat from the formation of base-pair stacks and the endothermic heats of breaking both base-pairs stacks in the hairpin stem and base–base stacking in the loop. The DSC unfolding curves of the reactant and product are shown in Fig. 1.3D, and the curves show transitions with negligible heat capacity effects between the initial and final states. *Hairpin* unfolded in a reversible monophasic transition with a T_M of 66.6 °C and enthalpy of 38.4 kcal/mol, indicating 100% formation of hairpin loop at temperatures below 35 °C. *HairpDup* unfolded in a reversible biphasic transition with T_Ms of 50.3 and 66.6 °C and ΔH_{cal}s of 36.1 and 38.4 kcal/mol, respectively, *HairpDup* is formed 100% at temperatures below 18 °C. The total enthalpy value is in agreement with the enthalpy estimated from nearest-neighbor parameters of 76.5 kcal/mol (Breslauer *et al.*, 1986; SantaLucia *et al.*, 1996). Furthermore, the similarity in T_M and ΔH_{cal} values of the second transition of *HairpDup* with the *Hairpin* transition suggests that the first transition of *HairpDup* corresponds to the unfolding of its duplex state into *Hairpin* & HairpCS and followed by the unfolding of *Hairpin* (Fig. 1.3D). Additional DSC experiments at different salt concentrations (data not shown) yielded ΔC_p values of 0.31 ± 0.06 kcal/°C-mol (Hairpin) and 0.57 ± 0.05 kcal/°C-mol (HairpDup), see Fig. 1.3E.

UV melting experiments for the hairpin reactant and the duplex product were carried out as a function of strand concentration (data not shown). The optical T_Ms for the hairpin remains constant, while the T_M of each product depends on strand concentration, indicating that the reactant forms an intramolecular hairpin loop, while the product forms a bimolecular duplex at low temperatures. We use these T_M dependencies on strand concentration to determine the actual heat contribution for the percent formation of *HairpDup* at 15 °C; however, this duplex forms 100% at 15 °C.

We use the DSC unfolding thermodynamic profiles of Table 1.3 to create a Hess cycle for this targeting reaction. This Hess cycle corresponds to temperatures equal to the T_Ms, which are extrapolated to the temperature of the ITC targeting reaction and corrected for both heat capacity effects and concentrations used in the ITC experiments. This procedure yielded a ΔH_{HC} value of −37.1 ± 5.4 kcal/mol for the formation of *HairpDup*, which is more exothermic than the ΔH_{ITC} value. To understand the measured exothermic heats obtained isothermally and from the DSC unfolding reactions, we need to consider additional hydration contributions. In the

ITC reactions, there are endothermic contributions from the release of electrostricted water molecules from the base-pair stacks of *Hairpin* (Gasan *et al.*, 1990), which are partially overridden by exothermic contributions from the release of structural water around the complementary strand and from the immobilization of electrostricted water by the duplex product (*HairpDup*; Marky and Kupke, 2000; Zieba *et al.*, 1991). In the DSC Hess cycle, these hydration contributions, release of electrostricted water molecule by *Hairpin* and the uptake of electrostricted water molecules by *HairpDup*, have similar magnitudes compared to those in the ITC reactions. By contrast, the removal of structural water from the complementary strand and the extent of its base–base stacking contributions contribute to a lesser extent due to their formation in DSC experiments at temperatures above their T_Ms.

In the following two targeting reactions of Fig. 1.3B and C, a complementary strand, *BILhairpCS*, is used to target both a hairpin loop with a bulge of three thymine residues (*Bhairpin*) and a hairpin loop with an internal loop of 6 thymine residues (*ILhairpin*). Each reaction yielded duplexes with 13 base-pairs and dangling ends of 12 (*BhairpDup*) and 15 (*ILhairpDup*) bases, respectively. The ITC titrations for these reactions were carried out at 20 °C (data not shown); *BILhairpCS* was placed in the reaction cell to diminish single-stranded stacking contributions, if any. After correcting each injection for the titrant dilution heat, the average heat of two to four injections for each titration yielded exothermic heats of −9.0 and −8.5 μcal, which after normalization by the concentration of the limiting reagent yielded ΔH_{ITC}s of −21.1 ± 2.0 kcal/mol (*BhairpDup*) and −20.7 ± 1.1 kcal/mol (*ILhairpDup*) for the formation of similar duplex products. The favorable reaction heats are due to a net compensation of exothermic heat measured from the formation of 13 base-pair stacks and endothermic heats from breaking 7 to 8 base-pair stacks of the hairpins and base–base stacking of the thymine loops.

The DSC unfolding curves of the reactants and products for each reaction are shown in Fig. 1.3F and H, and the resulting unfolding thermodynamic profiles are shown in Table 1.3. The DSC curves of the reactants and of the two duplex products show monophasic transitions with negligible heat capacity effects between the initial and final states. The reactants unfolded with T_Ms and ΔH_{cal}s of 57.1 °C and 75.1 kcal/mol (*Bhairpin*) and 57.1 °C and 86.1 kcal/mol (*ILhairpin*), respectively, indicative of stable hairpin loops that form completely at temperatures below 35 °C. The 11.0-kcal enthalpy difference suggests that the internal loop confers greater flexibility to the helical stem of *ILhairpin*, allowing better base-pair stacking of the base-pairs abating this internal loop. The product duplexes, *BhairpDup* and *ILhairpDup*, unfolded with T_Ms of 58.5 and 58.8 °C and ΔH_{cal}s of 92.8 and 111.6 kcal/mol, respectively, indicating that each is forming a 100% duplex at temperatures below 35 °C. Furthermore, the higher enthalpy (18.8 kcal) of *ILhairpDup* corresponds to the unfolding of 3 additional base-pair stacks. The additional DSC experiments at different salt concentrations (data not shown) generated the ΔH_{cal} versus T_M plots of Fig. 1.3G and I, and the slopes of the resulting lines yielded negligible heat capacity effects, ranging from -40 to 40 cal/°C-mol.

Additional UV melting experiments for each reactant and duplex product, as a function of strand concentration (data not shown), yielded similar T_Ms for the hairpin molecules, but the T_M of each product depended on strand concentration. This indicates that the reactants formed intramolecularly while each of the products is forming a bimolecular duplex at low temperatures.

We used the DSC unfolding enthalpies of the reactant and duplex product to create Hess cycles that correspond to the targeting reaction at temperatures equal to the T_Ms, which are extrapolated to the temperature of the ITC targeting reaction without correction for heat capacity effects because these values were negligible. This exercise yielded ΔH_{HC} values of -17.3 ± 6.0 kcal/mol (*ILhairpDup*) and -25.5 ± 6.0 kcal/mol (*ILhairpDup*), which are in good agreement with the exothermic enthalpies obtained in the ITC experiments. The net exothermic enthalpies for these targeting reactions correspond to a complete override of endothermic heat contributions by the exothermic heat contributions, which is due to a net gain of 5–6 base-pair stacks and to associated hydration changes in each reaction (Marky and Kupke, 2000; Zieba *et al.*, 1991).

3.4. Targeting a DNA pseudoknot

In this targeting reaction, the strand at the bottom of the *Pseudoknot* (Fig. 1.4A) is targeted with its complementary strand, *PseudoCS*, to form a long hairpin loop with a nick, *PseudoDup*. ITC titrations for this reaction were carried out at 5 °C (data not shown), after correcting for dilution heats, and taking the average heat of four injections and normalized by the concentration of the limiting reagent, yielded a ΔH_{ITC} of -60.0 ± 2.8 kcal/mol. This favorable heat is due to the net compensation of an exothermic heat for the formation of 9 base-pair stacks and the endothermic heats of both breaking five base–base stacks and base–base stacking of one of the thymine loops.

The DSC unfolding curves of *Pseudoknot* (reactant) and *PseudoDup* (product) are shown in Fig. 1.4D. Standard thermodynamic profiles for the unfolding of each reactant and product at 5 °C are shown in Table 1.4. The unfolding of both reactant and duplex product shows biphasic transitions, *Pseudoknot* unfolds with T_Ms of 40.0 and 69.8 °C and endothermic heats of 32.7 and 52.2 kcal/mol, while *PseudoDup* unfolds with T_Ms of 36.7 and 69.0 °C and endothermic heats of 85.5 and 45.8 kcal/mol. The curves show negligible heat capacity effects between the initial and final states. The similar T_Ms and ΔH_{cal}s for the second transition of each molecule

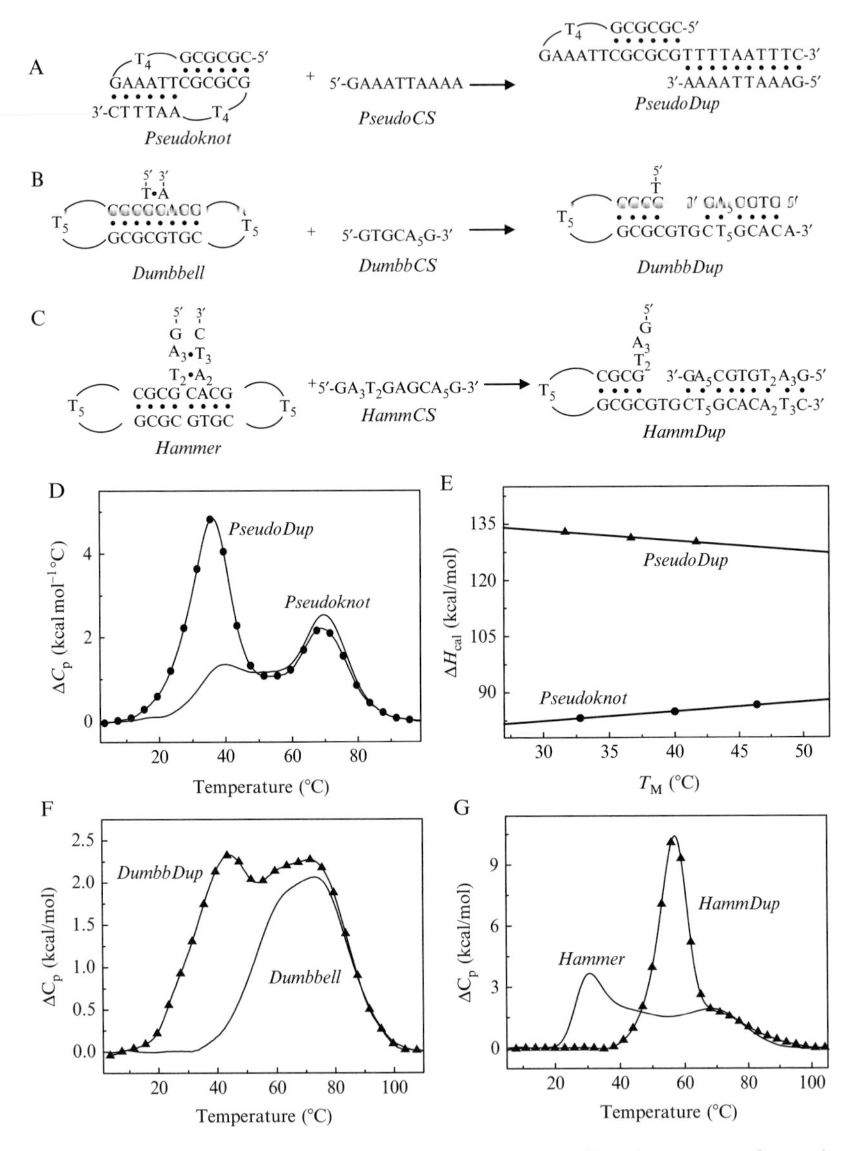

Figure 1.4 Targeting *Pseudoknot*, *Dumbbell*, and *Hammer*. (A–C) Cartoon of targeting reactions; (D, F, G) DSC curves of reactants and products; and (E) Determination of heat capacity effects, from the ΔH_{cal}-dependence on T_M, for the reactant and product of the targeting reaction of *Pseudoknot*. All experiments were carried out in 10 m*M* NaPi, 100 m*M* NaCl at pH 7.0.

indicate the unfolding of the nontargeted duplex stem of each molecule. These two molecules form 100% at the measuring temperature of 5 °C of the ITC experiments. Additional DSC scans for *Pseudoknot* and *PseudoDup*

Table 1.4 Thermodynamic unfolding profiles for targeting *Pseudoknot*, *Dumbbell*, and *Hammer*[a]

Molecule	Transition	T_M (°C)	ΔH_{cal} (kcal/mol)	$\Delta G°_T$ (kcal/mol)	$T\Delta S_{cal}$ (kcal/mol)	ΔC_p (kcal/°C-mol)
Targeting of *Pseudoknot*						
Pseudoknot	1st	40.0	32.7	2.6	30.1	
	2nd	69.8	52.2	8.3	43.9	
	Total		84.9	11.0	73.9	0.26 ± 0.01
PseudoDup	1st	36.7	85.5	6.0	79.5	
	2nd	69.0	45.8	7.2	38.6	
	Total		131.3	13.2	118.1	−0.27 ± 0.03
Targeting of *Dumbbell*						
Dumbbell	1st	61.4	38.6	6.5	32.1	
	2nd	7.4	35.4	7.3	28.1	
	Total		74.0	13.8	60.2	
DumbbDup	1st	43.1	63.0	7.6	55.4	
	2nd	62.1	35.3	6.0	29.3	
	3rd	78.3	30.5	6.4	24.1	
	Total		128.8	20.0	108.8	
Targeting of *Hammer*						
Hammer	1st	31.7	32.2	2.8	29.4	
	2nd	43.7	43.5	5.3	38.2	
	3rd	68.8	44.9	8.4	36.5	
	Total		120.6	16.5	104.1	
HammDup	1st	55.4	116.0	17.8	98.2	
	2nd	72.7	35.5	6.9	28.6	
	Total		151.5	24.7	126.8	

[a] The $\Delta G°_T$ and $T\Delta S_{cal}$ terms correspond to temperatures of 15 °C (*Pseudoknot* and *PseudoDup*) and 5 °C (all others). All experiments were carried out in 10 m*M* phosphate buffer at pH 7.0 and 100 m*M* NaCl.

at several salt concentrations yielded indirectly, from the slopes of the ΔH_{cal} versus T_M (Fig. 1.4E), ΔC_p values of 0.26 ± 0.01 kcal/°C-mol (*Pseudoknot*; Shikiya and Marky, 2005; Soto *et al.*, 2006) and −0.27 ± 0.03 kcal/°C-mol (*PseudoDup*). Additional UV melting experiments as a function of strand concentration (data not shown) show that the T_M remains constant (*Pseudoknot*) and increased (*PseudoDup*) with the increase in strand concentration, indicating their intramolecular and bimolecular formation at low temperatures.

Similar to the previous section, we created a Hess cycle with the DSC data that corresponds to the targeting reaction at the T_Ms, which is extrapolated to the temperature of the ITC targeting reaction, by correcting the

T_M for the strand concentration used in the ITC titration and for heat capacity effects. This exercise yielded a ΔH_{HC} of -57.3 ± 7.8 kcal/mol, which is in excellent agreement with the ΔH_{ITC} value of -60.0 ± 0.4. This net exothermic enthalpy corresponds to a complete override of the endothermic heat contributions, including the disruption of 5 base-pair stacks of the *Pseudoknot* and disruption of base–base stacking contributions of *PseudoCS*, by exothermic heat contributions that include the formation of 9 base-pair stacks in the duplex product (*PseudoDup*), the release of structural water from both reactants, and the immobilization of electrostricted water by the product duplex (Marky and Kupke, 2000; Zieba *et al.*, 1991).

3.5. Targeting of three-way junctions

In the following two reactions, we target intramolecular three-way junctions (Fig. 1.4B and C). In the reaction of Fig. 1.4B, we used a three-way junction that has two helical arms with 4 base-pairs and the third one with a single base-pair, and is designated *Dumbbell* for its similarity to the shape of a double hairpin loop. In Fig. 1.4C, we used a three-way junction similar in sequence to *Dumbbell,* but its third arm has 5 base pairs, and thus called *Hammer.* These two structures, *Dumbbell* and *Hammer,* are targeted with single strands, *DumbbCS* and *HammCS*, that are complementary to each of its right arms and right loops.

The experimental protocol employed for these two sets of reactions is to use DSC to measure unfolding thermodynamic profiles for each of the reactants and products of these two reactions. The resulting thermodynamic data are then used to create Hess Cycles to determine thermodynamic profiles for each targeting reaction with the assumption of negligible heat capacity effects (i.e., each thermodynamic parameter is not corrected for heat capacity effects), but extrapolated to a common temperature, at which 100% formation of reactant and product takes place. However, the molecularity of each reactant and product of these reactions is assessed by carrying out UV melting curves as a function of strand concentration.

The DSC unfolding curves of the reactant (*Dumbbell)* and product (*DumbbDup*) are shown in Fig. 1.4F and the resulting standard thermodynamic profiles in Table 1.4. The DSC curve of *Dumbbell* shows an intramolecular biphasic transition with T_Ms of 61.4 and 77.4 °C and ΔH_{cal}s of 38.6 and 35.4 kcal/mol, respectively. This indicates the melting of the right hand side helical domain (3′-ACACGT_5GTGC-5′) occurs followed by the melting of the left hand side helical domain (5′-TGCGCT_5GCGC-3′), Fig. 1.4B. The helical stem of the duplex product (*DumbbDup*) is made up of a bimolecular decameric duplex with a single-base dangling end and a 4 base-pair hairpin end loop interrupted by a row of four bases (Fig. 1.4B). The DSC curve of this product (*DumbbDup*) shows a triphasic transitions with negligible heat capacity effects (Fig. 1.4F). The complicated deconvolution procedure of these transitions, together with the additional UV melting curves (data not shown), indicates that the first transition with $T_M = 43.1$ °C and $\Delta H_{cal} = 63.0$ kcal/mol, corresponds to the melting of the decameric bimolecular duplex. Then, a partial dumbbell is formed at higher temperatures by refolding of the long dangling end, followed by the melting of the remaining domains. The T_Ms and associated enthalpies for the unfolding of these domains are shown in Table 1.4, where the total ΔH_{cal} of *DumbDup* is equal to 128.8 kcal/mol.

The DSC unfolding curves of the reactant (*Hammer*) and product (*HammDup*) are shown in Fig. 1.4G, and all show negligible heat capacity effects between the initial and final states. The resulting unfolding thermodynamic profiles are shown in Table 1.4. The DSC curve of *Hammer* (Fig. 1.4G) shows a biphasic transition with T_Ms of 31.7, 43.7, and 68.8 °C; and ΔH_{cal}s of 32.2, 43.5, and 44.9 kcal/mol, respectively, which indicates a sequential melting of its helical domains: 5′-GAAATT/ 5′-AATTTC, 5′-GTGCT_5GCAC, and 5′-GCGCT_5GCGC. The DSC curves of *HammDup* (Fig. 1.4D) shows a biphasic transition with T_Ms, and ΔH_{cal}s, of 55.4 °C and 116 kcal/mol (first transition), and 72.7 °C and 35.5 kcal/mol (second transition), corresponding to the unfolding of the hexadecameric duplex followed by melting of a hairpin loop with noncomplementary dangling ends.

We use the DSC unfolding thermodynamic profiles of the reactant and products to create Hess cycles that correspond to each targeting reaction at temperatures equal to the T_Ms, which are extrapolated to a temperature all species formed 100% helical structures, and with the assumption of negligible heat capacity effects. We obtained ΔH_{HC}s of -54.8 ± 7.4 and -30.9 ± 9.7 kcal/mol for the targeting reactions of *Dumbbell* and *Hammer*, respectively. These net exothermic enthalpies correspond to a complete override of the endothermic heat contributions, including the disruption of 5–6 base-pair stacks of *Dumbbell*, or 10–11 base-pair stacks of *Hammer*, and disruption of base–base stacking contributions of the targeting single strands, by exothermic heat contributions that include the formation of 8 base-pair stacks (*DumbbDup*), or 15 base-pair stacks (*HammDup*) in the duplex products, the release of structural water from both reactants, and the immobilization of electrostricted water by the product duplex (Marky and Kupke, 2000; Zieba *et al.*, 1991). Furthermore, close inspection of the structure of the duplex products (Fig. 1.4B and C) indicate that we should get similar ΔH_{HC} values for these two reactions; instead, we obtained a less favorable ΔH_{HC} value, by 23.9 kcal/mol, in the targeting reaction of *Hammer*. This difference may be explained in terms of *Hammer* forming a true three-way junction because it has 6 base-pairs in the third stem, while *Dumbbell* has only 1 base pair; as such, the junction site of *Hammer* has a higher hydration level and additional energy is needed to remove this water.

3.6. Thermodynamic profiles for the targeting reactions

Table 1.5 shows standard thermodynamic profiles for each targeting reaction. The first column of this table shows the ΔH_{ITC} values determined directly from the ITC titrations, while the second column lists the ΔH_{HC} values determined indirectly from Hess cycles using the unfolding thermodynamic profiles of the DSC experiments, as described earlier. These enthalpy values have been discussed in previous sections for each targeting reaction. The $T\Delta S_{HC}$ values of the third column of this table are determined from the DSC data in a similar way as the ΔH_{HC} terms at the temperatures of the ITC experiments, and also include heat capacity effects when appropriate. The $\Delta G^{\circ}{}_{HC}$ terms of the fourth column are calculated from the Gibbs equation: $\Delta G^{\circ}{}_{HC} = \Delta H_{HC} - T\Delta S_{HC}$, while the $\Delta G^{\circ}{}_{ITC}$ parameters for each targeting reaction (last column of Table 1.5) are determined from the following relationship, $\Delta G^{\circ}{}_{ITC} = \Delta G^{\circ}{}_{HC}\,(\Delta H_{ITC}/\Delta H_{HC})$ (Marky *et al.*, 1996; Rentzeperis *et al.*, 1994).

We obtained favorable $\Delta G^{\circ}{}_{ITC}$ values, ranging from −0.6 (targeting of *Pseudoknot*) to −10.7 kcal/mol (targeting of *G-Quadruplex*). Overall, the favorable free energy term of each targeting reaction resulted from a large compensation of a favorable enthalpy and unfavorable entropy

Table 1.5 Standard thermodynamic profiles of targeting reactions

ΔH_{ITC} (kcal/mol)	ΔH_{HC} (kcal/mol)	$T\Delta S_{HC}$ (kcal/mol)	$\Delta G^{\circ}{}_{HC}$ (kcal/mol)	$\Delta G^{\circ}{}_{ITC}$ (kcal/mol)
Targeting of *G-quadruplex*				
−84 ± 5	−112.4 ± 9.4	−98.1 ± 9.8	−14.3 ± 2.7	−10.7 ± 2.1
Targeting of *Triplex*				
−40 ± 2	−60.0 ± 7.0	−46.8 ± 9.2	−13.2 ± 1.6	−8.8 ± 2.2
Targeting of *Hairpin*				
−27.0 ± 1.1	−37.1 ± 5.4	−35.3 ± 5.4	−1.8 ± 0.7	−1.3 ± 0.5
Targeting of *Bhairpin*				
−21.1 ± 2.0	−17.7 ± 6.0	−15.3 ± 6.1	−2.4 ± 1.0	−2.9 ± 1.6
Targeting of *ILhairpin*				
−20.7 ± 1.5	−25.5 ± 7.1	−23.2 ± 7.2	−2.3 ± 1.1	−1.9 ± 0.9
Targeting of *Pseudoknot*				
−60.0 ± 0.4	−57.3 ± 7.8	−56.7 ± 7.8	−0.6 ± 0.6	−0.6 ± 0.6
Targeting of *Dumbbell*				
ND	−54.8 ± 7.4	−48.6 ± 7.6	−6.2 ± 1.7	ND
Targeting of *Hammer*				
ND	−30.9 ± 9.7	−22.7 ± 9.9	−8.2 ± 2.1	ND

All parameters measured in 10 m*M* Cs–HEPES, 100 m*M* KCl at pH 7.5 (*G-quarduplex* targeting), 10 m*M* NaPi, 200 m*M* NaCl at pH 7.0 (*Triplex* targeting), or 10 m*M* sodium phosphate buffer with 0.1 *M* NaCl, at pH 7 (all other experiments).

contributions (Table 1.4). The enthalpy contributions have been discussed extensively, while unfavorable entropy contributions include the bimolecular association of two reagents to form a stable duplex and the putative immobilization of electrostricted water molecules by the duplex product. The release of structural and electrostricted water by each of the reactants will make favorable entropy contributions.

In summary, our results show that the major driving force in each targeting reaction is the exothermic enthalpy contribution of product formation, which corresponds to the formation of base-pair stacks that involved the unpaired bases of the loops of each intramolecular DNA secondary structure investigated. In other words, the targeting complementary strand of each reaction is able to invade the particular DNA secondary structure, forming stable duplex products at low temperatures.

4. Conclusions

We have investigated several reactions, by using DNA single strands as reagents to target intramolecular DNA complexes. Specifically, we have used a combination of ITC, DSC, and spectroscopy techniques to determine standard thermodynamic profiles for the reaction of a variety of DNA secondary structures (G-quadruplex, triplex, hairpin loops, pseudoknot, and three-arm junctions) with their complementary strands. The enthalpies are measured directly in ITC titrations and are compared with those obtained indirectly from Hess cycles using the DSC unfolding data. All reactions investigated yielded favorable free energy terms, that is, each single strand is able to invade and disrupt the corresponding intramolecular DNA structure, resulting from the typical compensation of favorable enthalpy-unfavorable entropy contributions, these exothermic heat contributions are due primarily to the formation of additional base-pair stacks in the duplex products.

In summary, the results show the melting behavior of a variety of secondary DNA structures that can be used, together with nearest-neighbor parameters, in predicting the unfolding thermodynamics of a DNA molecule from knowledge of its sequence. Furthermore, the thermodynamic approach developed here can be used for the control of gene expression, by targeting the loops of the secondary structures formed by mRNA. In general, the larger the number of unpaired bases in a loop, the larger the number of base-pairs and base-pair stacks that are formed in the duplex product, yielding larger free energy terms. Similarly, the simultaneous targeting of several loops of the secondary/tertiary structure of mRNA will also yield favorable free energy terms that most likely will end up stopping transcription.

Furthermore, this investigation of nucleic acid targeting reactions has enabled us to develop a method, based on physicochemical principles, to examine the molecular forces that stabilized specific motifs in a variety of DNA secondary structures.

ACKNOWLEDGMENTS

This work was supported by Grants MCB-0315746 and MCB-0616005 from the National Science Foundation and a Shared Instrumentation Grant 1S10RR027205 from the National Institutes of Health.

REFERENCES

Beal, P. A., and Dervan, P. B. (1991). Second structural motif for recognition of DNA by oligonucleotide-directed triple-helix formation. *Science* **251,** 1360–1363.

Bock, L. C., Griffin, L. C., Latham, J. A., Vermaas, E. H., and Toole, J. J. (1992). Selection of single-stranded DNA molecules that bind and inhibit human thrombin. *Nature* **355,** 564–566.

Breslauer, K. J., Frank, R., Blocker, H., and Marky, L. A. (1986). Predicting DNA duplex stability from the base sequence. *Proc. Natl. Acad. Sci. USA* **83,** 3746–3750.

Brown, P. M., Madden, C. A., and Fox, K. R. (1998). Triple-helix formation at different positions on nucleosomal DNA. *Biochemistry* **37,** 16139–16151.

Cantor, C. R., and Schimmel, P. R. (1980). Biophysical Chemistry Part III: The Behavior of Biological Macromolecules. W. H. Freeman and Company, New York.

Cantor, C. R., Warshaw, M. M., and Shapiro, H. (1970). Oligonucleotide interactions. 3. Circular dichroism studies of the conformation of deoxyoligonucleotides. *Biopolymers* **9,** 1059–1077.

Crooke, S. T. (1999). Molecular mechanisms of action of antisense drugs. *BBA Gene. Struct. Expr.* **1489,** 31–43.

Firulli, A. B., Maibenco, D. C., and Kinniburgh, A. J. (1994). Triplex forming ability of a c-myc promoter element predicts promoter strength. *Arch. Biochem. Biophys.* **310,** 236–242.

Folini, M., Pennati, M., and Zaffaroni, N. (2002). Targeting human telomerase by antisense oligonucleotides and ribozymes. *Curr. Med. Chem. Anticancer Agents* **5,** 605–612.

Fox, K. R. (1990). Long (dA)n × (dT)n tracts can form intramolecular triplexes under superhelical stress. *Nucleic Acid. Res.* **18,** 5387–5391.

Fox, K. R. (2000). Targeting DNA with triplexes. *Curr. Med. Chem.* **7,** 17–37.

Gasan, A. I., Maleev, V. Y., and Semenov, M. A. (1990). Role of water in stabilizing the helical biomacromolecules DNA and collagen. *Stud. Biophys.* **136,** 171–178.

Gehring, K., Leroy, J.-L., and Gueron, A. (1993). A tetrameric, DNA structure with protonated cytosine•cytosine base pairs. *Nature* **363,** 561–565.

Ghildiyal, M., and Zamore, P. D. (2009). Small silencing RNAs: An expanding universe. *Nat. Rev. Genet.* **10,** 94–108.

Han, H., and Hurley, L. H. (2000). A potencial target for anti-cancer drug design. *Trends Pharmacol. Sci.* **21,** 136–142.

Helene, C. (1991). Rational design of sequence-specific oncogene inhibitors based on antisense and antigene oligonucleotides. *Eur. J. Cancer* **27,** 1466–1471.

Helene, C. (1994). Control of oncogene expression by antisense nucleic acids. *Eur. J. Cancer* **30A,** 1721–1726.

Huard, S., and Autexier, C. (2002). Targeting human telomerase in cancer therapy. *Curr. Med. Chem. Anticancer Agents* **2,** 577–587.

Juliano, R. L., Astriab-Fisher, A., and Falke, D. (2001). Macromolecular therapeutics: Emerging strategies for drug discovery in the postgenome era. *Mol. Interv.* **1,** 40–53.

Kankia, B. I., and Marky, L. A. (2001). Folding of the thrombin aptamer into a G-quadruplex with Sr^{2+}: Stability, heat, and hydration. *J. Am. Chem. Soc.* **123,** 10799–10804.

Koeppel, F., Riou, J.-F., Laoui, A., Mailliet, P., Arimondo, P. B., Labit, D., Petitgenet, O., Helene, C., and Mergny, J.-L. (2001). Ethidium derivatives bind to G-quartets, inhibit telomerase and act as fluorescent probes for quadruplexes. *Nucleic Acid. Res.* **29,** 1087–1096.

Lee, H.-T., Olsen, C. M., Waters, L., Sukup, H., and Marky, L. A. (2008). Thermodynamic contributions of the reactions of DNA intramolecular structures with their complementary strands. *Biochimie* **90,** 1052–1063.

Mahato, R. I., Cheng, K., and Guntaka, R. V. (2005). Modulation of gene expression by antisense and antigene oligodeoxynucleotides and small interfering RNA. *Expert Opin. Drug Deliv.* **2,** 3–28.

Maher, L. J., III, Wold, B., and Dervan, P. B. (1989). Inhibition of DNA binding proteins by oligonucleotide-directed triple helix formation. *Science* **245,** 725–730.

Malone, C. D., and Hannon, G. J. (2009). Small RNAs as guardians of the genome. *Cell* **136,** 656–668.

Marky, L. A., and Breslauer, K. J. (1987). Calculating thermodynamic data for transitions of any molecularity from equilibrium melting curves. *Biopolymers* **26,** 1601–1620.

Marky, L. A., and Kupke, D. W. (2000). Enthalpy–entropy compensations in nucleic acids: Contribution of electrostriction and structural hydration. *Methods Enzymol.* **323,** 419–441.

Marky, L. A., Blumenfeld, K. S., Kozlowski, S., and Breslauer, K. J. (1983). Salt-dependent conformational transitions in the self-complementary deoxydodecanucleotide d (CGCGAATTCGCG): Evidence for hairpin formation. *Biopolymers* **22,** 1247–1257.

Marky, L. A., Rentzeperis, D., Luneva, N. P., Cosman, M., Geacintov, N. E., and Kupke, D. W. (1996). Differential hydration thermodynamics of stereoisomeric DNA-benzo[a]pyrene adducts derived from diol epoxide enantiomers with different tumorigenic potentials. *J. Am. Chem. Soc.* **118,** 3804–3810.

Marky, L. A., Maiti, S., Olsen, C. M., Shikiya, R., Johnson, S. E., Kaushik, M., and Khutsishvili, I. (2007). Building blocks of nucleic acid nanostructures: Unfolding thermodynamics of intramolecular DNA complexes. *In* "Biomedical Applications of Nanotechnology," (V. Labhasetwar and D. Leslie-Pelecky, eds.), pp. 191–225. John Wiley & Sons, Inc, Hoboken, New Jersey.

Mills, M., Lacroix, L., Arimondo, P. B., Leroy, J.-L., Francois, J.-C., Klump, H., and Mergny, J.-L. (2002). Unusual DNA conformations: Implications for telomeres. *Curr. Med. Chem. Anticancer Agents* **2,** 627–644.

Moazed, D. (2009). Small RNAs in transcriptional gene silencing and genome defence. *Nature* **457,** 413–420.

Olsen, C. M., Gmeiner, W. H., and Marky, L. A. (2006). Unfolding of G-quadruplexes: Energetic, and ion and water contributions of G-quartet stacking. *J. Phys. Chem. B* **110,** 6962–6969.

Rando, R. F., Ojwang, J., Elbaggari, A., Reyes, G. R., Tinder, R., McGrath, M. S., and Hogan, M. E. (1995). Suppression of human immunodeficiency virus type 1 activity in vitro by oligonucleotides which form intramolecular tetrads. *J. Biol. Chem.* **270,** 1754–1760.

Rentzeperis, D., Kupke, D. W., and Marky, L. A. (1994). Differential hydration of dA.dT base pairs in parallel-stranded DNA relative to antiparallel DNA. *Biochemistry* **33,** 9588–9591.

Rich, A. (1993). DNA comes in many forms. *Gene* **135,** 99–109.

SantaLucia, J., Jr., Allawi, H. T., and Seneviratne, P. A. (1996). Improved nearest-neighbor parameters for predicting DNA duplex stability. *Biochemistry* **35,** 3555–3562.

Shikiya, R., and Marky, L. A. (2005). Calorimetric unfolding of intramolecular triplexes: Length dependence and incorporation of single AT→TA substitutions in the duplex domain. *J. Phys. Chem. B* **109,** 18177–18183.

Simonsson, T., Pecinka, P., and Kubista, M. (1998). DNA tetraplex formation in the control region of c-myc. *Nucleic Acid Res.* **26,** 1167–1172.

Soto, A. M., Loo, J., and Marky, L. A. (2002). Energetic contributions for the formation of TAT/TAT, TAT/CGC+, and CGC+/CGC+ base triplet stacks. *J. Am. Chem. Soc.* **124,** 14355–14363.

Soto, A. M., Rentzeperis, D., Shikiya, R., Alonso, M., and Marky, L. A. (2006). Substitution of deoxythymidine for deoxyuridine in DNA intramolecular triplexes: Unfolding energetics and ligand binding. *Biochemistry* **45,** 3051–3059.

Soyfer, V. N., and Potaman, V. N. (1996). Triple-Helical Nucleic Acids. Springer-Verlag, New York.

Vasquez, K. M., Dagle, J. M., Weeks, D. L., and Glazer, P. M. (2001). Chromosome targeting at short polypurine sites by cationic triplex-forming oligonucleotides. *J. Biol. Chem.* **276,** 38536–38541.

Wang, K. Y., Krawczyk, S. H., Bischofberger, N., Swaminathan, S., and Bolton, P. H. (1993). The tertiary structure of a DNA aptamer which binds to and inhibits thrombin determines activity. *Biochemistry* **32,** 11285–11292.

Wiseman, T., Williston, S., Brandts, J. F., and Lin, L.-N. (1989). Rapid measurement of binding constants and heats of binding using a new titration calorimeter. *Anal. Biochem.* **179,** 131–137.

Zahler, A. M., Williamson, J. R., Cech, T. R., and Prescott, D. M. (1991). Inhibition of telomerase by G-quartet DNA structures. *Nature* **350,** 718–720.

Zieba, K., Chu, T. M., Kupke, D. W., and Marky, L. A. (1991). Differential hydration of dA × dT base pairing and dA and dT bulges in deoxyoligonucleotides. *Biochemistry* **30,** 8018–8026.

CHAPTER TWO

Thermodynamics of Biological Processes

Hernan G. Garcia,* Jane Kondev,[†] Nigel Orme,[‡] Julie A. Theriot,[§] *and* Rob Phillips[¶]

Contents

Abstract

There is a long and rich tradition of using ideas from both equilibrium thermodynamics and its microscopic partner theory of equilibrium statistical mechanics. In this chapter, we provide some background on the origins of the seemingly unreasonable effectiveness of ideas from both thermodynamics and statistical mechanics in biology. After making a description of these foundational issues, we turn to a series of case studies primarily focused on binding that are intended to illustrate the broad biological reach of equilibrium thinking in biology. These case studies include ligand-gated ion channels, thermodynamic models of transcription, and recent applications to the problem of bacterial chemotaxis. As part of the description of these case studies, we explore a number of different uses of the famed Monod–Wyman–Changeux (MWC) model as a generic tool for providing a mathematical characterization of two-state systems. These case studies should provide a template for tailoring equilibrium ideas to other problems of biological interest.

* Department of Physics, California Institute of Technology, Pasadena, California, USA
[†] Department of Physics, Brandeis University Waltham, Massachusetts, USA
[‡] Garland Science Publishing, New York, USA
[§] Department of Biochemistry, Stanford University School of Medicine, Stanford, California, USA
[¶] Department of Applied Physics, California Institute of Technology, Pasadena, California, USA

Methods in Enzymology, Volume 492
ISSN 0076-6879, DOI: 10.1016/B978-0-12-381268-1.00014-8
© 2011 Elsevier Inc.
All rights reserved.

1. Introduction: Thermodynamics is Not Just for Dead Stuff

Thermodynamics has long been a key theory in biology, used in problems ranging from the interpretation of binding both *in vitro* and *in vivo* to the study of the conformations of DNA whether under the action of optical traps in well-characterized solutions or in the highly compacted state of the cellular interior. Despite this long tradition, there is often the sneaking suspicion that because thermodynamics (perhaps more properly referred to as thermostatics) is a theory of equilibrium that tells us how to reckon the "terminal privileged states" of systems (Callen, 1985), it is somehow irrelevant for thinking about the behavior of living cells which are demonstrably *not* in equilibrium. While the terminal state of a living system is death, there are many problems for which an equilibrium treatment is not only a good starting point, but may be the most appropriate tool for the problem of interest.

In a now classic article, Eugene Wigner spoke of the "unreasonable effectiveness of mathematics in the natural sciences," (Wigner, 1960), expressing surprise at the truth of Galileo's earlier assertion that "Mathematics is the language with which God has written the universe." In the time since Wigner's article, many others have taken liberties with his theme by noting the seemingly unreasonable effectiveness of other specific ideas in a much more general context than they were originally intended, and now it is our turn to add our names to the list. Indeed, the unreasonable effectiveness of equilibrium ideas for inherently out-of-equilibrium problems has already been developed by Astumian for specific cases such as a colloidal particle falling through water and a single molecule being stretched by an atomic force microscope (Astumian, 2007). This chapter complements that of Astumian by exploring the perhaps surprising effectiveness of equilibrium thermodynamics in thinking about a wide range of biological problems.

Our chapter has several goals. First, we describe the key theoretical foundations required for the application of equilibrium statistical mechanics models to problems spanning from ligand-gated ion channels to the action of enhancers in transcriptional regulation. In addition, we address conceptual issues related to the applicability of equilibrium concepts by using arguments about separation of time scales to determine when equilibrium ideas can be appropriately used in a living biological context, even though the cell as a whole is not in equilibrium. With these theoretical preliminaries in hand, we carry out a series of illustrative case studies from the last decade or so that show the broad reach of equilibrium ideas to a number of topics that are both timely and exciting. One of our main goals is to argue that equilibrium ideas are a good jumping-off point for thinking quantitatively about a range of problems in cell biology. In particular, they often lead to mathematical formulae that can be explicitly tested in biological experiments to arrive at a deeper understanding of a proposed mechanism. These ideas will be made explicit in the examples to follow.

2. States and Weights from the Boltzmann Rule

For all of the biological examples we wish to examine, the problem formulation plays out the same way. Our starting point is the notion of a "microstate," one of the many distinct ways that the microscopic objects making up our macroscopic system can be arranged. For example, if we are interested in the disposition of a fluorescently labeled DNA molecule on a surface, there are many different ways in which the molecule can lie down on the surface, as shown in Fig. 2.1. Each one of these conformations is a distinct microstate but they all share the common feature that the molecule is adsorbed on the surface (Maier and Radler, 1999). Similarly, if we have a collection of ligands in solution, both the positions and the momenta of the different ligands can be shuffled around without changing the overall concentration and temperature, for example. Again, each such arrangement corresponds to a different microstate. The job of statistical mechanics is to compute the relative probabilities of all the microstates consistent with the constraints imposed on the system. The constraints are

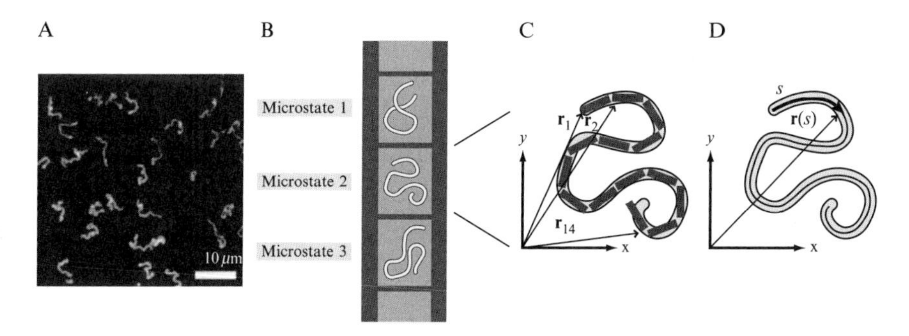

Figure 2.1 Microstates for DNA on a surface. (A) Fluorescence microscopy image of dye-labeled λ-phage DNA on a surface (Maier and Radler, 1999). Each configuration observed corresponds to a different microstate. (B) Schematic showing a series of different allowed microstates for a given DNA molecule on a surface. (C) Discrete representation of the microstate of the DNA molecule. The molecule is divided into a series of segments and there is a vector $\mathbf{r}_i$ which points to the ith segment. Each microstate is characterized by a different set of positions. (D) Continuous representation of the microstate of the DNA molecule. Each point on the molecule has its position defined by the vector $\mathbf{r}(s)$, where s is the arclength along the molecule.

defined by macroscopic variables like temperature, mean distance between the ends of the DNA, or the concentration of ligands in solution. For problems of biological interest, the challenge is to determine what set of microstates are biologically equivalent, and then to enumerate these microstates and calculate their probabilities. For example, a receptor in the presence of many molecules of ligand in solution may be considered "activated" if any one of the individual ligand molecules is bound, although these would all be considered distinct microstates. In practice, it would be tedious or impossible to actually enumerate the microstates for any real system, but the toolkit of statistical mechanics provides elegant methods to accurately estimate their numbers and probabilities, even for complex living systems.

For thinking about processes in the living world, one relevant constraint is the assumption of fixed temperature, which is equivalent to imposing the constraint of constant mean energy. For some biological systems, such as endothermic animals, this approximation is almost true, and in nearly all biological systems, the temperature changes very slowly compared to the rapid molecular transformations that we consider here. This is one example of the importance of the separation of time scales in the application of thermodynamics concepts to biological systems; as long as we can treat temperature as being nearly constant, we can vastly simplify the task of determining the probabilities of the microstates in the system. In this case, statistical mechanics provides us with an elegant and compact formula for the probabilities of all the microstates in the form of the celebrated Boltzmann formula, namely,

$$p_i = \frac{e^{-\beta E_i}}{Z}, \tag{2.1}$$

where E_i is the energy of the microstate i, $\beta = 1/k_B T$, k_B is the Boltzmann constant, and T is the temperature. The denominator in this expression is obtained by summing over the Boltzmann factors ($\exp(-\beta E_i)$) for each of the distinct microstates and is known as the partition function. The key intuition provided by this formula is that the probability of every microstate of the system is solely determined by its energy. For many biological experiments, it is easier to determine the probability of a state (e.g., the concentration of ligand-bound receptors) than to directly measure its energy. Within this framework, the two properties can be conveniently interconverted.

Perhaps the simplest problem of biological interest to which these ideas can be applied is that of a "two-state" ion channel like that shown in Fig. 2.2. In such models, it is assumed that the channel has only two states, closed and open, and the probabilities of these two states can be read off from the fraction of the time spent in each state, as is shown in Fig. 2.2A. There are several underlying assumptions explicit in this treatment, including the idea that the channel has no "memory" of how long it has been open or closed, and the

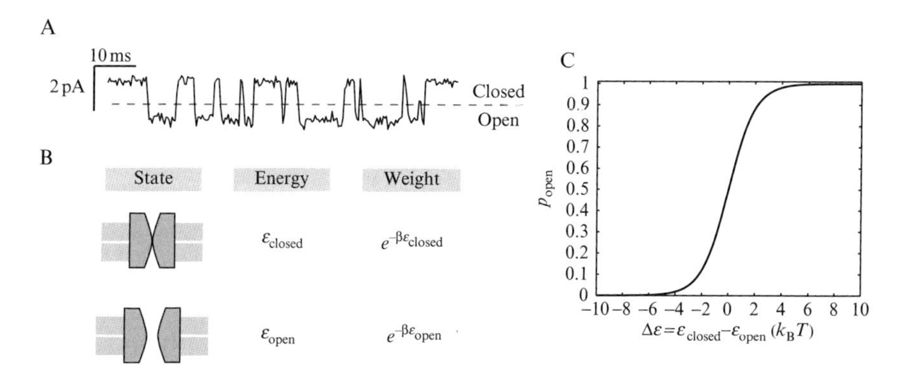

Figure 2.2 States and weights for ion channel dynamics. (A) Current trace showing how the channel transitions back and forth between the open and closed states (Keller *et al.*, 1986). By evaluating the fraction of time spent in either of these two states, we can compute the open (and closed) probability. (B) States of the two-state ion channel, the corresponding energies and the Boltzmann weights. (C) Plot of the probability of the channel being open as a function of the difference in energy between the open and closed states. This difference in energy can be tuned by the driving force such as ligands, voltages and tension on the membrane. The expression plotted corresponds to Eq. (2.3).

idea that all the channels in a population are functionally equivalent. In other words, the system is assumed to be ergodic, such that the average open probability for a single channel examined over time should be the same as the average fraction of channels in a population that happens to be open at any given instant. For cases where these assumptions are reasonable (or nearly reasonable), statistical mechanics tells us how to compute the probabilities of each of the states from their energies, or equivalently to compute their energies from their probabilities. In this chapter, we will repeatedly resort to the same cartoon depiction of the Boltzmann rule by showing a cartoon of the states and their corresponding Boltzmann weights which are obtained by exponentiating the energy of the relevant state, as shown in Eq. (2.1), and multiplying the Boltzmann factor by its associated multiplicity. For a channel like the one being considered here, the corresponding states and weights are shown in Fig. 2.2B. Using these ideas, we see that the probability of the open state is obtained as the ratio

$$p_{open} = \frac{e^{-\beta\varepsilon_{open}}}{e^{-\beta\varepsilon_{open}} + e^{-\beta\varepsilon_{closed}}}, \tag{2.2}$$

where ε_{open} and ε_{closed} are the energies of the open and closed states, respectively. This expression can be rewritten in the alternative fashion

$$p_{open} = \frac{1}{1 + e^{-\beta\Delta\varepsilon}}, \tag{2.3}$$

where $\Delta\varepsilon = \varepsilon_{closed} - \varepsilon_{open}$. The functional form introduced above is used widely in the fitting of opening-probability curves (Keller *et al.*, 1986; Perozo *et al.*, 2002; Zhong *et al.*, 1998) and is shown in Fig. 2.2C. Our interest here was simply to note the way in which this functional form arises completely naturally from the ideas of statistical mechanics.

Of course this is a deliberate oversimplification, as an ion channel that is opening and closing must go through a continuum of multiple structural states in between. However, inspection of the time trace in Fig. 2.2A reveals that the amount of time spent during these transitions is relatively brief compared to the time that the channel typically dwells in either the open or closed states, so for purposes of estimating probabilities, we may make the useful simplification that the system exists primarily in just these two states. Furthermore, we acknowledge that any one state, for example, "open," may in reality represent several or many structurally distinct substates that are equivalent as far as their biological function is concerned, that is, the amount of current that passes through them. One of the most useful properties of the thermodynamic framework for the analysis of biological systems is its flexibility with respect to the precision with which the states are defined; depending on the exact question being asked, the investigator can choose how finely to delineate the various states of the system. Overall, we argue that extremely simple models such as the two-state ion channel seem to fit experimental data unreasonably well, and furthermore provide extremely useful intuition as a starting point for thinking about highly complex systems.

There are many different kinds of ion channels, characterized not only by their selectivities for different ions, but also by the classes of driving forces that gate them (Hille, 2001). Regardless, from the two-state statistical mechanics perspective adopted here, the difference in gating mechanisms from one channel to the next is embodied in the dependence of $\Delta\varepsilon$ on the driving force, whether it is the voltage applied across the membrane, the concentration of some ligands, or the tension in the membrane. This is where the power and utility of the statistical mechanics approach becomes clear. All of the different environmental influences that may affect the opening and closing of the ion channel may be characterized with respect to their effect on the energy (or equivalently, on the probability) of the closed versus open state. So, the expectations for the behavior of a channel with multiple different ligands, or a channel affected by both ligand binding and voltage, can be described quantitatively within this framework, using energy as a universal currency. Formally, it is straightforward to predict quantitatively how a channel with multiple environmental influences is expected to respond when the several different factors operate independently of one another. If the factors such as ligand binding and voltage are not in fact independent, that will be revealed by the failure of the data (measurement of open probability as a function of these two variables) to fit the simple model, and the actual energy of the coupling between the factors can then be calculated.

3. Binding Reactions and Biological Thermodynamics

3.1. Thermodynamic models of binding

One of the poster children for the usefulness of equilibrium thermodynamics and its statistical mechanics partner ideas in biology is the study of binding reactions (Dill and Bromberg, 2003; Hill, 1985; Klotz, 1997). To illustrate our points, we focus on several key case studies. First, we use simple ideas about binding reactions to highlight a few key points about transcriptional regulation. With these ideas in hand, we turn to a class of models that have served as a centerpiece in the analysis of biological cooperativity, namely, the Monod–Wyman–Changeux models (MWC; Monod *et al.*, 1965) which, we will argue, serve in the same capacity in biology that the Ising model introduced to describe the magnetic properties of materials does in physics (Brush, 1967; Plischke and Bergersen, 2006). Both the MWC model and the Ising model make the extremely useful simplifications that, first, the individual elements within a complex system can exist only in a countable number of discrete states (rather than in a continuum), and that an individual element can sometimes change its state. For the simplest cases, such as the spins making up a magnet or ion channel opening, the number of discrete states is just two, but as we will see below, this same framework can be readily expanded to include more than two states.

As a biological case with very broad applicability, we start by considering binding problems in which several different molecular species can exist either separately or in complexes. As shown in Fig. 2.3A, the simplest receptor–ligand binding system can exist in one of two classes of states, or one of two macrostates. Either the receptor is unoccupied or occupied by a ligand molecule. However, for each of these macrostates, there are many different microscopic realizations of the system since the ligands can be distributed in many different ways throughout the solution. For simplicity, we introduce a model of the solution known as a "lattice model" in which the solution is divided into a huge number Ω of boxes and the configurations of the solute molecules are captured by their placement on these lattice sites. This idea is captured in the "multiplicity" column in Fig. 2.3A which tells us the number of distinct ways of arranging our L ligands in the lattice model of the solution adopted here. In reality, of course, the unbound ligands are not confined to boxes in the solution volume; they may exist at any location. However, the lattice model provides an unreasonably effective approximation to a continuous solution in the limit where the number of possible lattice positions is taken to be very large, and it greatly simplifies the statistical mechanics task of enumerating the microstates (Dill and Bromberg, 2003). To find the total statistical weight, we simply multiply the multiplicity of the two macrostates times their associated Boltzmann

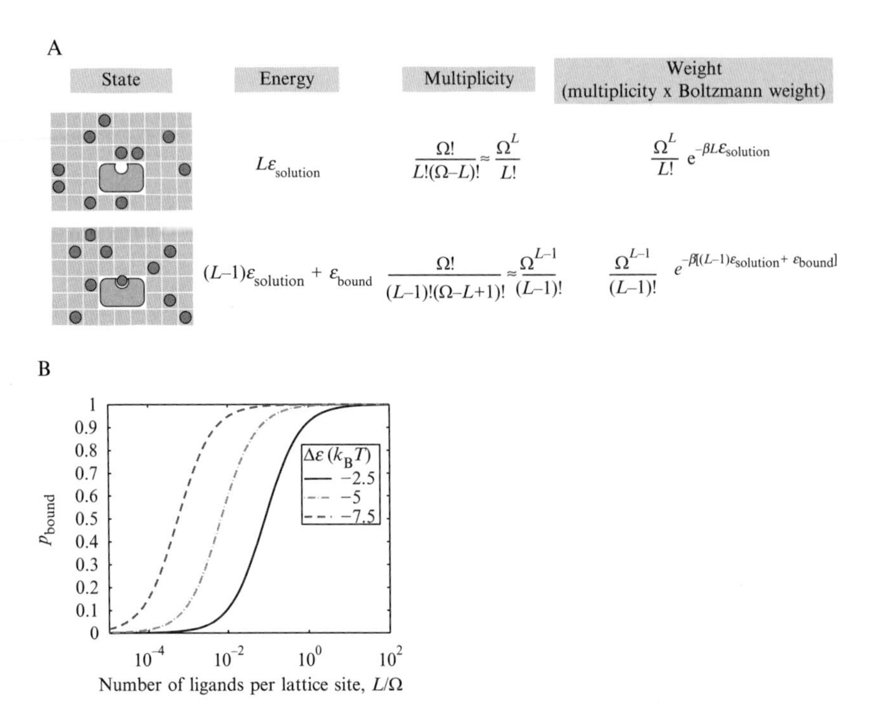

Figure 2.3 Statistical mechanics of receptor–ligand binding. (A) We consider a lattice model of L ligands in solution, represented as a lattice of Ω boxes, and one receptor. There are two broad classes of allowed states (i.e., two macrostates), those in which the receptor is empty and those in which the receptor is occupied by one of the ligands. To enumerate the microstates that make up each of the two macrostates, we count the number of ways that the ligands in solution can be distributed among the Ω boxes; this leads to the multiplicity factor. Furthermore, we assume that in the solution, the ligand has an energy $\varepsilon_{solution}$ and while bound to the receptor its energy drops to ε_{bound}, and these two energies define the Boltzmann factors associated with each of the two macrostates. (B) Plot of the probability of receptor occupancy as a function of the concentration of ligands as shown in Eq. (2.4) for three different choices of the difference in binding free energy between the solution and the receptor.

factors which depend upon their corresponding energies (i.e., upon the energy of binding ε_{bound} and the energy of being in solution $\varepsilon_{solution}$).

With the statistical weights in hand, we can now compute the probability of either of the two macrostates as its statistical weight divided by the sum of the statistical weights of all of the possible microstates. In particular, this leads to a formula for the probability of the receptor to be occupied by a ligand of the form

$$p_{bound}(L) = \frac{\frac{L}{\Omega} e^{-\beta\Delta\varepsilon}}{1 + \frac{L}{\Omega} e^{-\beta\Delta\varepsilon}}, \tag{2.4}$$

where $\Delta\varepsilon = \varepsilon_{\text{bound}} - \varepsilon_{\text{solution}}$ is the energy loss of the ligand upon binding to the receptor and we have assumed that the number of ligands L is much less than the size of the solution represented by the number of boxes in the lattice model, Ω. The factor L/Ω accounts for the loss in translational entropy of the ligand upon binding. As written, this equation describes the probability of receptor occupancy as a function of the number of ligands in our lattice model of solution. This probability is plotted in Fig. 2.3B as a function of several choices of $\Delta\varepsilon$. However, to make contact with concentrations, it is convenient to rewrite this expression by using the volume per elementary box in our lattice model (v) and occupied by ligand as a function of ligand concentration $[L]$. In particular, we can write the number of ligands L as $L = [L]\Omega v$, in which case the equation takes on a familiar form

$$p_{\text{bound}}([L]) = \frac{\frac{[L]}{K_{\text{d}}}}{1 + \frac{[L]}{K_{\text{d}}}}, \tag{2.5}$$

where $K_{\text{d}} = \frac{e^{\beta\Delta\varepsilon}}{v}$ is the equilibrium dissociation constant which provides the concentration at which the receptor has a probability of being occupied of 1/2.

In most interesting biological systems, the concentration of ligand will change over time (e.g., because of changes in cellular signaling), so the system is not truly in equilibrium. However, this is another instance where the separation of time scales is important. As long as the rate at which the ligand concentration changes is relatively slow compared to the individual rates of ligand binding and unbinding, the system can be considered to be nearly in equilibrium at each moment in time, with the probability of ligand binding simply adjusting as its concentration changes slowly.

Often in binding problems that are biologically interesting, the simple functional form defined above is not consistent with the data. This is usually the case when, for example, more than one ligand may bind to the same receptor simultaneously, or when ligand binding causes receptor dimerization. The general biochemical problems of understanding cooperativity and allostery have historically received a great deal of attention (Cui and Karplus, 2008). Below, we will argue that these more complex situations may also be analyzed usefully within this same formal framework. Indeed, the classic MWC model for allostery and cooperativity is a statistical mechanical model that considers molecules that intrinsically exist in a distribution of possible conformational states and assigns these different states different binding affinities (Cui and Karplus, 2008; Gunasekaran *et al.*, 2004). But first, with the basics of the statistical mechanics of single-ligand binding under our belt, we are now equipped to attack a specific problem of biological interest, the regulation of gene expression.

3.2. Thermodynamic models of transcription

Regulation is one of the great themes of biology. Few are left unimpressed after watching the ordered cell divisions and differentiation that attend embryonic development, which serves as a great reminder of what has been dubbed "the regulatory genome" (Davidson, 2006). The roots of regulatory biology are largely to be found in the study of prokaryotes (Ackers *et al.*, 1982; Jacob *et al.*, 2005; Ptashne and Gann, 2002), and these simple single-celled organisms continue to provide valuable insights into transcription and other processes of the central dogma of molecular biology (Buchler *et al.*, 2003; Michel, 2010; Wall *et al.*, 2004). One of our arguments is that the systems that were the early proving ground for our understanding of regulation, namely, questions centering on bacterial metabolism and the bacteriophage life cycle, can now be used as a test bed for a more stringent, systematic, and quantitative attack on questions in regulation. One of the earliest systematic uses of thermodynamic models for computing the properties of a regulatory network was carried out by Ackers and Shea on the decision-making apparatus in bacteriophage lambda (Ackers *et al.*, 1982). More recently, those efforts were generalized to consider the question of how various transcription factors by virtue of being present or absent from regulatory regions of the DNA can conspire to yield combinatorial control of the expression of a particular gene (Bintu *et al.*, 2005a,b; Buchler *et al.*, 2003). In the time since, these ideas have been used even more aggressively for an ever-increasing set of regulatory architectures (Dodd *et al.*, 2005; Fakhouri *et al.*, 2010; Giorgetti *et al.*, 2010; Kuhlman *et al.*, 2007).

To see the way in which these ideas play out most simply within the statistical mechanics framework, consider the case of repression of transcription by a transcription factor (repressor), as shown in Fig. 2.4. The idea is one of simple competition. The promoter can either be unoccupied, occupied by RNA polymerase, or occupied by repressor, but not by both simultaneously. The transcriptionally active state corresponds to that state in which RNA polymerase is bound to the promoter. In the thermodynamic models, all attention is focused on promoter occupancy, and it is assumed that the level of gene expression is proportional to the probability of promoter occupancy by RNA polymerase (Straney and Crothers, 1987). As with the examples worked out above for the two-state ion channel and the simple binding problem, we can compute the probability of interest by resorting to the states and weights diagram shown in Fig. 2.4 which tells us that the probability of promoter occupancy is given by

$$p_{\mathrm{bound}} = \frac{\frac{P}{N_{\mathrm{NS}}} e^{-\beta \Delta \varepsilon_{\mathrm{pd}}}}{1 + \frac{P}{N_{\mathrm{NS}}} e^{-\beta \Delta \varepsilon_{\mathrm{pd}}} + \frac{R}{N_{\mathrm{NS}}} e^{-\beta \Delta \varepsilon_{\mathrm{rd}}}}. \tag{2.6}$$

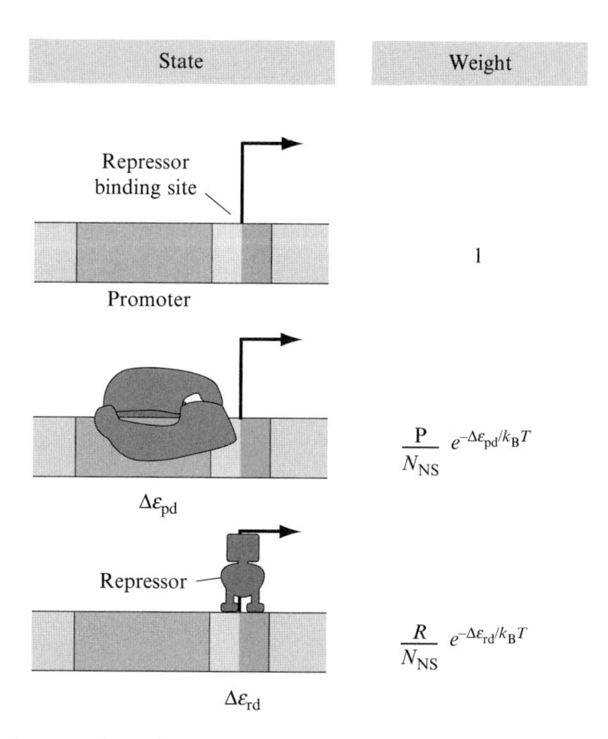

Figure 2.4 States and weights for simple repression. A promoter has a binding site for a repressor molecule which excludes the binding of RNA polymerase. The statistical weights of the different states depend upon the number of polymerases (P), the number of repressors (R), and their respective energies of binding to DNA, $\Delta\varepsilon_{pd}$, and $\Delta\varepsilon_{rd}$. To derive these weights, we use the same approach as that described for ligand–receptor binding, except now we assume that both polymerases and repressors when not bound to the promoter are distributed among N_{NS} sites on the bacterial genome (this is essentially the size of the genome). The energies in the Boltzmann factors are computed as the difference between the energy when repressor or polymerase is bound specifically to the promoter region of the DNA and when they are bound nonspecifically somewhere else on the genome (Bintu *et al.*, 2005b).

Here, the probability is expressed as a function of the number of polymerases (P), the number of repressors (R), the size of the genome N_{NS} in base pairs, and the relevant energy differences that characterize the binding of polymerase and repressor to promoter and operator DNA, $\Delta\varepsilon_{pd}$ and $\Delta\varepsilon_{rd}$, respectively. Details about how this formula is obtained in analogy to the probability of the ligand binding to a receptor from Eq. (2.4) are shown in the caption of Fig. 2.4.

From an experimental point of view, often the most convenient measurable quantity for carrying out the kind of quantitative dissection that is possible using thermodynamic models of gene expression is the fold-change, defined as the ratio of the level of expression in strains that harbor the repressor molecule to the level of expression in strains that do not. This definition can be generalized to an array of different regulatory architectures by always computing the ratio of the level of expression in the regulated strain to that in an unregulated strain. The prediction for the fold-change that follows from the thermodynamic model of simple repression described above is fold-change $= p_{\text{bound}}(R)/p_{\text{bound}}(R=0)$. For repression, the fold-change is always less than one, while for activation, the fold-change is greater than one. As shown in Fig. 2.5, several different bacterial promoters have had their fold-change systematically characterized, and we compare the measured value with the thermodynamic models that are appropriate for the particular promoter. Such experiments lead to knowledge of the parameters of the promoter architecture such as the relevant binding energies. Using these parameters, falsifiable predictions about the gene regulatory input–output relations can be generated (Bintu *et al.*, 2005a).

The idea to use models based on equilibrium ideas to describe the transcriptional output of a promoter might seem ill-conceived, given that transcription is an inherently out-of-equilibrium process with key steps like the elongation stage of transcription leading to mRNA production being essentially irreversible. Still, the key thing to keep in mind is what makes equilibrium ideas useful in these settings is always the separation of time scales. For example, even in the setting in which statistical mechanics and thermodynamics are typically taught, that of an ideal gas, the gas is thought of as being held in a container that is impermeable (i.e., molecules cannot

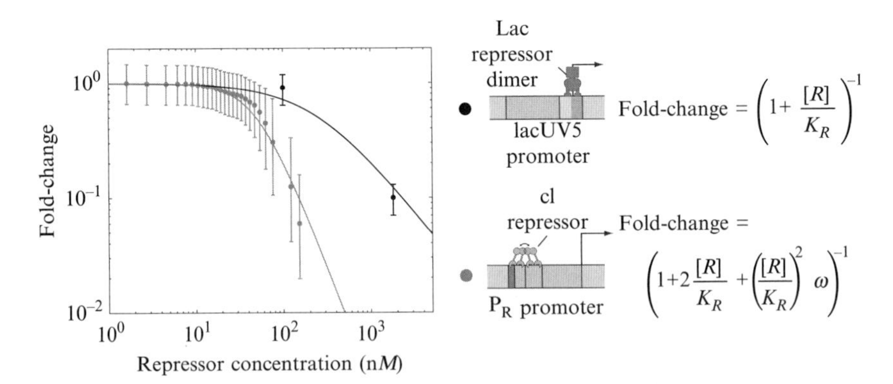

Figure 2.5 Fold-change in gene expression. Two different promoters have been characterized as a function of the number of repressors (Oehler *et al.*, 1994; Rosenfeld *et al.*, 2005). The thermodynamic models predict a precise dependence of the fold-change in gene expression on the concentration of repressors. The theoretical predictions are given by the curves, and the experimental results for two different promoters are given by the data points. These predictions were obtained using the reasoning outlined in Fig. 2.4, equation 2.5, and the weak promoter approximation described in Bintu *et al.* (2005a,b).

escape). In reality, no such container exists! Still, if the diffusion of the gas out of the container occurs on times scales that are much slower than the rate at which the gas explores the volume of the container (i.e., the time for a molecule to diffuse from one end of the container to the other), then we can consider the gas to be in equilibrium. Similarly, if transcription factor and RNA polymerase binding and falling off the DNA occur on time scales that are distinct from the time scales associated with initiation of transcription, we can treat the different states representing combinations of transcription factors bound to promoter DNA as being in equilibrium with each other. This is illustrated in Fig. 2.6. A more intuitive way of restating this

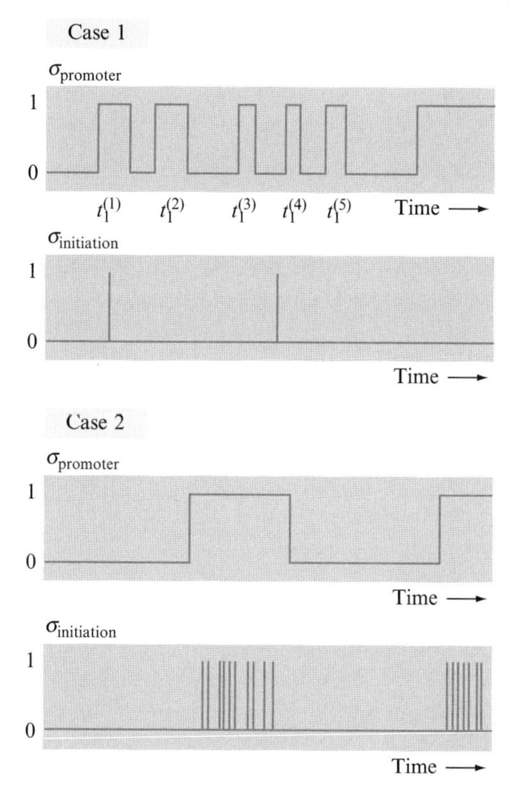

Figure 2.6 Transcriptional time series for several different classes of rate constants. The schematic emphasizes two scenarios in which thermodynamic models of gene regulation are valid. In case 1, the promoter switches fast on the time scale defined by transcription initiation, while in case 2, the opposite limit is illustrated. In both limits, the steady-state number of transcripts (mRNA degrades due to the action of RNases and by dilution after cell division) is proportional to the fraction of time the promoter is in the active state (state 1), which can be computed using equilibrium techniques. We are grateful to Alvaro Sanchez for his articulation of the ideas embodied in this figure.

conclusion is that the rate of transcription should depend on the concentration and activity of the transcription factors, a proposition that is likely to be widely accepted. Here, we have simply developed the formal underpinnings of this assertion.

3.3. The unreasonable effectiveness of MWC models

In the world of statistical mechanics, the Ising model has celebrity status and can be argued to be one of the most useful conceptual frameworks in all of physics. One of the arguments we want to make here is for a similar status for the MWC model in the context of biology (Monod *et al.*, 1965). The biological essence of the MWC philosophy is that many of the molecules of life, or complexes consisting of many molecules, can exist in several different functional states (e.g., inactive and active), and their propensity to bind ligands is different in those states. For a protein that is activated by ligand binding, the simplest picture is that the free energy of the inactive state is intrinsically lower, making it more likely in the absence of ligands. However, if the binding energy for ligands is greater when the molecule is in the *active* state, then the presence of ligands can shift the equilibrium toward this state. What this means in turn is that as ligands are titrated in, the active state will ultimately be the thermodynamic winner. More generally, the same kind of enumeration of discrete states can be applied to any other reversible biological transformation such as protein phosphorylation and dephosphorylation, and transport into or out of a subcellular compartment. There are many important and nuanced features of this idea, some of which will be made mathematically explicit in the case studies to be given in the remainder of the chapter.

3.3.1. MWC and hemoglobin: Where it all began

The MWC model in its various forms has been applied in many different contexts. The most famous example and a story told many times before concerns the application of these ideas to the binding of oxygen to hemoglobin. Because hemoglobin can bind four separate oxygen molecules, there are at least five distinct states of occupancy: empty, single-, double-, triple-, and quadruple occupancy (we are glossing over the possible distinctions among the substates within these states; e.g., a single hemoglobin tetramer with two bound oxygens may carry those oxygens either on the two alpha chains, the two beta chains, or one of each). One of the most important experimental findings about these binding probabilities is the existence of cooperativity: one way of couching it is the idea that the K_d for adding the next ligand depends upon how many ligands are already present. In this situation, the simple binding curves such as those shown in Fig. 2.3B fit the experimental data very poorly. In this case, people often resort to a richer binding curve known as a Hill function, which is a generalization of the functional form shown in Eq. (2.5) to the case where the ratio $[L]/K_d$ in the numerator and the denominator is raised to the power n,

$$p_{\mathrm{bound}}([L]) = \frac{\left(\frac{[L]}{K_d}\right)^n}{1 + \left(\frac{[L]}{K_d}\right)^n}. \tag{2.7}$$

The parameter n is the so-called Hill coefficient and is usually associated with the degree of cooperativity. For the hemoglobin case, the cooperativity concept was developed by Linus Pauling in 1935 specifically as a way to explain the nontrivial shape of the observed binding curve (Pauling, 1935). In this framework, the binding of one oxygen molecule to hemoglobin alters its affinity for the subsequent binding of another oxygen molecule to another site. While conceptually attractive and very useful for fitting experimental data, the Pauling model for cooperativity and subsequent elaborations of it (Koshland *et al.*, 1966) require an explicit accounting for how each ligand affects the energetics of subsequent binding events. This formulation becomes increasingly unwieldy if other kinds of interactions are also considered. For example, the metabolic byproduct 2,3-bisphosphoglycerate (2,3-BPG) is found at high concentrations in red blood cells and binds to a site on the hemoglobin tetramer far from the heme groups, substantially decreasing the affinity of hemoglobin for oxygen as part of the blood-based oxygen delivery system in mammals (Benesch and Benesch, 1967). Incorporation of 2,3-BPG into a Pauling-style model for hemoglobin (or, similarly, incorporation of the Bohr effect, etc.) requires a proliferation of coupling terms describing how the binding of each ligand affects the affinity for every other possible ligand (Phillips *et al.*, 2009a).

The MWC view of the cooperativity problem is fundamentally different. The original MWC model took the approach of assuming that hemoglobin itself could exist in only two distinct structural states: in one, the binding of oxygen to all the sites is weak, while in the other, it is strong; there is also an energy penalty to be paid when switching from the state in which oxygen is bound weakly to the one in which it is bound more strongly. The cooperativity in this case arises from the fact that the penalty for binding one, two, three, or four oxygen molecules tightly is the same regardless of the number of molecules. In other words, the presence of one or more bound ligands simply alters the probability of the protein being in each of the two structural states (or in the language of statistical mechanics, alters the energy difference $\Delta\varepsilon$ between the two; Monod *et al.*, 1965). Inclusion of 2,3-BPG in this framework is straightforward; binding of 2,3-BPG also alters the population distribution between the states, lowering the relative energy of the weak oxygen-binding state, and therefore driving the population of hemoglobin molecules in that direction. For this first-order model, the ligands can all be assumed to stabilize or destabilize each possible protein structural state independently, and the effect of combining the various different ligands can be predicted by calculating the linear combination of all of the binding energies with respect to the state probabilities. Though the hemoglobin example was historically foundational, we believe that the MWC framework for biological statistical mechanics can be even more usefully applied to an unreasonably broad range of biological problems by virtue of its intrinsic ability to describe systems that exist primarily in a countable number of discrete functional states.

3.3.2. MWC and ligand-gated ion channels: Cooperative gating

The general applicability of the MWC philosophy is perhaps best illustrated with the example of ion channels. This time our discussion is based on an ion channel that is gated by the binding of ligands. Even though it is an oversimplification, we continue with the picture of ion channels that have only two allowed conformational states, an open state which permits the flow of ions and a closed stated which forbids any ionic current. Further, imagine an ion channel like the nicotinic acetylcholine receptor that has two binding sites for ligands, meaning that there are four possible states of occupancy when the channel is in a given state: unoccupied by ligand, occupied by ligand on site 1, occupied by ligand on site 2, and occupied by ligands on both sites 1 and 2. This is a reasonable first description of the acetylcholine receptor involved in the neuromuscular junction, which is also one of the best-studied ligand-gated channels, though detailed studies show that a faithful interpretation of these channels requires more than this simplest of models provides (Colquhoun and Sivilotti, 2004). The interesting twist that results from exploiting the MWC framework is that the binding energy for the ligands is different in the open and the closed state. All of these eventualities are shown in Fig. 2.7A.

If we make the simplifying assumption that the binding energy for the two different sites is identical, then the statistical weights of the different states can be written in the simple form shown in Fig. 2.7A. The outcome of this model is that the open probability as a function of ligand concentration has the simple but subtle form

$$p_{\text{open}} = \frac{e^{-\beta\varepsilon_{\text{open}}}\left(1+\frac{[L]}{K_{\text{d}}^{(\text{o})}}\right)^2}{e^{-\beta\varepsilon_{\text{open}}}\left(1+\frac{[L]}{K_{\text{d}}^{(\text{o})}}\right)^2 + e^{-\beta\varepsilon_{\text{closed}}}\left(1+\frac{[L]}{K_{\text{d}}^{(\text{c})}}\right)^2}. \tag{2.8}$$

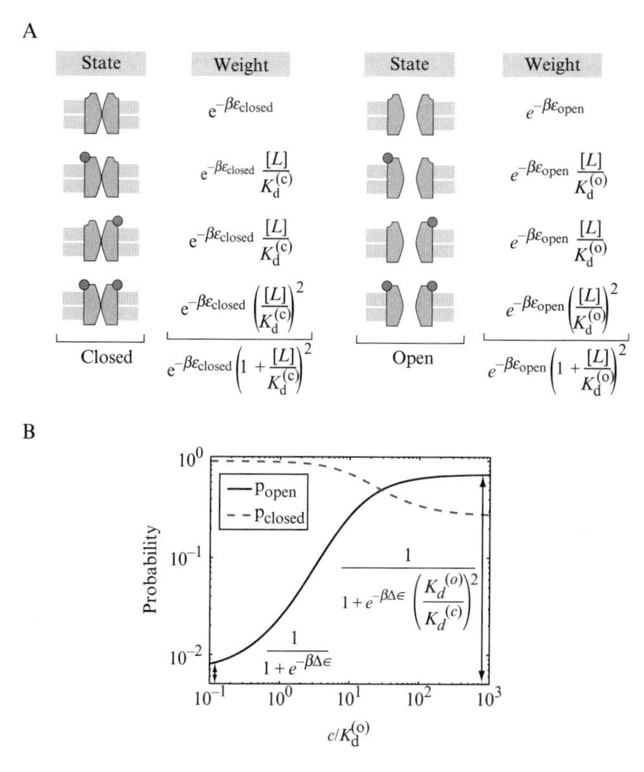

Figure 2.7 MWC model of ligand-gated ion channel. The channel is presumed to exist in one of two states, closed and open. The binding affinities of the ligands for the two binding sites on the channel are the same and they depend upon whether the channel is closed or open. This dependence leads to cooperative binding of the ligands. (A) States and weights for a toy model of a ligand-gated ion channel with two binding sites for the gating ligand. (B) Probability of the two states of the channel (open and closed) as a function of the gating ligand. Notice how in the absence of ligand the probability of the channel being open is the same as that calculated in Eq. (2.3), while the presence of ligand biases the channel toward the open state. The parameters used are $\Delta\varepsilon = \varepsilon_{\text{open}} - \varepsilon_{\text{closed}} = 5k_BT$ and $K_d^{(c)}/K_d^{(o)} = 20$.

The parameters that come into play here include the energies of the open and closed states, namely, $\varepsilon_{\text{open}}$ and $\varepsilon_{\text{closed}}$, and the dissociation constants for the ligand when in the open and closed states, namely, $K_d^{(o)}$ and $K_d^{(c)}$, while the concentration of the ligands themselves is given by $[L]$. Note that this functional form bears some resemblance to that worked out earlier for the simple two-state ion channel, but as a result of the fact that the concentration-dependent terms come in a quadratic fashion, the dependence of the open probability on the ligand concentration is sharper than revealed in our earlier model. This sharpness can be explored by looking at the way that the open probability changes with concentration.

Not surprisingly and just as in the case of hemoglobin, more careful studies of the dynamics of ligand-gated channels reveal behavior that is more nuanced than that captured in the simplest MWC model (Colquhoun and Sivilotti, 2004). Nevertheless, the simple treatment represents a very good first approximation to describing the system that can be used to build intuition and refine the precision of the quantitative questions that can be brought to bear. Within the same statistical mechanics framework, more sophisticated models can be constructed by including more precisely defined structural states and including the possibility for energetic coupling between the two ligand-binding sites (Colquhoun and Sivilotti, 2004).

3.3.3. MWC and chemotaxis: Cooperativity in signal detection

One of the most beloved microscopy videos in the history of modern biology was taken by David Rogers and shows the purposeful motion of a neutrophil as it chases down a bacterium, *Staphylococcus aureus*. This compelling directed motion, a few frames of which are shown in Fig. 2.8, captures people's imaginations because at first blush one cannot avoid a sense of amazement that so many different processes can be so exquisitely synchronized on such short time scales. Indeed, similar rich and complex behavior of the single-celled *Paramecium* led some to wonder whether they were capable of some form of primitive thought (Greenspan, 2006). One of the captivating features of the Rogers video is that the neutrophil "knows" which way to go in order to track down its prey, revealing a specific example of the widespread phenomenon of chemotaxis. Though eukaryotic chemotaxis is a field unto itself, the study of chemotaxis in bacteria is, in many ways, the fundamental paradigm of signal transduction and has also been fruitfully viewed through the prism of equilibrium statistical mechanics (Berg, 2004).

The motion of a bacterium such as *E. coli* is characterized by "runs" and "tumbles" in which the bacterium moves forward in a nearly straight path, reorients in the tumbling process, and then heads off in a new direction (Berg, 2000). Bacterial chemotaxis refers to the way in which bacteria will bias the frequency of their tumbles in the presence of a gradient of chemoattractants (Cluzel *et al.*, 2000). At the molecular level, this behavior is mediated by surface-bound chemoreceptors and cytoplasmic response regulators that communicate with the flagellar rotary apparatus (Falke *et al.*, 1997). To illustrate how equilibrium statistical mechanics has been used to study chemotaxis, we consider the simplified scenario shown in Fig. 2.9. This watered-down version of the chemotaxis process centers on membrane-bound receptors that can bind soluble chemoattractants in the surrounding medium. The receptor communicates the presence of chemoattractants in the external milieu by modifying response regulators within the cell through phosphorylation. More precisely, from the standpoint of the statistical mechanics approach advocated here, the receptor can

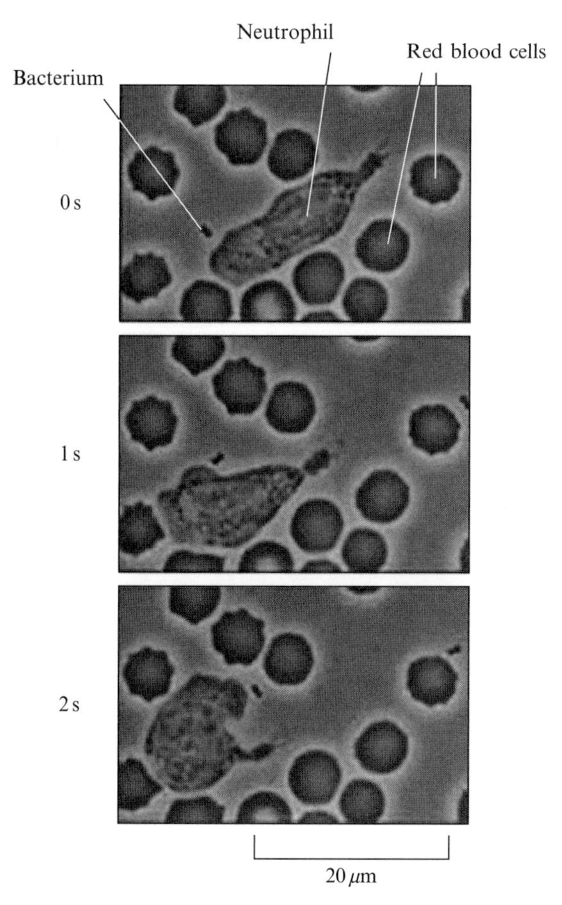

Figure 2.8 Snapshots from the Rogers video showing the directed motion of a neutrophil. Three different frames from the video separated by one-second time intervals reveal that the cell has made a sharp right turn in its pursuit of the bacterium (video by David Rogers, Venderbilt University and digital capture by Tom Stossel, Brigham and Women's Hospital, Harvard Medical School).

be either in an inactive or an active state, with only the active state able to perform the posttranslational modification of the response regulator. The balance of the active and inactive states of the receptor is determined, in turn, by whether or not the receptor is occupied by a ligand. Just as the balance between the open and closed states of the ligand-gated channel is altered by the presence of a ligand, here, the kinase activity of the receptor is tuned by ligand binding.

To compute the probability that a given receptor is activated and hence that the frequency of tumbles is altered, we resort to precisely the same states

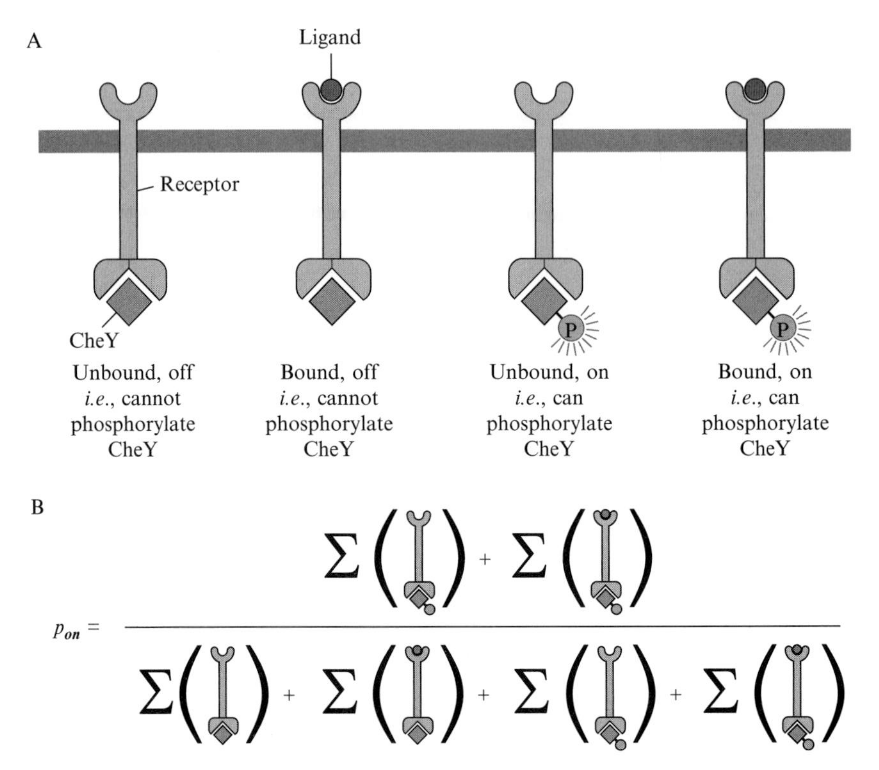

Figure 2.9 States of a single chemoreceptor. (A) The chemoreceptor can exist in four distinct states, characterized by whether or not it is occupied by a ligand and by whether it is in the active state or not. (B) The probability of being in the active (on) state is obtained by summing over the statistical weights of the states where the receptor is active and normalizing by summing over the statistical weights of all states.

and weights philosophy already favored throughout the chapter. We begin with the simplest model of an isolated chemoreceptor, as shown in Fig. 2.9. In this case, the states and weights are shown in the figure and reflect the four eventualities that can be realized: the receptor is either inactive or active and ligand-bound or not. When the ligand is bound, the entropy of the ligands in solution is changed and there is an additional binding energy. This results in probability of being active of the form

$$p_{\text{active}} = \frac{e^{-\beta\varepsilon_{\text{active}}}\left(1+\frac{[L]}{K_{\text{d}}^{(\text{active})}}\right)}{e^{-\beta\varepsilon_{\text{active}}}\left(1+\frac{[L]}{K_{\text{d}}^{(\text{active})}}\right)+e^{-\beta\varepsilon_{\text{inactive}}}\left(1+\frac{[L]}{K_{\text{d}}^{(\text{inactive})}}\right)}, \tag{2.9}$$

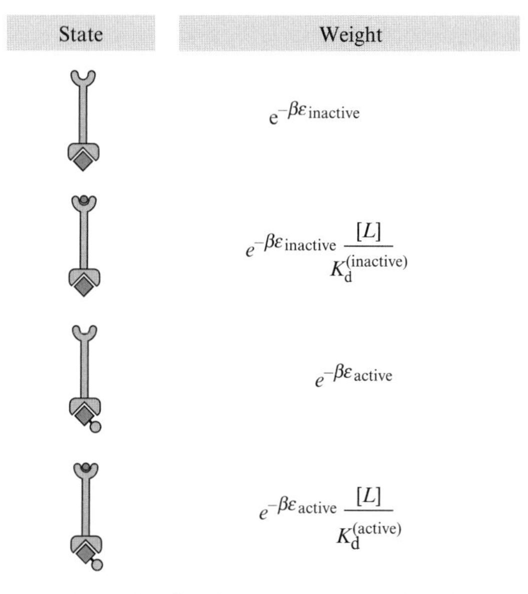

Figure 2.10 States and weights for chemoreceptors in the chemotaxis process. This figure shows the states and weights for a single receptor.

where we have introduced the energies $\varepsilon_{\text{active}}$ and $\varepsilon_{\text{inactive}}$ to capture the energy of the receptor in the active and inactive states, respectively, and $K_d^{(\text{active})}$ and $K_d^{(\text{inactive})}$ to capture the equilibrium dissociation constant for the ligand to bind the receptor when in the active and inactive states, respectively. The states and weights corresponding to this model are shown in Fig. 2.10.

One of the most important outcomes of systematic quantitative experimentation on bacterial chemotaxis is the recognition that the behavior is much more cooperative than indicated by the simple formula derived above (Sourjik and Berg, 2002). The first level of sophistication beyond the naïve model written above is to incorporate the idea that chemoreceptors exist in clusters (Mello and Tu, 2005). In this case, as shown in Fig. 2.11A, the various weights conspire to yield an expression for the probability of the active state as a function of the concentration of chemoattractant, namely,

$$p_{\text{active}} = \frac{e^{-\beta\varepsilon_{\text{active}}}\left(1+\frac{[L]}{K_d^{(\text{active})}}\right)^N}{e^{-\beta\varepsilon_{\text{active}}}\left(1+\frac{[L]}{K_d^{(\text{active})}}\right)^N + e^{-\beta\varepsilon_{\text{inactive}}}\left(1+\frac{[L]}{K_d^{(\text{inactive})}}\right)^N}. \tag{2.10}$$

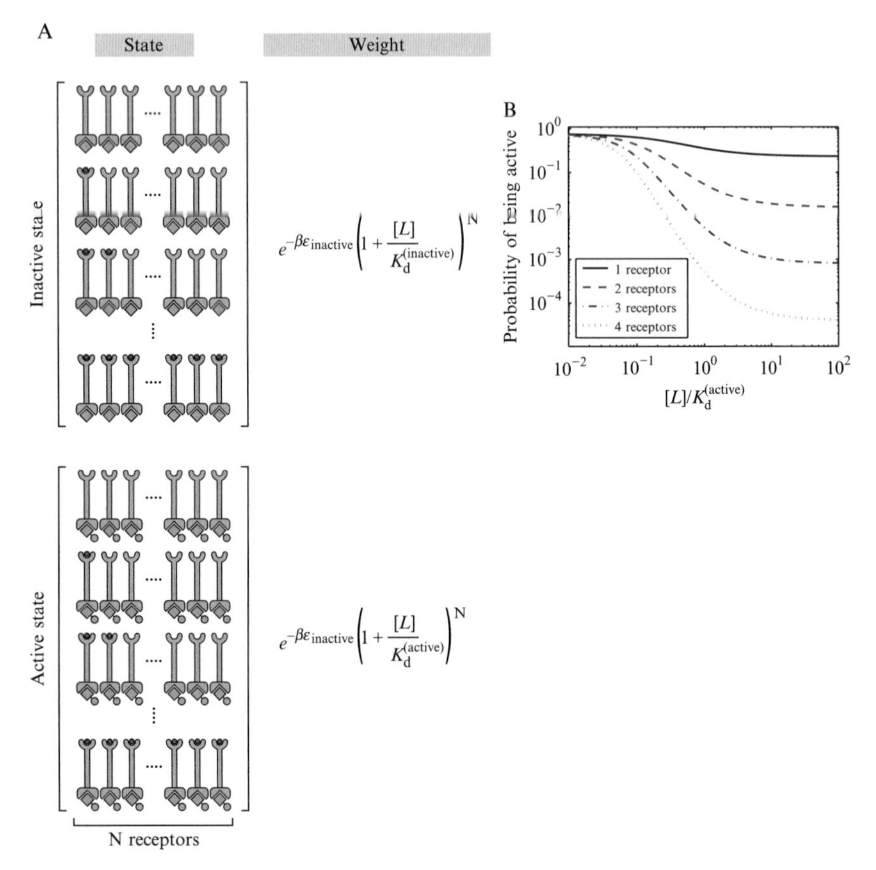

Figure 2.11 MWC model for bacterial chemotaxis. The chemotactic receptors are modeled to exist in clusters with N distinct receptors. Collectively, these receptors can either be in the off or on state, where when on they are able to phosphorylate their downstream response regulator. (A) States and weights diagram for the receptors. (B) Plot of the probability of the active state as a function of the concentration of chemoattractant for different numbers of receptors within a cluster. The parameters used for the plots are $\Delta\varepsilon = \varepsilon_{active} - \varepsilon_{inactive} = -2k_B T$ and $K_d^{(inactive)}/K_d^{(active)} = 1/20$.

In this scenario, N individual receptor molecules within a cluster are envisioned as acting as a unit, where the entire cluster can interconvert between the active and inactive states. This translates into the sharpness of the transition from inactive to active shown in the plot in Fig. 2.11B. Conceptually, the cooperativity for ligand-based activation of the clusters of receptor molecules can be treated in much the same way as the cooperativity for oxygen binding in the MWC model for hemoglobin.

Structurally, the chemotaxis receptors can, in fact, be seen in trimeric clusters on the bacterial surface (Briegel *et al.*, 2009; Shimizu *et al.*, 2000), in support of the validity of this treatment. In fact, this picture is itself only the starting point of a much more sophisticated set of models which acknowledge the collective action of many such receptors as the trimers are arranged in structurally connected networks. Such models even account for the possibility that different receptor types can interact, thus explaining the intriguing experimental observation that the presence of a ligand for one type of chemotaxis receptor can alter the apparent sensitivity of the bacteria to ligands for other receptor types. These models accomplish this without the need to postulate the existence of any unidentified signaling pathways that would enable this kind of crosspathway communication (Keymer *et al.*, 2006; Mello and Tu, 2005). Yet, a further complication in the chemotaxis signaling system is the fact that receptors can be reversibly methylated at several sites in response to continuous stimulation, allowing adaptation over a wide range of ligand concentrations. Within the MWC framework, these posttranslational modifications can also be incorporated as effectively independent "ligands" that alter the probability that the receptors will be either active or inactive, by altering the relative stability of the two states. A statistical mechanics model based on these ideas for modifying the population distribution of simple two-state receptors is unreasonably well-able to reproduce experimental data over a broad range of conditions, including the prediction of system behavior for mutants where methylation is either constitutively on or off at any of several of the possible modification sites (Keymer *et al.*, 2006).

3.3.4. MWC and eukaryotic transcriptional regulation: From nucleosomes to enhancers

A less familiar example of the use of MWC-like models is to binding problems involving DNA and its binding partners. In particular, in a recent set of papers, it was suggested that by analogy to the inactive and active states of a protein, DNA could be either inaccessible or accessible to binding by transcription factors (Mirny, 2010; Raveh-Sadka *et al.*, 2009). One concrete mechanism for how that idea might be realized in a biological system is that the DNA could either be wrapped up in nucleosomes (inaccessible) or open for interaction with other factors. In Fig. 2.12A, we show a schematic of the states and weights for this case, with the "ligands" in this case now being DNA-binding proteins such as transcription factors which bind to some enhancer. For the concrete case shown in the figure, inspired by an enhancer in *Drosophila*, we consider an enhancer region containing seven binding sites, all of which have the same affinity for the transcription factor of interest (though this simplification is not at all crucial).

The idea embodied in the figure is once again that embodied in an MWC model. This means that the system can exist in two overall states

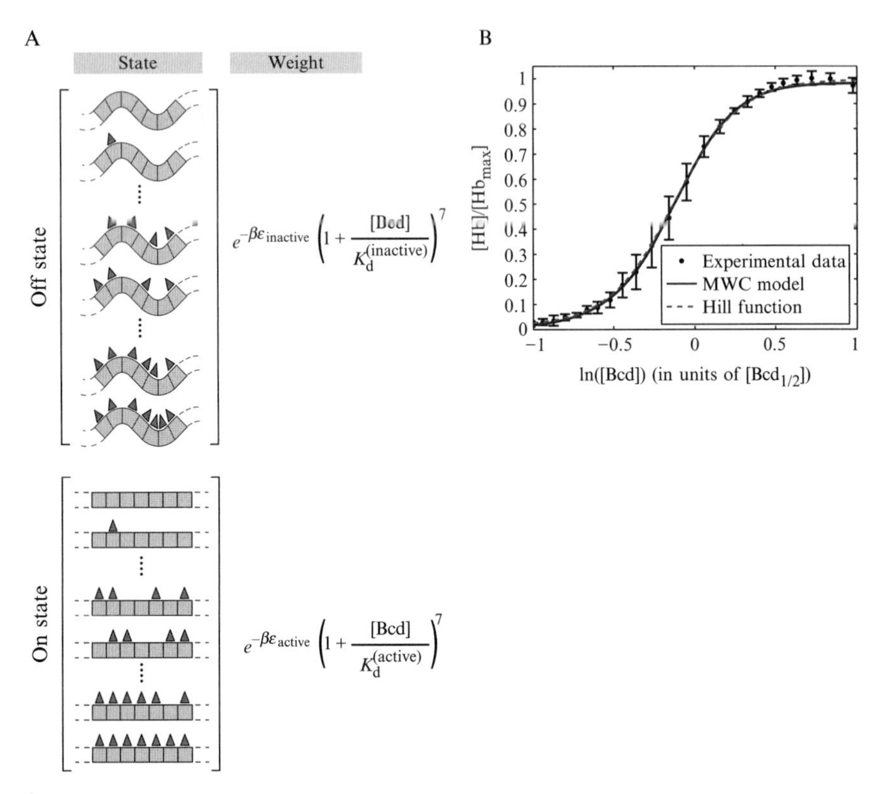

Figure 2.12 MWC model for eukaryotic action at a distance. The regulatory region of the DNA is pictured here to have seven distinct binding sites. The organization of the DNA itself is further posited to exist in two different states. Following the MWC philosophy, the binding energy of the transcription factors (pictured here as triangles) depends upon whether the DNA is in the "closed" or "open" state. (A) States and weights corresponding to the model. (B) Data for the normalized level of Hunchback protein as a function of the level of Bicoid protein (Gregor *et al.*, 2007) measured in units of $[\mathrm{Bcd}_{1/2}]$, the concentration of Bicoid protein for which $[\mathrm{Hb}]/[\mathrm{Hb}_{\max}] = 1/2$, overlaid with a fit to a Hill function $[\mathrm{Hb}]/[\mathrm{Hb}_{\max}] = b(1 + f([\mathrm{Bcd}]/K_d)^5)/(1 + ([\mathrm{Bcd}]/K_d)^5)$ and the MWC model shown in Eq. (2.11). The parameters of the Hill function are $b = 0.01$, $f = 99$, and $K_d = 0.88$. For the MWC model, we took the limit $K_d^{(\mathrm{inactive})} \to +\infty$ and used the parameters $K_d^{(\mathrm{active})} = 0.3$, and $\Delta\varepsilon = \varepsilon_{\mathrm{active}} - \varepsilon_{\mathrm{inactive}} = 9.5\ \mathrm{k}_B T$.

(accessible and inaccessible) and that the affinity of the relevant ligands for their target sites depends upon which of the two overall conformational states the system is in. An equivalent way of stating this is that the relative population distribution and therefore relative stability for each of the two DNA conformational states is influenced by the binding of the ligands. For the particular example shown here, we were loosely inspired by the binding of the transcription factor Bicoid in its role as an activator of a second gene known as Hunchback, two genes that play a specific role in the much larger process of development in the *Drosophila* embryo (Gilbert, 2010). For simplicity, we assume that each of the seven distinct bicoid target sites has the same binding energy and that there is no cooperativity in the sense that the binding of one protein does not alter the binding energy of a second molecule of the bicoid protein to one of the other sites. As a result, the partition function can be evaluated simply in the closed form shown here and results in the level of Hunchback activation given by

$$[\mathrm{Hb}] = [\mathrm{Hb_{max}}]\frac{e^{-\beta\varepsilon_{\mathrm{active}}}\left(1+\frac{[\mathrm{Bcd}]}{K_{\mathrm{d}}^{(\mathrm{active})}}\right)^{7}}{e^{-\beta\varepsilon_{\mathrm{active}}}\left(1+\frac{[\mathrm{Bcd}]}{K_{\mathrm{d}}^{(\mathrm{active})}}\right)^{7}+e^{-\beta\varepsilon_{\mathrm{inactive}}}\left(1+\frac{[\mathrm{Bcd}]}{K_{\mathrm{d}}^{(\mathrm{inactive})}}\right)^{7}}, \quad (2.11)$$

where [Bcd] and [Hb] are the Bicoid and Hunchback concentrations, respectively. The data for the relationship between bicoid binding and hunchback expression has been explored in a recent paper (Gregor *et al.*, 2007). Empirically, the authors of that study found that the expression of Hunchback can be fit to a Hill function that depends upon the concentration of Bicoid. An example of both the Hill function approach favored in that study and the MWC functional form described here are shown in Fig. 2.12B. At this point, the quantitative dissection of developmentally important enhancers in eukaryotes is still in the very early stages, and there is a huge amount still to be done both in carrying out experiments that are at once quantitative and revealing and in finding the right set of "knobs" that can be tuned in both these experiments and the models that are developed in response. Our discussion is meant simply to illustrate the types of questions that are currently being considered and the way that simple thermodynamics are beginning to be used to answer those questions (Fakhouri *et al.*, 2010).

3.3.5. The biological reach of MWC models

Of the nearly 5700 citations at the time of this writing of the original paper by Monod, Wyman, and Changeux (Monod *et al.*, 1965), many are concerned with the limits and validity of this class of models and how they can be used to reflect on a broad class of biological problems with special interest in the fitting of some class of data. Our intent here has mainly focused instead on what such models assume about the molecules they describe and how to use simple ideas from equilibrium statistical mechanics to compute the MWC expressions for binding probability.

It is important to realize that in all of the case studies set forth here, the key point is to illustrate the style of analysis and not the claim that the particular models are the final word on the subject in question. For example, our treatment of the ligand-gated ion channel, while a useful starting point, has been found to miss certain detailed features of the gating properties of these channels. Similarly, our introduction to the MWC approach for bacterial chemotaxis has swept many of the key nuances for this problem under the rug. For example, to really capture the detailed behavior of these systems requires positing a heterogeneous clustering of the different types of chemoreceptors. As concerns transcriptional activation in eukaryotic enhancers, the use of models like that presented here is in its infancy and may end up not being the right picture at all. The key reason for promoting these models is that they provide quantitative hypotheses about the processes of interest which can be used a starting point for developing experiments that test them. As is often the case for the application of simplified analytical models to biological systems, their most useful role can be to help the investigator determine what information is missing. To a first approximation, experimental data that are extremely well-fit by MWC models may be reasonably assumed to operate more-or-less as discrete state systems, where the relevant separation of time scales has rendered the equilibrium assumption of statistical mechanics to be close to correct. In such cases, no further complexifications of the mechanism need be postulated to explain the phenomenon at hand, at least within the limits of the available data which is well fit by the simple model. In the more interesting and perhaps more common case where the simplest statistical mechanics models reveal systematic differences from the data, new kinds of experiments may suggest themselves that will account for the discrepancies and reveal more insight into the workings of the system. Thus, a careful comparison of theory and experiment can serve to uncover quantitative details of the mechanism, whether it be gene regulation, ion-channel gating, or detection of chemoattractant.

4. The Unreasonable Effectiveness of Random-Walk Models

So far, our emphasis has been almost exclusively on binding problems. However, our argument that equilibrium ideas have a broad reach in the biological setting transcends these applications. To demonstrate that point, we close with a brief discussion of the power of such thinking in the context of random-walk models in general and their uses for thinking about polymer problems in particular. The random-walk model touches on topics ranging from evolution to economics, from materials science to biology (Rudnick and Gaspari, 2004). For our purposes, we reflect on the random-walk model in its capacity as the first approach one is likely to try when thinking about the equilibrium disposition of polymers, including those referred to by Crick as the two great polymer languages, namely, nucleic acids and proteins. Though the particular case study we address here concerns proteins that harbor tethered receptor–ligand pairs, the same underlying ideas can be applied just as well to nucleic acids for thinking about the ubiquitous process of DNA looping in transcriptional regulation, for example, (Garcia *et al.*, 2007; Rippe, 2001).

There is a vast literature on the use of models from equilibrium statistical mechanics to explore the properties of biological polymers (de Gennes, 1979; Grosberg and Khokhlov, 1997). As usual, the idea is to figure out what the collection of allowed microstates is for the biological polymer of interest (an example for the conformations of DNA was given in Fig. 2.1). Perhaps the simplest example imagines the polymer of interest in much the same way we would think of a chain of interlinked paper clips. In particular, we treat the polymer as a chain of N segments, each of which has length a. We then posit that each and every configuration has the same energy (and hence the same Boltzmann factor) and thus, the problem of finding the probability of different configurations becomes one of counting their degeneracies. For example, those macrostates, characterized by a particular end–end distance which can be realized in the most different ways are the most likely. These ideas and their generalizations have been used to consider many interesting problems (Phillips *et al.*, 2009b). One of the most celebrated examples that we will not elaborate on here concerns the use of these ideas in the setting of single-molecule biophysics where it has now become routine to manipulate individual proteins and nucleic acids. Indeed, the force-extension properties of these biological polymers are so well described by ideas of equilibrium polymer physics that stretching individual DNA molecules has become a way to *calibrate* various single-molecule apparatus such as optical and magnetic traps.

To get a sense of how these ideas from polymer physics insinuate themselves into biological binding problems hence building upon the earlier parts of the chapter, we consider the simple competition between a tethered ligand–receptor pair and soluble competitor ligands. This kind of motif exists in a number of signaling proteins and has also been the basis of fascinating recent experiments in synthetic biology (Dueber *et al.*, 2003). In particular, the toy model introduced here mimics a synthetic receptor–ligand pair in which the actin cytoskeletal regulatory protein, N-WASP, has been modified to include a single PDZ domain, thus allowing N-WASP activity to be artificially brought under the influence of the PDZ ligand. Furthermore, a copy of the ligand is also attached to the modified N-WASP, with both ligand and receptor domains attached by flexible unstructured protein domains that serve as tethers. As shown in Fig. 2.13A, there are three distinct classes of states available to the system. In the first state, the tethered ligand and receptor are bound to each other. In the second state, the receptor is unoccupied. In the third state, one of the

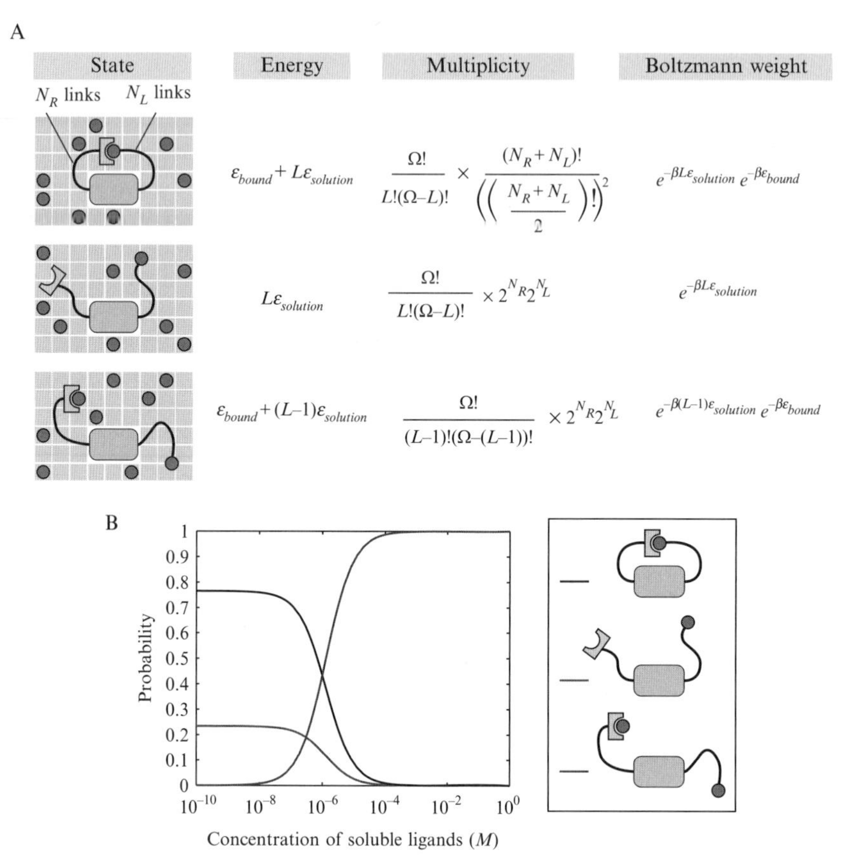

Figure 2.13 Random-walk models of tethered ligand–receptor pairs. (A) Schematic of the tethered ligand–receptor pair and the associated statistical weights. The multiplicities for the polymer are computed using a one-dimensional toy model of the tether. N_R and N_L represent the number of right- and left-pointing segments out of the total $N = N_R + N_L$. The rest of the parameters are defined as in Fig. 2.3. For more realistic calculations, see Van Valen *et al.* (2009). (B) Concentration dependence of the probabilities of the different states that can be realized by the tethered ligand–receptor pair. The parameters used in the plot are $\Delta\varepsilon = \varepsilon_{bound} - \varepsilon_{solution} = -15\ K_BT$ and a probability of looping given by the one-dimensional random walk of 10^{-6}. The volume of the elementary box v has been chosen to be approximately 1.67 nm^3. (See Color Insert.)

soluble ligands is bound to the receptor. The question we are interested in addressing is the relative probability of the two different bound states and how they depend upon the concentration of soluble ligands.

The intuitive argument is that the probability that the receptor will be occupied by a ligand is a result of the competition between the tethered ligand and its soluble partners. As the concentration of the soluble ligands is increased, it becomes increasingly likely that they will form a partnership with the tethered receptor. To explore the nature of this competition, we compute the ratio of the probabilities for the free and tethered ligands. For the purposes of the model shown in Fig. 2.13A, we treat the tether using the simplest one-dimensional model of a random walk since all we are trying to demonstrate is the concept, as opposed to the quantitative details. What this means really is that we evaluate the entropic cost of loop formation using a one-dimensional model which makes it a simple counting exercise to determine the fraction of conformations which close on themselves. Stated simply, if we think of each monomer in the polymer as pointing left or right, then loop formation in this context requires that the number of right and left-pointing monomers be the same. The key point is that in the closed conformation, the two tethers have many fewer conformations available to them in comparison with the case when they are no longer linked, and each side is free to flop around on its own. The result of this competition as a function of the soluble ligand concentration is shown in Fig. 2.13B, and is consistent with our intuition in the sense that in the high concentration limit, the receptor is saturated by soluble ligands. The specific concentration at which the tethered ligand and the soluble ligands have the same probability of being bound to the receptor depends upon the looping probability. As the relative flexibility and length of the tethers are varied experimentally, the quantitative predictions of this simple model can be rigorously tested (Dueber *et al.*, 2003).

5. Conclusions

Thermodynamics is unreasonably effective in the biological setting, but effective it is. As noted by Einstein in his autobiography, "A theory is the more impressive the greater the simplicity of its premises is, the more different kinds of things it relates, and the more extended its area of applicability. Therefore the deep impression which classical thermodynamics made upon me. It is the only physical theory of universal content concerning which I am convinced that, within the framework of the applicability of its basic concepts, it will never be overthrown." (Schilpp, 1970).

Equilibrium thermodynamic ideas and their statistical mechanics partner concepts pervade not only the *in vitro* domain of traditional biochemical binding reactions, but also permeate our thinking for more biologically relevant *in vivo* examples ranging from gene regulation to signaling networks to the physical limits on biological detection (Bialek and Setayeshgar, 2005, 2008). In this chapter, we have tried to articulate some of the fundamentals of equilibrium models for a variety of different problems. Our analysis has focused more on the conceptual underpinnings that on the specific and detailed ways that biological data is greeted by these kinds of models. Here, we have described a few of our favorite examples of the confrontation of the models and corresponding experiments, and many more can be found elsewhere (Bintu *et al.*, 2005a,b; Hill, 1985; Keymer *et al.*, 2006; Klotz, 1997; Mello and Tu, 2005; Phillips *et al.*, 2009c). We have argued that conceptually part of the reason for the effectiveness of equilibrium ideas in the biological setting is likely a matter of separation of time scales, and the most unreasonably effective simplification underlying the MWC and statistical mechanical treatment of these problems, that they exist primarily in a countable number of interconvertible functional states rather than as a squishy continuum.

In his book "How the Mind Works," Steven Pinker notes "The linguist Noam Chomsky once suggested that our ignorance can be divided into problems and mysteries. When we face a problem, we may not know its solution, but we have insight, increasing knowledge, and an inkling of what we are looking for. When we face a mystery, however, we can only stare in wonder and bewilderment, not knowing what an explanation would even look like. I wrote this book because dozens of mysteries of the mind have recently been upgraded to problems. Every idea in the book may turn out to be wrong, but that would be progress, because our old ideas were too vapid to be wrong." (Pinker, 2009). In our view, one of the most important reasons for the potency of the quantitative slant which equilibrium models are but one example of is that they are a tool for generating specific and detailed hypotheses which are a step along the way to turning mysteries into problems and which give us an opportunity to design experiments that can tell us whether we are wrong. The rigorous framework of statistical mechanics provides no space for being vapid.

ACKNOWLEDGMENTS

We are grateful to a number of people for giving us guidance in thinking about this problem and/or providing data: Tom Kuhlman, Terry Hwa, Ulrich Gerland, Leonid Mirny, Henry Lester, Doug Rees, and Dennis Dougherty. H. G. and R. P. are also extremely grateful to the NIH for support through the NIH Director's Pioneer Award (DP1 OD000217), RO1 GM085286, and RO1 GM085286-01S. J. K. acknowledges the support of the National Science Foundation through grant DMR-0706458. J. A. T. was supported by the National Institutes of Health and the Howard Hughes Medical Institute.

REFERENCES

The list of references provided here is meant to be representative rather than comprehensive and are meant as an entry point into the literature for interested readers. The literature on each of the topics described in the text is vast and the few references cited here are largely those we have found are a convenient starting point.

Ackers, G. K., Johnson, A. D., and Shea, M. A. (1982). Quantitative model for gene regulation by lambda phage repressor. *Proc. Natl. Acad. Sci. USA* **79,** 1129–1133.

Astumian, R. D. (2007). Coupled transport at the nanoscale: The unreasonable effectiveness of equilibrium theory. *Proc. Natl. Acad. Sci. USA* **104,** 3–4.

Benesch, R., and Benesch, R. E. (1967). The effect of organic phosphates from the human erythrocyte on the allosteric properties of hemoglobin. *Biochem. Biophys. Res. Commun.* **26,** 162–167.

Berg, H. C. (2000). Motile behavior of bacteria. *Phys. Today* **53,** 24–29.

Berg, H. C. (2004). *E. coli* in Motion. Springer, New York.

Bialek, W., and Setayeshgar, S. (2005). Physical limits to biochemical signaling. *Proc. Natl. Acad. Sci. USA* **102,** 10040–10045.

Bialek, W., and Setayeshgar, S. (2008). Cooperativity, sensitivity, and noise in biochemical signaling. *Phys. Rev. Lett.* **100,** 258101.

Bintu, L., Buchler, N. E., Garcia, H. G., Gerland, U., Hwa, T., Kondev, J., Kuhlman, T., and Phillips, R. (2005a). Transcriptional regulation by the numbers: Applications. *Curr. Opin. Genet. Dev.* **15,** 125–135.

Bintu, L., Buchler, N. E., Garcia, H. G., Gerland, U., Hwa, T., Kondev, J., and Phillips, R. (2005b). Transcriptional regulation by the numbers: Models. *Curr. Opin. Genet. Dev.* **15,** 116–124.

Briegel, A., Ortega, D. R., Tocheva, E. I., Wuichet, K., Li, Z., Chen, S., Muller, A., Iancu, C. V., Murphy, G. E., Dobro, M. J., Zhulin, I. B., and Jensen, G. J. (2009). Universal architecture of bacterial chemoreceptor arrays. *Proc. Natl. Acad. Sci. USA* **106,** 17181–17186.

Brush, S. (1967). History of the Lenz-Ising model. *Rev. Mod. Phys.* **39,** 883.

Buchler, N. E., Gerland, U., and Hwa, T. (2003). On schemes of combinatorial transcription logic. *Proc. Natl. Acad. Sci. USA* **100,** 5136–5141.

Callen, H. B. (1985). Thermodynamics and an Introduction to Thermostatistics. 2nd edn. Wiley, New York.

Cluzel, P., Surette, M., and Leibler, S. (2000). An ultrasensitive bacterial motor revealed by monitoring signaling proteins in single cells. *Science* **287,** 1652–1655.

Colquhoun, D., and Sivilotti, L. G. (2004). Function and structure in glycine receptors and some of their relatives. *Trends Neurosci.* **27,** 337–344.

Cui, Q., and Karplus, M. (2008). Allostery and cooperativity revisited. *Protein Sci.* **17,** 1295–1307.

Davidson, E. H. (2006). The Regulatory Genome: Gene Regulatory Networks in Development and Evolution. Academic, Burlington, MA; San Diego.

Dill, K. A., and Bromberg, S. (2003). Molecular Driving Forces: Statistical Thermodynamics in Chemistry and Biology. Garland Science, New York.

Dodd, I. B., Shearwin, K. E., and Egan, J. B. (2005). Revisited gene regulation in bacteriophage lambda. *Curr. Opin. Genet. Dev.* **15,** 145–152.

Dueber, J. E., Yeh, B. J., Chak, K., and Lim, W. A. (2003). Reprogramming control of an allosteric signaling switch through modular recombination. *Science* **301,** 1904–1908.

Fakhouri, W. D., Ay, A., Sayal, R., Dresch, J., Dayringer, E., and Arnosti, D. N. (2010). Deciphering a transcriptional regulatory code: Modeling short-range repression in the *Drosophila* embryo. *Mol. Syst. Biol.* **6,** 341.

Falke, J. J., Bass, R. B., Butler, S. L., Chervitz, S. A., and Danielson, M. A. (1997). The two-component signaling pathway of bacterial chemotaxis: A molecular view of signal transduction by receptors, kinases, and adaptation enzymes. *Annu. Rev. Cell Dev. Biol.* **13,** 457–512.

Garcia, H. G., Grayson, P., Han, L., Inamdar, M., Kondev, J., Nelson, P. C., Phillips, R., Widom, J., and Wiggins, P. A. (2007). Biological consequences of tightly bent DNA: The other life of a macromolecular celebrity. *Biopolymers* **85,** 115–130.

de Gennes, P. G. (1979). Scaling Concepts in Polymer Physics. Cornell University Press, Ithaca, N.Y.

Gilbert, S. F. (2010). Developmental Biology. 9th edn. Sinauer Associates, Sunderland, MA.

Giorgetti, L., Siggers, T., Tiana, G., Caprara, G., Notarbartolo, S., Corona, T., Pasparakis, M., Milani, P., Bulyk, M. L., and Natoli, G. (2010). Noncooperative interactions between transcription factors and clustered DNA binding sites enable graded transcriptional responses to environmental inputs. *Mol. Cell* **37,** 418–428.

Greenspan, R. J. (2006). An Introduction to Nervous Systems. Cold Spring Harbor Laboratory Press, Cold Spring Harbor, NY.

Gregor, T., Tank, D. W., Wieschaus, E. F., and Bialek, W. (2007). Probing the limits to positional information. *Cell* **130,** 153–164.

Grosberg, A. I. U., and Khokhlov, A. R. (1997). Giant Molecules: Here, and There, and Everywhere. Academic Press, San Diego.

Gunasekaran, K., Ma, B. Y., and Nussinov, R. (2004). Is allostery an intrinsic property of all dynamic proteins? *Proteins* **57,** 433–443.

Hill, T. L. (1985). Cooperativity Theory in Biochemistry: Steady-State and Equilibrium Systems. Springer-Verlag, New York.

Hille, B. (2001). Ion Channels of Excitable Membranes. 3rd edn. Sinauer, Sunderland, MA.

Jacob, F., Perrin, D., Sanchez, C., Monod, J., and Edelstein, S. (2005). The operon: A group of genes with expression coordinated by an operator. *C. R. Biol.* **328,** 514–520*C. R. Acad. Sci. Paris* 250 (1960) 1727–1729.

Keller, B. U., Hartshorne, R. P., Talvenheimo, J. A., Catterall, W. A., and Montal, M. (1986). Sodium channels in planar lipid bilayers. Channel gating kinetics of purified sodium channels modified by batrachotoxin. *J. Gen. Physiol.* **88,** 1–23.

Keymer, J. E., Endres, R. G., Skoge, M., Meir, Y., and Wingreen, N. S. (2006). Chemosensing in *Escherichia coli*: Two regimes of two-state receptors. *Proc. Natl. Acad. Sci. USA* **103,** 1786–1791.

Klotz, I. M. (1997). Ligand–Receptor Energetics: A Guide for the Perplexed. John Wiley & Sons, New York.

Koshland, D. E., Jr., Nemethy, G., and Filmer, D. (1966). Comparison of experimental binding data and theoretical models in proteins containing subunits. *Biochemistry* **5,** 365–385.

Kuhlman, T., Zhang, Z., Saier, M. H., Jr., and Hwa, T. (2007). Combinatorial transcriptional control of the lactose operon of *Escherichia coli. Proc. Natl. Acad. Sci. USA* **104,** 6043–6048.

Maier, B., and Radler, J. O. (1999). Conformation and self-diffusion of single DNA molecules confined to two dimensions. *Phys. Rev. Lett.* **82,** 1911–1914.

Mello, B. A., and Tu, Y. (2005). An allosteric model for heterogeneous receptor complexes: Understanding bacterial chemotaxis responses to multiple stimuli. *Proc. Natl. Acad. Sci. USA* **102,** 17354–17359.

Michel, D. (2010). How transcription factors can adjust the gene expression floodgates. *Prog. Biophys. Mol. Biol.* **102,** 16–37.

Mirny, L. A. (2010). Nucleosome-mediated cooperativity between transcription factors. *Proc. Natl. Acad. Sci. USA* **107**(52), 22534–22539.

Monod, J., Wyman, J., and Changeux, J. P. (1965). On the nature of allosteric transitions: A plausible model. *J. Mol. Biol.* **12,** 88–118.

Oehler, S., Amouyal, M., Kolkhof, P., von Wilcken-Bergmann, B., and Müller-Hill, B. (1994). Quality and position of the three *lac* operators of *E. coli* define efficiency of repression. *EMBO J.* **13,** 3348–3355.

Pauling, L. (1935). The oxygen equilibrium of hemoglobin and its structural interpretation. *Proc. Natl. Acad. Sci. USA* **21,** 186–191.

Perozo, E., Kloda, A., Cortes, D. M., and Martinac, B. (2002). Physical principles underlying the transduction of bilayer deformation forces during mechanosensitive channel gating. *Nat. Struct. Biol.* **9,** 696–703.

Phillips, R., Kondev, J., and Theriot, J. (2009a). Physical Biology of the Cell, Chap. 7. Garland Science, New York.

Phillips, R., Kondev, J., and Theriot, J. (2009b). Physical Biology of the Cell, Chap. 8. Garland Science, New York.

Phillips, R., Ursell, T., Wiggins, P., and Sens, P. (2009c). Emerging roles for lipids in shaping membrane-protein function. *Nature* **459,** 379–385.

Pinker, S. (2009). How the Mind Works. Norton Pbk. Ed. Norton, New York.

Plischke, M., and Bergersen, B. (2006). Equilibrium Statistical Physics. 3rd edn. World Scientific, Hackensack, NJ.

Ptashne, M., and Gann, A. (2002). Genes and Signals. Cold Spring Harbor Laboratory Press, New York.

Raveh-Sadka, T., Levo, M., and Segal, E. (2009). Incorporating nucleosomes into thermodynamic models of transcription regulation. *Genome Res.* **19,** 1480–1496.

Rippe, K. (2001). Making contacts on a nucleic acid polymer. *Trends Biochem. Sci.* **26,** 733–740.

Rosenfeld, N., Young, J. W., Alon, U., Swain, P. S., and Elowitz, M. B. (2005). Gene regulation at the single-cell level. *Science* **307,** 1962–1965.

Rudnick, J. A., and Gaspari, G. D. (2004). Elements of the Random Walk: An Introduction for Advanced Students and Researchers. Cambridge University Press, Cambridge; New York.

Schilpp, P. A. (1970). Autobiographical Notes in Albert Einstein: Philosopher–Scientist, the Library of Living Philosophers, Vol. 7, 3rd edn. Open Court, Chicago, IL.

Shimizu, T. S., Le Novere, N., Levin, M. D., Beavil, A. J., Sutton, B. J., and Bray, D. (2000). Molecular model of a lattice of signalling proteins involved in bacterial chemotaxis. *Nat. Cell Biol.* **2,** 792–796.

Sourjik, V., and Berg, H. C. (2002). Receptor sensitivity in bacterial chemotaxis. *Proc. Natl. Acad. Sci. USA* **99,** 123–127.

Straney, S. B., and Crothers, D. M. (1987). Kinetics of the stages of transcription initiation at the *Escherichia coli* lac UV5 promoter. *Biochemistry* **26,** 5063–5070.

Van Valen, D., Haataja, M., and Phillips, R. (2009). Biochemistry on a leash: The roles of tether length and geometry in signal integration proteins. *Biophys. J.* **96,** 1275–1292.

Wall, M. E., Hlavacek, W. S., and Savageau, M. A. (2004). Design of gene circuits: Lessons from bacteria. *Nat. Rev. Genet.* **5,** 34–42.

Wigner, E. P. (1960). The Unreasonable Effectiveness of Mathematics in the Natural Sciences. *Commun. Pure Appl. Math.* **13,** 1–14.

Zhong, W., Gallivan, J. P., Zhang, Y., Li, L., Lester, H. A., and Dougherty, D. A. (1998). From *ab initio* quantum mechanics to molecular neurobiology: A cation-pi binding site in the nicotinic receptor. *Proc. Natl. Acad. Sci. USA* **95,** 12088–12093.

CHAPTER THREE

Protein Stability in the Presence of Cosolutes

Luis Marcelo F. Holthauzen,* Matthew Auton,† Mikhail Sinev,‡ *and* Jörg Rösgen‡

Contents

* Sealy Center for Structural Biology and Molecular Biophysics, University of Texas Medical Branch, Galveston, Texas, USA
† Department of Medicine, Cardiovascular Research, Baylor College of Medicine, Houston, Texas, USA
‡ Department of Biochemistry and Molecular Biology, Penn State College of Medicine, Hershey, Pennsylvania, USA

Methods in Enzymology, Volume 492
ISSN 0076-6879, DOI: 10.1016/B978-0-12-381268-1.00015-X
© 2011 Elsevier Inc.
All rights reserved.

Abstract

Protein scientists have long used cosolutes to study protein stability. While denaturants, such as urea, have been employed for a long time, the attention became focused more recently on protein stabilizers, including osmolytes. Here, we provide practical experimental instructions for the use of both stabilizing and denaturing osmolytes with proteins, as well as data evaluation strategies. We focus on protein stability in the presence of cosolutes and their mixtures at constant and variable temperature.

1. Introduction

Proper folding of proteins is of crucial importance for the survival of living organisms. It was recognized a century ago that it is possible to denature proteins, rendering them inactive (Hopkins, 1930). A more general denaturing effect of urea on muscle tissue was noted even earlier (Limbourg, 1887). More recently, naturally occurring organic osmolytes have been identified that force proteins to fold, rather than unfold (Baskakov and Bolen, 1998). The importance of considering protein stabilizers in addition to denaturants has become especially clear based on the wealth of proteins that are either completely unfolded under "native" conditions or have at least disordered domains (Dyson and Wright, 2005; Uversky *et al.*, 2000). Often, activation of these proteins involves a folding transition as part of their function (Uversky *et al.*, 2000).

In the following, we discuss several example cases that illustrate how to investigate protein stability as a function of osmolyte concentration at both constant and variable temperature. We start with the most basic case and proceed to more involved ones. Within cases, the general experimental basics are covered first, followed by the specifics of the given example.

We then continue with simple data evaluation considerations and conclude each case with more advanced data analysis concepts. Four sections (Sections 2–5) are presented after this pattern. The last section of this chapter covers the mathematical background needed for deriving equations used throughout this chapter. In this section, we also present an alternative representation for protein stability that circumvents serious problems arising from the classical approach. The chapter focuses on monomeric, two-state protein folding. But at the end, we also provide additional information on the stability of oligomeric proteins and non-two-state folding.

1.1. Protein stability

Small proteins tend to exhibit all-or-none transitions between their native and unfolded forms in equilibrium (Lumry *et al.*, 1966). This holds particularly for equilibrium conditions, while in kinetic measurements there may be folding intermediates, or temporary off-pathway misfolding events (Bedard *et al.*, 2008). Such all-or-none transitions are characterized by a shift in population, where fully native proteins are in equilibrium with fully denatured proteins. The stability of a protein is defined as the difference in Gibbs free energy between the denatured (D) and native (N) state:

$$\Delta G^{\circ} = -RT \ln K = -RT \ln \frac{[\mathrm{D}]}{[\mathrm{N}]}, \tag{3.1}$$

where R is the gas constant (8.314 J/mol K) and T absolute temperature in K. When there is an excess of native state ([D]/[N] < 1), the stability ΔG° is positive and it is negative when the denatured state predominates ([D]/[N] > 1). The midpoint of the transition is characterized by zero stability because [D]/[N] $= 1$ in that case and $\ln 1 = 0$. Equation (3.1) can be inverted to obtain the equilibrium constant

$$K = \exp\left(-\frac{\Delta G^{\circ}}{RT}\right). \tag{3.2}$$

The fraction of denatured proteins is

$$f_{\mathrm{D}} = \frac{[\mathrm{D}]}{[\mathrm{N}] + [\mathrm{D}]} = \frac{K}{1+K}. \tag{3.3}$$

In general, we do not observe the fraction of unfolded protein directly in an experiment, but rather a signal

$$S = S_{\mathrm{N}} + (S_{\mathrm{D}} - S_{\mathrm{N}}) \frac{K}{1+K}, \tag{3.4}$$

where S_N and S_D are the signals of the native and denatured protein, respectively. This is a fundamental equation used throughout this chapter, and it will be discussed in detail how the equilibrium "constant" K varies with osmolyte concentration and temperature. Depending on the experimental method, also S_N and S_D could depend on these parameters. Examples are given below of how to handle such situations. A side remark: Eq. (3.4) assumes that the signal is proportional to the populations of native and denatured proteins, as is usually the case. Yet, for some experimental observables, the assumption may not hold, as explained (e.g., Eftink, 1994).

Figure 3.1A shows a typical transition curve of a protein that is denatured by increasing concentrations of urea. The data have been normalized according to the native and denatured signals, so that Eq. (3.4) is reduced to the fraction of denatured proteins (Eq. (3.3)). The curve can be subdivided into three regions: the native region between zero and about two molar urea, a denatured region beyond about 4.5 *M* urea, and the transition zone in between. Among the most easily identified features of the curve is the midpoint concentration c_m, where half of the proteins are native, and half denatured (in the given case, $c_m \approx 3.3\ M$). Note in Fig. 3.1B that the protein stability is zero at c_m. Another feature that is nearly as easily determined is the width of the transition. Shown in Fig. 3.1 is the estimated range from about 96% native to about 4% native. Such 96% to 4% range corresponds to a change in $\Delta G°$ of about 16 kJ/mol because according to Eq. (3.2), we have $\Delta G°(96\%N) - \Delta G°(4\%N) = -RT \ln(0.04/0.96) + RT \ln(0.96/0.04) = 16$ kJ/mol at room temperature (~300 K). Since the width of the transition in the current case spans about 2 *M* urea, there is a slope of $\Delta G°$ of about –8 kJ/mol *M* of urea. This slope of the protein stability with osmolyte concentration

$$\left(\frac{\partial \Delta G°}{\partial c}\right) = m \tag{3.5}$$

is called the *m*-value (Greene and Pace, 1974), a property that will be frequently encountered throughout this chapter.

2. Isothermal Folding/Unfolding of Protein in the Presence of Stabilizing/Denaturing Osmolyte

2.1. General outline of how to measure isothermal protein unfolding

Methods: Fluorescence and circular dichroism (CD) spectroscopies are the most common among the spectroscopic methods, and so we will focus primarily on these. Less common methods are available as well, such as pulse

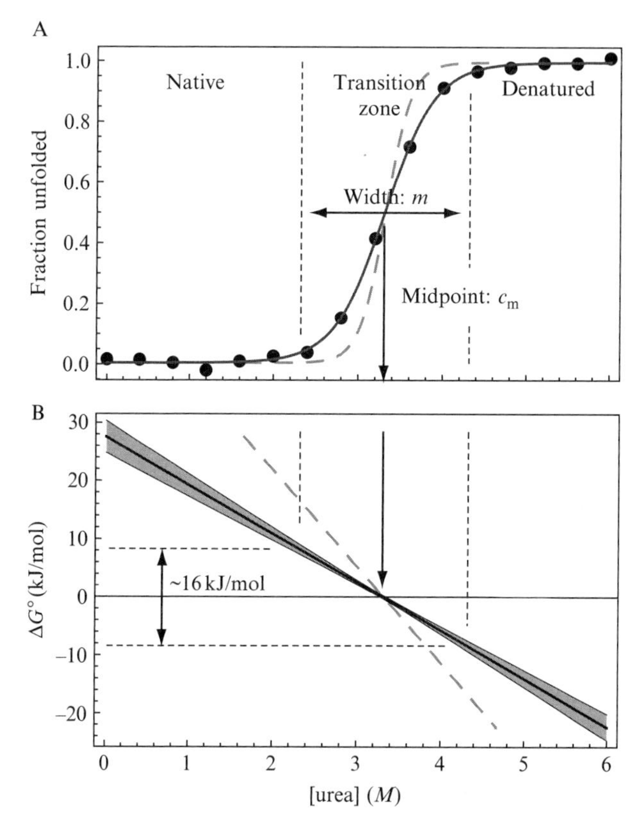

Figure 3.1 Typical solvent-induced protein unfolding: (A) experimental points of fraction unfolded versus urea concentration with curve fit (discussed in Section 2.3). Major features are highlighted. (B) Protein stability derived from the given data. Error limits (two standard deviations of fitting error) are given in gray. The transition region from 4% denatured to 4% native corresponds to a stability decrease of about 16 kJ/mol (see text). The dashed gray lines illustrate that the transition narrows to half width if the slope m of the protein stability versus urea is doubled.

proteolysis (Kim *et al.*, 2009; Park and Marqusee, 2004), isothermal titration calorimetry (Pfeil and Privalov, 1976), tyrosine exposure measurements by second derivative UV spectroscopy (Ragone *et al.*, 1984), viscosity (Neurath and Saum, 1939) and density measurements (Crouch and Kupke, 1977; Skerjanc *et al.*, 1970), NMR (Bai *et al.*, 1994), and dynamic light scattering (Ferreon and Bolen, 2004).

Whatever experimental method is used, it is important to have a good temperature control of the sample, because both kinetics and equilibria do in general depend on temperature.

Purity of the protein: The needed purity of the protein can greatly vary, but normally it is best to have the protein as pure as possible. Only in exceptional cases are impurities tolerable (Kim *et al.*, 2009; Park and Marqusee, 2004). In general, spectroscopic methods are used to monitor the folding state of the protein, and impurity may severely interfere with proper data collection. Often, a purity of 95% is sufficient, but the tolerable limit will always depend on the specific method. It should be kept in mind that the sample purity may also suffer from covalent damage to the protein under investigation (Manning *et al.*, 2010).

Concentration of the protein: The amount of protein needed depends on several factors. It is best to measure the protein signal with the method of choice in advance, so that a realistic estimate of the total sample consumption can be made ahead of time. Typically, solutions for the final measurements will be prepared by diluting protein stock into other solutions. Therefore, the stock should be about a factor of 10 more concentrated than the final desired concentration. It should be made sure that the protein is soluble at such concentrations. Otherwise, alternative approaches for solution preparation may be needed (see below under "Solution preparation" in this section for alternatives).

The sample consumption in CD spectroscopy is comparably predictable. As a rule of thumb, a good secondary structure signal around 220 nm is obtained at a protein concentration of 1 mg/mL in a 0.1 mm pathlength cell. The concentration can be scaled down if the pathlength is increased in proportion. It is desirable to have a protein concentration that yields a range of at least 10 mdeg signal change between the native and the denatured state. But the total signal should not exceed ±100 mdeg. Table 3.1 lists typical protein concentrations needed for CD. The actually needed values should be in the same order of magnitude, but may deviate depending on the particular protein.

In choosing a pathlength and protein concentration according to Table 3.1, it is important to consider that the absorption of the buffer, salts, and other additives in the solution may interfere with the measurement, particularly high osmolyte concentrations. In such cases, cuvettes with small pathlengths should be used. Generally, cuvettes up to 1 cm can be used with highly concentrated urea down to about 220 nm. Interference

Table 3.1 Approximate protein concentrations needed for CD spectroscopy using cuvettes of various pathlengths

Pathlength	0.01 mm	0.1 mm	1 mm	1 cm
Secondary structure (~220 nm)	10 mg/mL	1 mg/mL	0.1 mg/mL	0.01 mg/mL
Tertiary structure (~280 nm)	–	100 mg/mL	10 mg/mL	1 mg/mL

of the solution absorption with the signal can be recognized by the resulting excessively high photomultiplier high voltage. Consult the manual of your instrument in what voltage range you can expect usable data.

The tertiary structure of proteins is observed in the near UV region where aromatic amino acid residues absorb. The signal is often about 100-fold weaker in the near UV than it is in the far UV, and thus the concentration, the pathlength, or both have to be adjusted accordingly.

Protein fluorescence is normally based on tryptophan or tyrosine. The signal can be quite variable depending on the context of these residues in the protein. Expect to need concentrations roughly between 1 and 10 μM. In principle, cuvettes with small volumes and pathlengths can be used. However, it may be necessary to use cuvettes with 1-cm pathlength to accommodate a stirring bar. Such approach removes photobleached protein out of the light path. Whether or not photobleaching is an issue with the investigated protein should be tested in advance by observing the time dependence of the signal of fully native and fully denatured protein.

Osmolyte stock: The osmolyte stock should have as little impurities as possible, especially those that interfere with the measurements or the protein behavior. It is a good practice to first dissolve the osmolyte in pure water, treat this solution with granulated activated carbon, and filter the solution with a 0.2 μM filter. Note that some osmolyte solutions can be quite viscous at elevated concentration. In those cases, it may be necessary to choose a filter with larger pore size or, if feasible, use a lower osmolyte stock concentration. A repetition of the cleaning procedure may be necessary when larger amounts of impurities are present.

The final buffered osmolyte solution is prepared by mixing a concentrated buffer stock (including salts, etc.) with the osmolyte stock. For example, mix one part of a 10× buffer stock with nine parts of the osmolyte stock.

If the pH of the buffer changes slightly upon mixing, it can be directly adjusted. However, if the pH change is large, either there may still be significant amounts of impurities present, or the solution pH may be too close to a protonation pK_a of the osmolyte (see Table 3.2 for pK_a values of osmolytes). Osmolytes alter their behavior when they are in protonation states that normally do not occur in their physiological context (Singh *et al.*, 2005). Such altered behavior may lead to undesired results due to a change in the charge of the osmolyte.

Normally, the pH-meter reading is assumed to already reflect the proper pH, that is, the *activity* of protons (Garcia-Mira and Sanchez-Ruiz, 2001; Nozaki and Tanford, 1967). In the unlikely case that the actual *concentration* of protons is desired, complicated corrections may be necessary (Garcia-Mira and Sanchez-Ruiz, 2001).

The concentration of the buffered osmolyte solution can be determined by the refractive index of the solution relative to plain 1× buffer (ΔN).

Table 3.2 Osmolyte pK_a values

Osmolyte	pK_a	Source
Glycine betaine	1.83	Lide (2004)
Sarcosine	2.21	Lide (2004)
	10.1	Lide (2004)
Proline	1.95	Lide (2004)
	10.64	Lide (2004)
Urea	−0.1	Average of following two:
	(0.1)	Lide (2004)
	(−0.3)	Wen and Brooker (1993)
TMAO	4.65	Average of following three:
	(4.56)	Lin and Timasheff (1994)
	(4.75)	Qu and Bolen (2003)
	(4.65)	Lide (2004)

Data for the trimethylamine N-oxide (TMAO) concentration as a function of ΔN was given by Wang and Bolen (1997).

$$[\text{TMAO}] = -0.0038 + 103.3151(\Delta N) - 259.43(\Delta N)^2. \tag{3.6}$$

Pace (1986) listed the values for urea

$$[\text{Urea}] = 117.66(\Delta N) + 29.753(\Delta N)^2 + 185.56(\Delta N)^3, \tag{3.7}$$

and guanidinium chloride

$$[\text{GdmCl}] = 57.147(\Delta N) + 38.68(\Delta N)^2 - 91.60(\Delta N)^3. \tag{3.8}$$

We shall also add the equation for sarcosine

$$[\text{Sarcosine}] = 69.333(\Delta N) + 83.378(\Delta N)^2. \tag{3.9}$$

When either refractive index data or a refractometer is unavailable, it is possible to prepare the stock from thoroughly dried osmolyte by weight and calculate the concentration from the weighed masses of osmolyte and water, and the mass of added buffer stock. This leads to the molality (moles per kg of water) to be known, rather than the molarity (moles per liter of solution). If the density of the solution ρ is known, the molarities c_i of each component can be directly calculated from the molalities m_i and the molar masses M_i

$$c_i = \frac{m_i \rho}{1 + \sum_k m_k M_k / 1000}. \quad (3.10)$$

The sum is over all components of the solution, except for the water. If the density is not known, the molarity of an osmolyte O can be calculated to a good approximation by (Rösgen, 2007; Rösgen *et al.*, 2004)

$$c_\mathrm{O} = \frac{m_\mathrm{O} \rho_\mathrm{W}}{1 + \sum_k m_k \rho_\mathrm{W} / c_{\max,k}}, \quad (3.11)$$

where ρ_W is the density of pure water (0.99705 kg/L at 25 °C), and $c_{\max,k} = \rho_k / M_k$ (ρ_k is the density of the neat substance *k*). For example, crystalline urea has the density $\rho_\mathrm{Urea} = 1323.0$ g/L, the molecular weight $M_\mathrm{Urea} = 60.06$ g/mol, and thus $c_{\max,\mathrm{Urea}} = 22.03$ mol/L. Values for ρ_k and M_k can be found in Lide (2004) or other editions of the same book, and values for $c_{\max,k}$ are compiled in various places (Harries and Rösgen, 2008; Rösgen, 2007; Rösgen *et al.*, 2004). Using Eq. (3.11), it is important to remember that it applies to compounds that have a small, if any, concentration dependence of their partial molar volume. This is normally true for osmolytes, but not for other molecules like salts, which can exhibit considerable electrostriction effects on the solution density.

Some osmolytes can be stored in aqueous solution at room temperature for prolonged times. However, some are unstable in solution. This is in particular true for urea, which decomposes when dissolved in water (Hagel *et al.*, 1971; Lin *et al.*, 2004). Since proteins can be covalently modified by the decomposition product cyanate (Gerding *et al.*, 1971; Stark *et al.*, 1960), it is best to use only freshly prepared and cleaned urea solutions and limit use of this solution to 1 day. It also may be acceptable to use urea stock that was deep frozen directly after preparation and cleaning. TMAO also suffers from storage. When exposed to light, even the solid osmolyte decomposes to trimethylamine and oxygen. This reaction can be detected by the resulting fishy smell.

Sample preparation (manual approaches): There are two major ways of preparing the final solutions that are ready for the measurements: either (1) the protein is added after all solutions are prepared or (2) it is added to both the buffer and the osmolyte stock before the osmolyte concentration series is prepared.

In the classical strategy (1), which is discussed in this paragraph, a series of tubes (typically 1–2 mL) is prepared with increasing concentrations of osmolyte. The solution in each tube is prepared by mixing the desired amounts of 1× buffer and buffered osmolyte solution. Finally, equal amounts of concentrated protein stock in 1× buffer are added to each tube. These final solutions are equilibrated overnight at controlled temperature before the measurement is done. For a reversibility check, an identical set of solutions is prepared, but the protein added in the final step should be in buffered osmolyte, rather than in the plain 1× buffer. Both sets of experiments (the titration and its reversibility check) should yield the same results when the reaction is well equilibrated (and reversible) within the incubation time. Equilibration overnight is often sufficient, but in extreme cases several weeks may be needed (Rosengarth *et al.*, 1999). Figure 3.2A illustrates the reversibility check. The gray points from the unfolding experiment approach the white points from the refolding experiment, and they fall on the same line when the system is in equilibrium (black points). If the available amount of protein is limited, the reversibility series can be replaced by a single point that corresponds to the midpoint of the equilibrium. Naturally, this experiment can only be done after the initial unfolding curve has been recorded. Recording the progress as shown in Fig. 3.2 can also be used to monitor both kinetics and equilibrium parameters simultaneously (Erilov *et al.*, 2007).

As a side remark, the folding and unfolding kinetics are normally slowest at the midpoint of the equilibrium curve when denaturants are used. Similarly, use of stabilizers in forcing unstable proteins into the native state should lead to a kinetics that is slowest at the midpoint (Chang and Oas, 2010; Russo *et al.*, 2003).

Strategy (2) works similarly as strategy (1). But here, two solutions at their final protein concentration are prepared, one without osmolyte, and one with the maximal osmolyte concentration. These solutions are mixed in different ratios to obtain the various samples at the final osmolyte

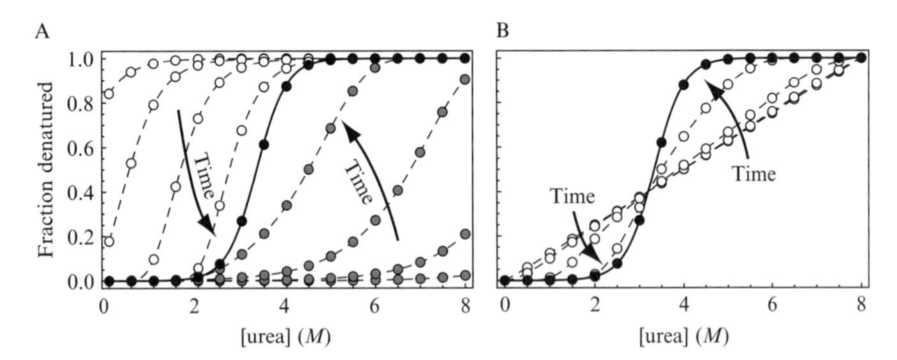

Figure 3.2 Approach to equilibrium: different pipetting strategies involve adding the protein after (A) or before (B) mixing the 1× buffer and osmolyte stock solutions. Black circles indicate the equilibrium curve. The gray/white circles in panel A track the signal as a function of time in the case when native/denatured protein was added. The time is spaced logarithmically. (Panel B) When both buffer and 8 *M* urea contain already protein, the initial fraction of denatured protein equals the fraction of urea solution mixed with buffer (the diagonal line). With time, the signal at high urea relaxes up to the denatured state, and at low urea down to the native one.

concentrations. A reversibility series is not possible in this setup, and it should be replaced with a kinetic folding or unfolding trace in the middle of the transition, because this is where the equilibration is slowest. The kinetics should be well finished within the equilibration time that was used in the titration experiment. Figure 3.2B shows how the signal approaches its equilibrium values within strategy (2). This strategy is especially advisable when the protein is not much more soluble than the desired final concentration. In that case, two separate aliquots of the protein can be directly dialyzed against the zero molar solution and the solution at the highest osmolyte concentration to obtain the two concentration extremes. Such procedure avoids the necessity of using protein concentrations that are higher than the final experimental concentration. This is in contrast to strategy (1), where highly concentrated protein is diluted to its final concentration.

About 10 concentration points is the absolute minimum for the isothermal measurement of solvent-dependent protein stability. Protein should be available in sufficient amounts for number of planned sample points including reversibility points.

Sample preparation (automatic titration): Stites *et al.* (1995) suggested an alternative to the solution preparation strategies we just described, minimizing sample consumption and spent effort. In this approach, one cuvette filling of protein solution serves as the basis for the entire experimental series. This solution is titrated with an identical one that contains osmolyte in addition, and excess solution is automatically removed. The spectroscopic signal is recorded after each addition of osmolyte solution.

While this strategy requires special equipment, if pursued in the originally proposed automated setup (Stites *et al.*, 1995), the titration can also be performed by hand. After each addition of osmolyte there should be sufficient waiting time before the final signal is recorded, because the protein needs some time to reach equilibrium. How long is sufficient? Equilibration is normally slow within the region around the midpoint of the equilibrium folding transition and fast outside that region. Rather than choosing a constant equilibration time for all osmolyte concentrations (Stites *et al.*, 1995), it is therefore advisable to monitor the kinetic signal at each concentration to determine the minimal needed waiting time (Harms *et al.*, 2008). The next titration step can be performed as soon as it is clear that equilibrium has been reached. Such approach has the additional benefit that both kinetic and equilibrium data are collected simultaneously.

This method should work well as long as the relaxation time for the folding equilibrium does not exceed a couple of dozen minutes. Slower folding would make this approach impractical.

Forced folding versus unfolding: This section was mainly geared toward the more common approach of unfolding a protein by addition of a denaturing osmolyte. However, it is also possible to do the reverse, that is, force an unfolded protein into the native state by addition of a stabilizing osmolyte (Baskakov and Bolen, 1998). There is a series of osmolytes with different capacities toward force-folding proteins (Auton and Bolen, 2005; Auton *et al.*, 2007b). One problem that may occur with protein stabilizers is that they often also tend to decrease the solubility of proteins, thus precipitating them (Bolen, 2004). Such problems may be overcome by mixing the precipitating stabilizer, such as TMAO, with a solubilizer, such as proline. A mixture of four parts of TMAO with one part of proline has been successfully used (Kumar *et al.*, 2001).

2.2. Case 1: Adenylate kinase unfolding by urea

The unfolding of *Escherichia coli* adenylate kinase (Müller *et al.*, 1996; Yan and Tsai, 1999) with urea serves as a practical example of an isothermal titration of a protein with a denaturing osmolyte. We choose here a procedure mimicking the automatic titration discussed above.

The protein is dialyzed against buffer (10 mM PIPES pH 7.7, 0.1 m*M* EDTA, and 150 m*M* NaCl) at a concentration of > 100 μ*M* (> 2.4 mg/mL, $M_r \approx 24$ kDa). The urea stock is prepared by dissolution into water up to the solubility limit. This solution is treated with activated carbon, incubated with mixed bed ion exchange resin, and finally filtrated through 0.2 μ*M* pore size filters.

Concentrated buffer was mixed with the urea stock to the correct buffer concentration, and the protein was diluted to a final concentration of 2 μ*M* into both this buffered urea stock and the plain buffer. In all, 2.5 mL of each solution was used for the measurements. A difference of 0.0760 in refractive index between urea-free and urea containing protein solution was measured with an Abbe refractometer, corresponding to 9.2 *M* according to Eq. (3.7).

The degree of folding is monitored by CD spectroscopy in a Jasco J-720 Polarimeter. A 1-cm cuvette is best for the experimental setup of exchanging small volumes of the solution in the cuvette. The photomultiplier high voltage at 223 nm (within the region of α-helical signal) is appropriate in our system: it is 345 V for protein in buffer without urea, and 378 V with urea. These values should be determined before setting up the experiment to make sure the signal will be usable. In all, 2.2 mL of protein in buffer is transferred into the cuvette containing a stirring bar. The signal is recorded as a kinetic scan (lowest trace in Fig. 3.3A). Afterward, 120 μL is removed from the cuvette and replaced with 120 μL of buffered protein in 9.2 *M* urea. The stirring of the solution should be sufficiently strong to allow for a quick and complete mixing. Again, a kinetic scan is recorded until the signal becomes stable. The sequence of volume replacement and data collection is repeated until past the transition, well into the region of the denatured state baseline. The individual data sets are shown in Fig. 3.3A.

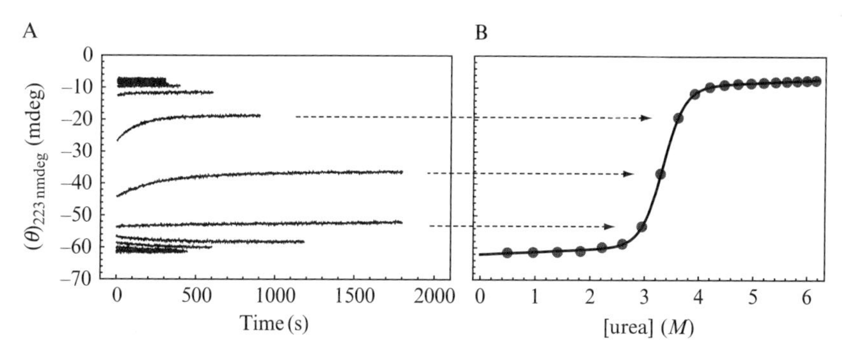

Figure 3.3 Titration of adenylate kinase with urea: (A) individual kinetic traces obtained after each exchange step of 120 μL of cuvette content against the same volume of protein in the buffered urea stock. (B) Equilibrium curve resulting from the kinetic end points of the traces shown in panel A. Dashed arrows illustrate this process for three points in the transition region.

It is important to do the pipetting steps accurately. It may be desirable to monitor the pipetting by weight if the density of the solutions is known. Given that the pipetting was performed correctly, the urea concentration for the individual measurements is given by

$$c_{\mathrm{osmolyte}} = c_{\mathrm{stock}} - c_{\mathrm{stock}}\left(1 - \frac{v_{\mathrm{exchange}}}{V_{\mathrm{cell}}}\right)^{n}, \tag{3.12}$$

where c_{stock} is the concentration of the urea stock solution, v_{exchange} is the exchanged volume (120 μL in our case), and V_{cell} is the volume of the solution in the cuvette (2.2 mL in our case). The point before adding the first urea solution corresponds to $n = 0$, and each addition increases n by 1. Figure 3.3B uses Eq. (3.12) to plot the final signals from Fig. 3.3A as a function of urea concentration.

What can be seen directly: Figure 3.3 shows several features of adenylate kinase without resorting to complicated data analysis. First, the kinetics can clearly be seen to be slow (Fig. 3.3A, center), but not too slow to perform this type of experiment with titration by solution exchange. Second, the equilibrium curve in Fig. 3.3B instantly reveals that the midpoint of the unfolding reaction c_{m} is a little below 3.5 *M* urea. A quick determination is also possible for the slope of the stability with urea concentration, the *m*-value (Eq. (3.5)). To determine this slope, we need to know the width of the transition and the amount of stability change within this concentration range. The width (between about 96% and 4% native) is around 1.3 *M* urea. Just before Eq. (3.5), it was explained that the region between 4% denatured and 4% native corresponds to a decrease in stability by 16 kJ/mol. So, the slope *m* of the stability as a function of urea is about $-16/1.3$ kJ/mol $M \approx -12$ kJ/mol M.

These initial estimates are valuable not only for a quick evaluation of the protein's properties but also for streamlining the following fitting procedure.

2.3. Data treatment

2.3.1. Curve fitting: Basics

Both free and commercial software is available for the data analysis. Common examples for free software are R (http://www.r-project.org/) and *gnuplot* (http://www.gnuplot.info/). Popular commercial software includes *Origin* (http://www.originlab.com/), *SigmaPlot* (http://www.sigmaplot.com/), *Matlab* (http://www.mathworks.com/), and *Mathematica* (http://www.wolfram.com/). Here, we use the latter one.

The data from Section 2.2 can be directly fit to a combination of Eqs. (3.4), (3.2), and (3.15), resulting in the equation (Santoro and Bolen, 1988)

$$S=\left[S_{\mathrm{N,m}}+\mathrm{d}S_{\mathrm{N,m}}(c-c_{\mathrm{m}})\right]+\left[S_{\mathrm{D,m}}-S_{\mathrm{N,m}}+\left(\mathrm{d}S_{\mathrm{D,m}}-\mathrm{d}S_{\mathrm{N,m}}\right)(c-c_{\mathrm{m}})\right]$$

$$\frac{\exp\left(-\frac{m(c-c_{\mathrm{m}})}{RT}\right)}{1+\exp\left(-\frac{m(c-c_{\mathrm{m}})}{RT}\right)}, \tag{3.13}$$

where $S_{\mathrm{N,m}}$ and $S_{\mathrm{D,m}}$ are the signals of native and denatured state at the midpoint concentration c_{m}, $\mathrm{d}S_{\mathrm{N,m}}$ and $\mathrm{d}S_{\mathrm{D,m}}$ are the slopes of the signals with concentration, and m is the m-value. R and T were defined in Eq. (3.1). Fig. 3.3B shows the fit as a continuous line along with the data. The resulting fit parameters are given in Table 3.3. In the initial fit, we noticed that the slopes of the signals of the native and denatured states are equal (cf. also Fig. 3.3B), so we decided to use a single-slope $\mathrm{d}S$ in place of the two slopes of both states.

The table shows the best estimate for each parameter, its standard error (SE), and the 95% confidence interval for each parameter. This piece of

Table 3.3 Fit result for Case 1

Parameter	Unit	Estimate	SE	Confidence interval
$S_{\mathrm{N,m}}$	mdeg	−62.09	0.14	(−62.40, −61.89)
$S_{\mathrm{D,m}}$	mdeg	−12.37	0.45	(−13.32, −11.42)
m	kJ/mol M	−12.6	0.2	(−13.0, −12.2)
c_{m}	M	3.328	0.003	(3.323, 3.334)
$\mathrm{d}S_{\mathrm{m}}$	mdeg/M	0.81	0.08	(0.63, 0.99)

information is not often given and serves as a control whether the error is distributed symmetrically to both sides of the parameter estimate, and then the 95% confidence interval covers a range of about $\pm$2SEs around the parameter estimate. In the cases reported in this chapter, the errors are symmetric. However, when nonsymmetrical errors are found, it is best practice to find an equation that provides parameters with more a symmetrical error distribution.

2.3.2. Curve fitting: Looking beyond the obvious

The general expression for describing protein denaturation is given by Eq. (3.4). A central aspect of this equation is the equilibrium constant of unfolding K, which is determined by the stability ΔG° of the protein (Eq. (3.2)). We will therefore focus now on the optimal representation of ΔG°.

Three alternative representations of the protein stability: The protein stability has been found to linearly depend on the concentration of many additives, with urea being the best characterized (Courtenay *et al.*, 2000; Ferreon and Bolen, 2004; Greene and Pace, 1974; Holthauzen and Bolen, 2007; Makhatadze, 1999; Mello and Barrick, 2003; Santoro and Bolen, 1988; Timasheff and Xie, 2003). The slope m of the stability is called the "m-value." It appears that, in general, naturally occurring osmolytes share this linearity of ΔG°—they have a constant m. There are various ways to express this so-called linear extrapolation model (LEM) in terms of equations. Usually, researchers are interested in knowing the stability of a protein in the absence of urea $\Delta G^\circ(0\ M)$. Then the most practical stability equation is:

$$\Delta G^\circ = \Delta G^\circ(0M) + mc \tag{3.14}$$

While this equation is correct, it may not always be the best choice to use because of (1) the ease of estimating the parameters prior to the curve fitting and (2) the errors in the parameters. These limitations are not too serious in the given case of just a single independent variable, a single equilibrium, and a linear dependence of the stability on the independent variable. However, severe issues can arise in more complicated cases. But because discussing these issues is more easily done in the current case, we present them here. For this purpose, we first show two alternatives to Eq. (3.14).

The following equation is very similar to Eq. (3.14), but it focuses on the midpoint concentration of the denaturation transition $c_m = -\Delta G^\circ(0\ M)/m$:

$$\Delta G^\circ = m(c - c_m) \tag{3.15}$$

Both equations (3.14) and (3.15) have the following general form

$$\Delta G^\circ = \Delta G^\circ(c_{ref}) + m(c - c_{ref}), \tag{3.16}$$

where $c_{ref} = 0$ in Eq. (3.14) and $c_{ref} = c_m$ in Eq. (3.15). Note that at the midpoint of the transition c_m, we have $\Delta G°(c_m) = 0$.

We already saw at the end of Section 2.2 that it is very easy and straightforward to estimate both parameters, c_m and the m-value, in Eq. (3.15). Under circumstances as simple as here, it is just one additional step to derive $\Delta G°(0\ M) = -c_m m$ for Eq. (3.14) from these two parameters. However, under more complicated circumstances, this may not always be as easy, and then equations of the type of Eq. (3.15) may be preferred over Eqs. (3.14) and (3.16).

After discussing the first point raised about Eq. (3.14), we turn now to the second point. How does use of the three equations ((3.14), (3.15), (3.16)) relate to the errors in the parameters? To answer this question, we need to understand their strengths and weaknesses under realistic experimental circumstances. The point is that Eq. (3.15) provides the most certain parameter evaluations, and that use of the other equations leads to larger error estimates as discussed in the following paragraphs. These larger estimates can lead to wrong and artificially large propagated errors, even if in themselves the error estimates were originally correct. This becomes an issue, when error estimates are needed at a different concentration than 0 M (if Eq. (3.14) is used), c_m (if Eq. (3.15) is used), or in general c_{ref} (if Eq. (3.16) is used). The easiest way to get around the issues described in the rest of this Section (2.3.2) is to repeatedly fit Eq. (3.16) to the data, each time with c_{ref} set to one of the concentrations for which error estimates are needed.

The uncertainty in the parameter estimates is smallest when using Eq. (3.15): We generated hundreds of sets of example data that are given in Fig. 3.4A. The noise in the data is represented by its standard deviation σ_{noise} and ranges from 0% to about 10% of the total signal amplitude, well covering typical conditions. Panels B–D show fitting results as a function of the noise level in the data, and how the choice of the fitting equation influences the uncertainty of the resulting parameters. When Eq. (3.14) is used for the fitting, both parameters $\Delta G°(0\ M)$ and m get rapidly more uncertain as σ_{noise} increases (Fig. 3.4B and C). The estimates of these parameters quickly deviate by 100% and more from their true values (26.4 kJ/mol and −8 kJ/mol M) as the noise in the data approaches 10%. Also, the fitting errors (indicated by the error bars) for each parameter estimate become large. We are switching now to the results from Eq. (3.15), which are given in Fig. 3.4C and D. One of the parameters, the m-value, occurs in both equations, and there is no difference here. But notice in contrast how tight the estimates of the midpoint concentration c_m cluster around the true value of 3.3 M, deviating by hardly 10%.

Coupling between parameters: The finding of the previous paragraph is consistent with intuition. In Fig. 3.4A, it is straightforward to visually determine the midpoint of the transition (c_m) even when the data are subject to a large amount of noise. However, the width of the transition (m) can

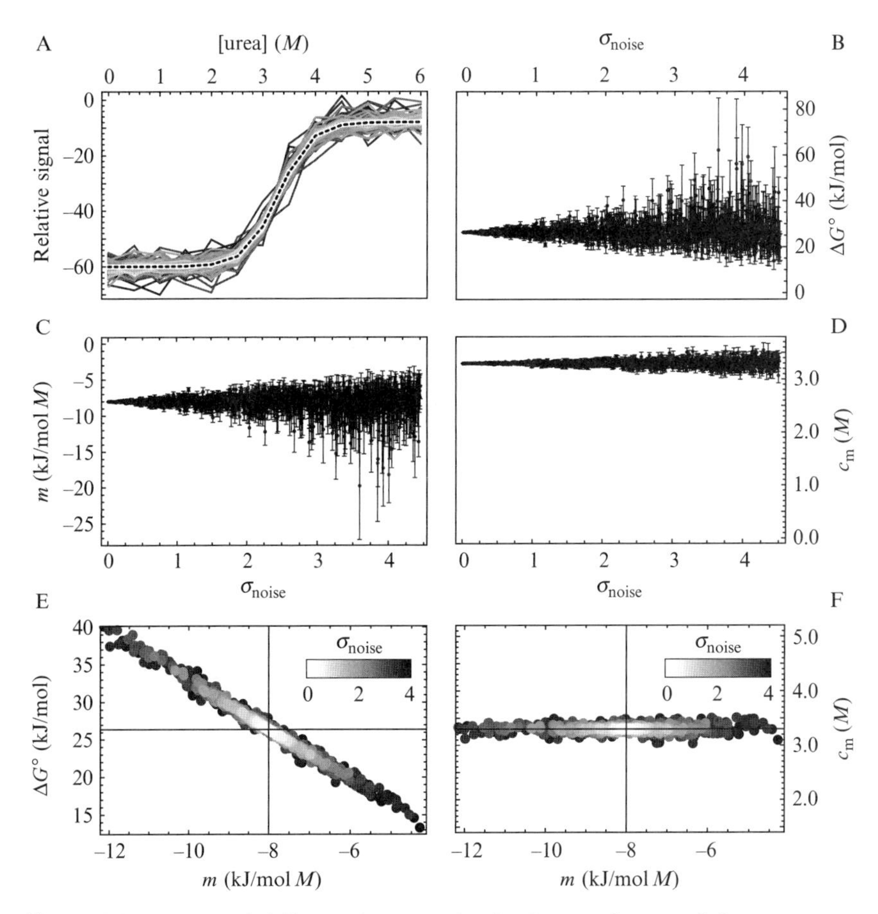

Figure 3.4 Impact of different degrees of noise in raw data on fitting parameter uncertainty: (A) Unfolding data as a function of urea concentration at variable noise level (standard deviation σ_{noise} of random noise increases from 0% [dashed line] to 10% of the total signal amplitude). (B and C) Fitting results for $\Delta G°(0\ M)$ and m using Eq. (3.14) in Eq. (3.4). (C and D) Fitting results for m and c_m using Eq. (3.15) in Eq. (3.4). (E) High degree of coupling between fitting parameters m and $\Delta G°(0\ M)$ (using Eq. (3.14) in Eq. (3.4)). (F) Low degree of coupling between fitting parameters m and c_m (using Eq. (3.15) in Eq. (3.4)).

easily get distorted by noise and may thus lead to estimates of m that are rather different from its true value. These two parameters, c_m and m, determine the position and shape of the transition. The stability $\Delta G°$ $(0\ M)$ is only a derived quantity equaling $-c_m\ m$. So, if c_m is certain, and m uncertain, then any increase in m must lead to a decrease in $\Delta G°(0\ M)$, and vice versa. It is therefore not surprising to find the errors in $\Delta G°(0\ M)$ coupled to the errors in m, as seen in Fig. 3.4B and C. Indeed, plotting these two fitting parameters against each other (Fig. 3.4E) reveals the strong degree of coupling: negative deviations in m lead to positive ones in the stability $\Delta G^\circ(0\ M)$, and vice versa. Figure 3.4F shows that, in contrast, the midpoint c_m and width m of the transition are independent fitting parameters.

Propagation of errors: The strong coupling between $\Delta G^\circ(0\ M)$ and m can lead to a gross overestimation of errors when they are calculated for concentrations different from the reference point 0 M. We will not get further into the details of error propagation and just give the equations for calculating the propagated error for Eq. (3.14)

$$\mathrm{Var}[\Delta G^\circ] = 1^2\mathrm{Var}[\Delta G^\circ(0\,M)] + c^2\mathrm{Var}[m] + 2c\mathrm{Cov}[\Delta G^\circ(0\,M), m]. \quad (3.17)$$

The variances (squared SEs) for the parameters were already listed in Table 3.3. But the covariance $\mathrm{Cov}[\Delta G^\circ(0\ M), m]$ equals the product of the SEs in $\Delta G^\circ(0\ M)$ and m, times their correlation coefficient. This coefficient can be read from the covariance matrix that the fitting software should provide as part of the fitting result. In our case, there is a strong (−99.9%) anti-correlation. Without taking into account this significant covariance between the two parameters, the errors can be grossly misestimated for conditions far away from the reference condition (0 M). For example, the last term in Eq. (3.17) is nearly completely compensating the other two terms at c_m in the given case:

$$\begin{aligned}\mathrm{Var}[\Delta G^\circ] &= (661\,\mathrm{J/mol})^2 + (3.328\,M)^2(198\,\mathrm{J/mol\ M})^2 \\ &\quad +2(3.328\,M)(-0.999)\times 661\,\mathrm{J/mol}\times 198\,\mathrm{J/mol\,M} \\ &= 436,921(\mathrm{J/mol})^2 + 434,207(\mathrm{J/mol})^2 - 870,253(\mathrm{J/mol})^2 \\ &= 875(\mathrm{J/mol})^2 \approx (30\,\mathrm{J/mol})^2.\end{aligned} \quad (3.18)$$

Based on the discussions in this section, we see that, although Eqs. (3.14)–(3.16) are fully equivalent, the choice of the equation can make a difference in the ease of their application.

3. Isothermal Protein (Un)Folding in the Presence of Osmolyte Mixtures

3.1. General outline how to use osmolyte mixtures

In nature, osmolytes often occur in mixtures. One of the best studied examples is the osmolyte cocktail in the human kidney (Garcia-Perez and Burg, 1991). Here, about half a dozen osmolytes counter the deleterious effects of high urea and salt concentrations. Urea is known to destabilize proteins; in the example case below, we investigate how the osmolyte sarcosine counteracts this effect. The highest impact of urea is expected with proteins that have an intrinsic low stability, that is, those that are already at zero molar urea in the transition region from native to denatured. Then, the population of the denatured state will significantly change even at low urea concentrations. This is in contrast to the example discussed in Section 2, where the transition region is at elevated urea concentration (Fig. 3.1). Similarly, a partially unstable protein will also show a response to stabilizers even at low osmolyte concentration. As a consequence, the intrinsic stability of a protein determines what kind of experiment can reasonably be done, viz. folding or unfolding or both. A fully folded protein is best studied with denaturants, a fully denatured one with stabilizers, and a partially unfolded one can be studied with both. Additional strategies are discussed in Section 5.

One important question when studying folding in the presence of mixed osmolytes is whether the effects of the osmolytes are additive, or whether synergies/interferences between the additives occur. Below, we describe a way to test whether osmolytes act additively. We work at the melting temperature T_m as a strategy to poise the unfolding equilibrium to its midpoint. This method works particularly well for proteins that have low values for T_m, that is, those which are of low stability. If the T_m is too high, this method may not work because the protein has a higher risk of unfolding irreversibly.

Methods: In general, the same principles apply for the methods, purity, concentrations, etc., as discussed in Section 2.1. However, there are more limitations with regard to the methods that can be applied—in particular, when a protein is investigated that is already partially unfolded in the absence of denaturant, as in our current example below. Such circumstances complicate the identification of the baseline of the native state because this state is only populated in a subset of experiments (when sufficient stabilizer is added). For this reason, it is advisable to use an experimental method in which the baselines of the native and denatured states are very stable and do not depend much on osmolyte concentration. Fluorescence is a method in which baselines are often osmolyte or temperature dependent, and it may take considerable extra effort to reliably define them (Wu and Bolen, 2006). To avoid such complications, we use CD. Here, the native state should yield a comparably constant signal. The denatured state signal may vary with osmolyte addition (Holthauzen *et al.*, 2010), but this is not an issue in our case. This is because all denaturation curves have a sufficiently large region in which the denatured state is populated, and all folding curves have such a short region of denatured state baseline that its slope does not affect the results too much.

Solution preparation: Again, the same principles apply as in Section 2.1, with the difference that the buffered osmolyte stock contains an additional osmolyte at a fixed concentration, and the plain buffer solution also contains the same amount of the additional "fixed" osmolyte. Full titrations are then repeated at various concentrations of this osmolyte. In cases where possible, it is best to repeat the series with exchanged roles between the "fixed" and "variable" osmolytes. However, in cases where the protein is either mostly native (or mostly denatured) in the absence of additive, only one set of experiments makes sense: titrations with denaturant at variable stabilizer concentration when the protein is relatively stable, or vice versa when the protein is mostly denatured to begin with.

3.2. Case 2: Nank isothermal folding/unfolding in urea–sarcosine mixtures

As an example of how to determine the additivity between a pair of stabilizing/denaturing osmolytes, we discuss below a protein called "Nank4-7★," a marginally stable (T_m = 34 °C) 15 kDa truncated version of the ankyrin domain of the *Drosophila* notch receptor (Mello and Barrick, 2004).

Method: In short, the basic idea is to do a series of denaturant unfolding of the protein in the presence of increasing concentrations of the protecting/stabilizing osmolyte in the background, and a series of forced folding of the protein in the presence of increasing concentrations of the denaturant (Holthauzen and Bolen, 2007). In that way, we will have several concentrations of the *variable* osmolyte for each concentration of the *fixed* osmolyte.

Sample preparation: The osmolyte solutions were treated with activated charcoal and filtered through a 0.2 μ*M* filter, slight deviations in pH adjusted, and their concentrations measured refractometrically (see Eq. (3.9)). The protein stock was centrifuged to remove aggregates, and afterwards the protein concentration determined by the Edelhoch method (Pace *et al.*, 1995).

For the concentration series, two stocks must be prepared: (1) the desired concentration of the *fixed* osmolyte in buffer (the *fixed* osmolyte stock) and (2) the same concentration of *fixed* osmolyte plus the maximum concentration of the *variable* osmolyte in buffer (the *variable* osmolyte stock).

The method that yields the least error in terms of variation in protein concentration when preparing samples is having protein added to both *fixed* osmolyte stock and the *variable* osmolyte stock before preparing the concentration series (strategy 2 from Section 2.1: solution preparation). Both stock solutions are then mixed to several ratios to yield the denaturing or force-folding curve. The first point should consist of pure *fixed* osmolyte stock and the last point of pure *variable* osmolyte stock.

We measured data for about 14 points for the variable osmolyte in the presence of four different concentrations of the fixed osmolyte (in a total of about 112 experimental points). Ideally, the experiment should be repeated independently at least once with fresh solutions so that the results can be validated.

Optimization of experimental conditions: To get the maximal response of the protein equilibrium to even lower concentrations of both the denaturant urea and the stabilizer sarcosine, we tune the equilibrium to be at 50% native and 50% denatured by working at the melting temperature T_m.

In order to further optimize the experimental conditions, we first perform thermal denaturations of the protein in question as a function of pH and [NaCl]. We select conditions where the midpoint temperature, T_m, of the transition changes as little as possible with these two variables so that small deviations during the mixed osmolyte experiment will not affect the results. In addition, it is necessary to determine how long the protein must be equilibrated at the experimental temperature before any measurement can be made. This can be easily done by performing kinetic experiments under conditions where the reaction is expected to be slowest (at the midpoint concentration where native and denatured populations are equal). When in doubt, a kinetic scan can be used for all data points as described in Section 2.2. For Nank4-7*, we determined that ideal conditions (which we use subsequently) are 34 °C, pH 7.0, 200 m*M* NaCl, with a preequilibration time of 10 min in the instrument prior to data acquisition. Under these conditions, the stability of Nank4-7* does not depend on pH, the dependence on salt concentration is moderate, and the reaction is at its midpoint. The signal of each sample is averaged for 5 min.

Regarding the CD signal, we have to consider that Nank4-7* is rich in α-helices so that selecting a wavelength of 222 nm should give the largest signal change upon folding/unfolding. However, as mentioned before, osmolytes absorb strongly at lower wavelengths in the far UV region, and the osmolytes of interest must be tested before the experiment can be performed. In the case of Nank4-7*, where we study the pair sarcosine and urea, 228 nm was the lowest wavelength with low signal-to-noise ratio up to 4 *M* urea plus 1.5 *M* sarcosine and an acceptable ratio with 4 *M* sarcosine plus 1.5 *M* urea.

3.3. Data treatment with mixed osmolytes: Independent analysis of the individual curves

There are multiple ways to determine *m*-values from the data. Comparison of the various *m*-values then shows whether or not the osmolytes act additively.

Deriving m-values from each curve: The profile obtained from this series of experiments (eight curves in total, four for each fixed concentration of the denaturant and four for the protecting/stabilizing osmolyte) are shown in Fig. 3.5A–B. Note that the sarcosine concentration axis is inverted. Fitting each curve individually is problematic in the cases where either the native or denatured state baseline is never reached. This can be seen in Fig. 3.5C, where the fitting results with error limits are given for individual fits of each

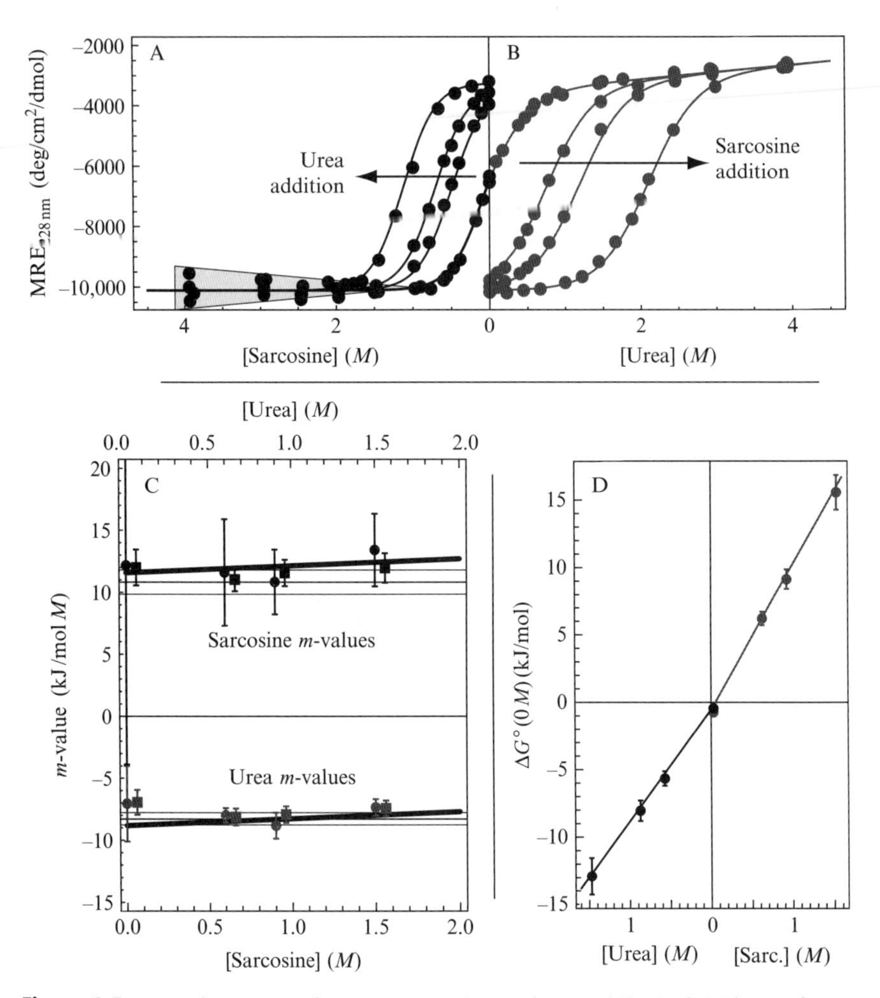

Figure 3.5 Test for synergy between sarcosine and urea: (A) Nank4-7* was force-folded with sarcosine at increasing background urea concentrations (0, 0.6, 0.9, and 1.5 *M*), and the mean residue ellipticity monitored. Note the decrease in the signal-to-noise level as the concentration of the highly absorbing sarcosine increases, as indicated by the gray wedge. (B) Same as in A, but with the roles of urea and sarcosine switched. (C) *m*-values derived by various methods from panels A and B. Bold solid lines: global fit; circles: individual fits of each of the eight transition curves; squares: individual fits with shared common baselines (concentration axis is offset for better visibility); and straight thin lines: *m*-values with error limits derived from panel D. (D) Stability, ΔG° (0 *M*), at zero molar variable osmolyte from each of the eight individual fits with shared baseline of the data in panels A and B. Data derived from experiments with variable urea or sarcosine are shown in gray or black, respectively.

curve (circles). The high level of uncertainty of the m-values is striking. However, the error limits can be reduced by fitting the curves simultaneously with shared baselines. In this fitting strategy, each curve has its own m-value and $\Delta G^\circ(0\ M)$, but the values for slopes and intercepts for the baselines are shared. The results are shown in Fig. 3.5C (squares; slightly offset to the right for visibility). Both kinds of fits use Eq. (3.13). All m-values thus obtained are referred to as *direct* m-values.

Deriving m-values from extrapolated stabilities: From the fits described in the previous paragraph, we obtain the protein stabilities at 0 M of the variable osmolyte, $\Delta G^\circ(0\ M)$, at 0, 0.6, 0.9, and 1.5 M of the constant osmolyte (Fig. 3.5D). The slopes of these curves ($m_S = \partial \Delta G^\circ(0\ M)/\partial\,[\text{sarcosine}]$ and $m_U = \partial \Delta G^\circ(0\ M)/\partial\,[\text{urea}]$) yield two more m-values for protein stability in urea and sarcosine (referred here as *indirect* m-values). These m-values (plus minus fitting error) are given in Fig. 3.5C as thin continuous lines. Note that the data in the left half of Fig. 3.5D are obtained from fitting the individual four curves in Fig. 3.5A and those in the right half from Fig. 3.5B (baselines shared in all eight fits).

Comparison of the various m-values: Comparison of the *indirect* m-values with the *direct* m-values gives information on the additivity of the pair of osmolytes used. If the values are identical within experimental error, then the pair of the analyzed osmolytes has effects on the protein that are additive within error. Consequently, the osmolytes can then be used in combination without complication to study folding and denaturation in mixed systems. However, the *direct* and *indirect* m-values may not agree. In that case, we would typically expect that the *direct* m-values exhibit a concentration dependence, and that the *indirect* m-values match those *direct* m-values that come from curves at 0 M background osmolyte concentration.

From Fig. 3.5C, we see that in the case of the pair urea and sarcosine, their effects are independent of one another within error, that is, the effect of sarcosine on the stability of Nank4-7* (its m-value) is the same regardless of the presence of urea and *vice versa*. While this shows that there is no strong synergy between these two osmolytes, there could still be an effect that is smaller than the error of the methods employed so far. A further reduction of the uncertainty can be achieved by the phase diagram method (Fig. 3.6), or by the global analysis described in Section 3.4.

Detecting synergy by the phase diagram method: A very simple way to test for synergy is given by the phase diagram method, as mentioned in Sections 6.4 and 6.5. A protein phase diagram is a plot of phase separation lines, which are 50% population lines, for example, midpoint concentrations c_m of one osmolyte as a function of background concentrations of the other osmolyte (Fig. 3.6). In case of the present figure, where two osmolyte concentrations are plotted against each other, synergy causes a curvature of the phase separation line. Equation (3.74) quantifies this behavior. It can be rewritten to

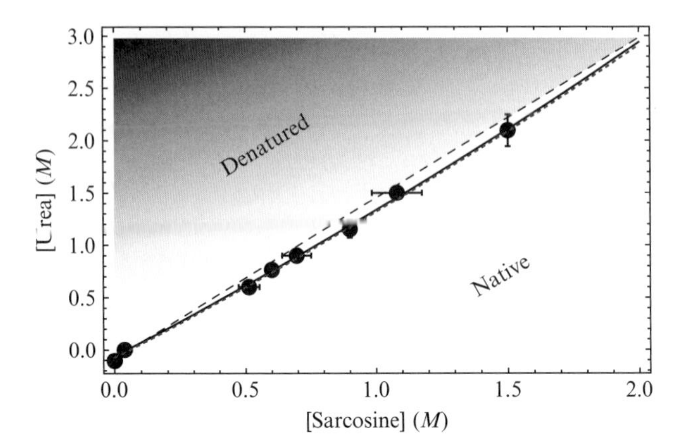

Figure 3.6 Use of the phase diagram method to test for synergy between sarcosine and urea: synergy leads to curvature in this plot. The points are the midpoint concentrations of the eight transition curves shown in Fig. 3.5A–B: urea denaturation curves at four sarcosine concentrations (vertical error bars); and sarcosine folding curves at four urea concentrations (horizontal error bars). The lines are drawn according to Eq. (3.19), using as parameters the average of the *m*-value points in Fig. 3.5C along with the 0 *M* intercept of ΔG° in panel D (black dashed line), or the results of a fit to Eq. (3.19) (continuous line), or the results from the global fit in Section 3.4 (dotted gray line). In the region above the lines, the protein is predominantly denatured and below predominantly native.

$$c_U = \frac{\left(\Delta G^\circ_{ref}/m_U\right) + c_S\left(m_S/m_U\right)}{1 + c_S\left(m_{US}/m_U\right)}, \tag{3.19}$$

where ΔG°_{ref} is the protein stability in the absence of osmolyte, m_U and m_S are the *m*-values of the two osmolytes (urea and sarcosine), $m_{US} = (\partial m_U/\partial c_S) = (\partial m_S/\partial c_U)$ is the synergy between the osmolytes, and c_U and c_S are the molar concentrations.

In the absence of synergy, m_{US} is zero and the denominator of Eq. (3.19) is unity. Then the remaining equation is simply linear, as shown by the dashed line in Fig. 3.6. This line was calculated using parameters from the individual fits described above, which resulted in no significant signs of synergy. However, the line does not exactly go through the data points. Therefore, it is necessary to do a curve fit. A fit of the points without further assumptions cannot resolve all four parameters because they occur in three groups. Thus, a three-parameter equation of the form $c_U = (P_1 + P_2c_S) / (1 + P_3c_S)$ was used, resulting in the continuous line in Fig. 3.6. The line is clearly curved, but this curvature is barely significant within the given error limits.

Overall, the current Section 3.3 shows that it can be difficult to detect synergy with classical approaches, and only the phase diagram method gave initial signs of synergy. A global analysis of the data is more difficult to perform, but superior in terms of the results as we shall see in the following section. Interestingly, it is the simplest data evaluation technique in the current section (the phase diagram method), which yields the same result with respect to synergy as the global data analysis does (gray dotted line in Fig. 3.6).

3.4. Global analysis

For the global analysis, we need to modify Eq. (3.13). First, we see in Fig. 3.5A that the signal of the native state does not seem to depend on the concentration of sarcosine, and it is reasonable to assume that the same holds for urea. The denatured state, however, clearly exhibits a urea-dependent signal (Fig. 3.5B). Such dependence is due to expansion of the denatured state in urea, and the opposite behavior (contraction of the denatured state) is expected of the stabilizer sarcosine (Holthauzen *et al.*, 2010). We therefore modify Eq. (3.13) to

$$S = S_{N,\mathrm{ref}} + \left[S_{D,\mathrm{ref}} - S_{N,\mathrm{ref}} + c_1 \mathrm{d}S_{\mathrm{D},1,\mathrm{ref}} + c_2 \mathrm{d}S_{\mathrm{D},2,\mathrm{ref}}\right] \frac{\exp(-\Delta G^\circ(c_1, c_2)/RT)}{1 + \exp(-\Delta G^\circ(c_1, c_2)/RT)}, \qquad (3.20)$$

where the signals of the native and denatured state, $S_{\mathrm{N,ref}}$ and $S_{\mathrm{D,ref}}$, are constants, as well as the slope of the denatured state with respect to osmolyte 1 and 2, $\mathrm{d}S_{\mathrm{D},1,\mathrm{ref}}$ and $\mathrm{d}S_{\mathrm{D},2,\mathrm{ref}}$. If the signal of the native state depends on the concentrations of the osmolytes, then $S_{\mathrm{N,ref}}$ should be replaced with the term $S_{\mathrm{N,ref}} + c_1 \mathrm{d}S_{\mathrm{N},1,\mathrm{ref}} + c_2 \mathrm{d}S_{\mathrm{N},2,\mathrm{ref}}$. The stability equation that we are using for a mixture of two osmolytes

$$\begin{aligned} \Delta G^\circ(c_1, c_2) &= \Delta G^\circ\left(c_{1,\mathrm{ref}}, c_{2,\mathrm{ref}}\right) + m_1\left(c_1 - c_{1,\mathrm{ref}}\right) + m_2\left(c_2 - c_{2,\mathrm{ref}}\right) \\ &\quad + m_{1,2}\left(c_1 - c_{1,\mathrm{ref}}\right)\left(c_2 - c_{2,\mathrm{ref}}\right) \\ &= \Delta G^\circ(0M, 0M) + m_1 c_1 + m_2 c_2 + m_{1,2} c_1 c_2 \end{aligned} \qquad (3.21)$$

contains a synergy term $m_{1,2}c_1c_2$ that leads to m-values that depend on the concentration of the additional osmolyte. For simplicity, we use as reference conditions 0 M of either osmolyte.

The fit of Eqs. (3.20) and (3.21) to the data is shown in Fig. 3.5A and B. The resulting fitting parameters are given in Table 3.4. A comparison of the m-values from this fit (bold lines in Fig. 3.5C) with the ones derived in the previous section (symbols and thin lines in Fig. 3.5C) shows that they are generally consistent with each other. However, the uncertainty in the global fit is smaller than in the other approaches, and so it is possible to detect a

Table 3.4 Global fit results for Case 2: Nank4-7*

Parameter	Unit	Estimate	SE	Confidence interval
m_U	kJ/mol M	−9.0	0.3	(−9.6, −8.5)
m_S	kJ/mol M	11.7	0.3	(11.1, 12.3)
$m_{U,S}$	kJ/mol M^2	0.59	0.14	(0.31, 0.87)
$S_{N,ref}$	deg cm^2/dmol	−10,105	28	(−10,160, −10,050)
$S_{D,ref}$	deg cm^2/dmol	−3670	85	(−3840, −3500)
$dS_{D,U,ref}$	deg cm^2/dmol M	293	40	(214, 372)
$dS_{D,S,ref}$	deg cm^2/dmol M	−170	83	(−330, −6)
$\Delta G°(0\ M, 0\ M)$	J/mol	−710	130	(−460, −970)

slight synergistic effect between the osmolytes (slope of the m-value as a function of the other osmolyte). This coupling between urea and sarcosine, $m_{U,S}$, is small, but significant (see Table 3.4). Figure 3.5 also reveals why in the absence of a global fit the coupling was not detected previously (Holthauzen and Bolen, 2007): the fitting errors are about of the same magnitude as the degree of synergy.

Another feature seen in the fitting results (Table 3.4) is the dependence of the denatured state signal on the concentrations of the osmolytes, $dS_{D,U,ref}$ and $dS_{D,S,ref}$. Even visible inspection made obvious that $dS_{D,U,ref}$ is significant and positive. However, no stretch of denatured state baseline is available for the denatured state as a function of sarcosine concentration. The results demonstrate that nevertheless the dependence of the denatured state signal on sarcosine is still detectable and has the expected sign.

On a final note for this case, we see that we are slightly off T_m at 34 °C, as seen by the small, but significant value of $\Delta G°(0\ M, 0\ M)$ at this temperature (see Table 3.4). The population of the denatured state is between 54% and 59%, rather than 50%.

4. Osmolyte-Induced Unfolding at Variable Temperature

4.1. General outline

There are various reasons why researchers are interested in the temperature dependence of protein stability. One group of reasons is of technical nature—for example, by temperature variation, it is possible to populate low-abundant protein states that would be difficult to study otherwise. Another group of reasons is concerned with the variation of environmental temperatures that normally occur not only with poikilothermal organisms but also within homothermal organisms (think about the variations in skin temperature in a chilly winter breeze compared to sauna conditions). Quantification of the thermal stability of small proteins tends to be relatively straightforward in many cases. However, complications may arise for proteins that are larger, have high melting temperatures, or are composed of multiple domains. Irreversible processes can occur when such proteins are thermally unfolded, including the formation of protein aggregates and precipitates. These complications can prevent an adequate analysis of the thermodynamic stability of proteins and protein domains and their interactions. What is needed is a means (or method) to determine these properties under reversible conditions that circumvent such problems of protein solubility. Urea denaturation is a good alternative to thermal unfolding because urea solubilizes proteins (Bolen, 2004) and therefore allows the analysis of protein stability under reversible or near-reversible conditions even in cases where thermal scans are problematic.

Urea denaturation curves can also be used to better define thermodynamic properties that were obtained by thermal denaturation (Pace and Laurents, 1989). It is normally straightforward to extract the enthalpy of unfolding (the slope of the stability with temperature) from thermal unfolding curves. However, it is more difficult to determine the heat capacity of unfolding (the curvature of the stability), and the additional stability points at lower temperature provide valuable information to define this curvature.

Methods: The general approach is to repeat experiments of the type discussed in Section 2 at various temperatures. The needed equilibration time may greatly vary as the temperature is changed (Chen *et al.*, 1989), normally leading to quicker experiments at high temperature, and slower ones at low temperature. The upper temperature limit for reversible solvent denaturation is dependent on the velocity of the irreversible step that occurs at elevated temperatures. This limit may be significantly below the actually observable thermal transition range under unfavorable circumstances. In particular, the duration of the experiment has an impact on the temperature at which irreversible denaturation is first observed (Sanchez-Ruiz *et al.*, 1988). The longer the protein is exposed to the final experimental temperature, the higher the likelihood that aggregation occurs.

The lower temperature limit is normally given by the freezing point of the solution. Urea lowers the freezing point of the solution by about 2 °C/*M*. Thus, if, for example, an experimental temperature of −5 °C is desired in a urea unfolding experiment, the lowest concentration point should not be below 2.5 *M* (better a little higher). Consult freezing point tables for reference (Lide, 2004). If temperatures below the freezing point are desired, note that it is in principle possible to work with supercooled solutions. However, the experimental time is more limited, because spontaneous nucleation of ice can occur, especially when dust particles are in the solution, the temperature is far below the freezing point, or the solution volume is large. When the solution freezes, the sample cell may burst either instantly (when supercooled far below freezing point) or after a short time during which it may be possible to bring the solution above the freezing point and prevent damage.

Data evaluation can become involved when data are collected in several dimensions (temperature and urea concentration in our case). When such complications occur, then it is often possible to still derive many thermodynamic parameters with relative ease by the phase diagram method (Ferreon *et al.*, 2007; Rösgen and Hinz, 2003). In this method, the data are reduced to the most essential features by plotting phase separation lines, that is, lines of 50% population of protein states. In the current case, this is just one line, because only two states exist in the discussed folding/unfolding equilibria. Specific features of these lines then permit extraction of essential thermodynamic information with relative ease. The method is particularly valuable when there are more than two states or more than 2D. In the cases discussed below, we discuss several phase diagrams to illustrate some principles. More details about the method were discussed by Rösgen and Hinz (2003) and Ferreon *et al.* (2007).

4.2. Case 3: von Willebrand factor: two-state unfolding of the collagen-binding A3 domain

The von Willebrand factor contains a series of domains that can be investigated separately and exhibit various degrees of complexity in their folding transitions. The von Willebrand factor platelet GP1bα-binding A1 domain unfolds through a stable intermediate (Auton *et al.*, 2009). Such extra complication of an additional unfolding transition can be taken into account (see Section 6.6). However, for the sake of simplicity, we will present the analysis for the von Willebrand factor collagen-binding A3 domain, which unfolds in a two-state manner (Auton *et al.*, 2010). This ~23 kDa protein has a single disulfide bond and adopts a classic alpha/beta Rossmann fold (pdb = 1AO3) (Bienkowska *et al.*, 1997). Thermal denaturation is irreversible in the absence of urea, due to aggregation at high temperature, as demonstrated by differential scanning calorimetry (Auton *et al.*, 2010). However, urea denaturation at temperatures less than 50 °C is reversible (Auton *et al.*, 2007a).

The A3 domain construct used here involves von Willebrand factor residues S_{1667}–G_{1874} and was expressed in *E. coli* as a fusion protein containing the 6×His tag at the N-terminus (Auton *et al.*, 2007a; Bienkowska *et al.*, 1997). The concentration was determined by UV spectroscopy at 280 nm using the method of Pace (Pace *et al.*, 1995). Urea denaturation of the protein was monitored by CD spectroscopy at 222 nm at a final protein concentration of 10 μ*M*. The urea concentration was varied by preparing a series of tubes (200 μL) according to the classical strategy (1) described above (Section 2.1). The samples were mixed briefly by short vortex pulses and then incubated in a water bath at a defined temperature of 5, 15, 25, 35, and 45 °C for a minimum of 3 h. Equilibrium was verified by monitoring the ellipticity which was found to be constant as a function of time. To minimize thermal perturbation, the 1-mm pathlength cuvettes were also incubated in the water bath for 3 min prior to sample addition. The ellipticity data were averaged for 2 min. Reversibility was verified by a single point approach (cf. Section 2.1, sample preparation) in which the protein was first incubated at ~7 *M* urea, diluted to the urea concentration at the midpoint of the transition and incubated, followed by a measurement of the ellipticity.

4.3. Data treatment: Independent analysis of the individual curves

The denaturation curves at each temperature were first analyzed by fitting to Eq. (3.13), with linear pre- and postdenaturation baselines, for the midpoint of the transition (c_m) and the *m*-value. $\Delta G°(0\ M)$ was calculated from Eq. (3.15) as $\Delta G°(0\ M) = -c_m\ m$. The initial fit revealed that the slopes of the native and denatured state signals with urea concentration are equal within error. We therefore fit the curves using but a single-slope dS for both native and denatured state baseline, and we share this slope as well as the native state signal among all individual fits. The results are shown in Table 3.5. To eliminate baseline problems in the subsequent global fits (Section 4.4), we normalized the data to the fraction of native molecules, as shown in Fig. 3.7. This figure also shows the *m*-values (Fig. 3.7B), stabilities at 0 *M* urea (Fig. 3.7C), and c_m values (Fig. 3.7D) as a function of temperature.

Both the midpoint concentrations, c_m, and the stabilities at 0 *M* urea, $\Delta G°(0\ M)$, show that the stability of the A3 domain has a maximum within the range of experimental temperatures. Such maximum around room temperature is a normal feature in protein stability curves (Becktel and

Table 3.5 Individual fit results for Case 3: VWF A3 domain

Parameter	Unit	5 °C	15 °C	25 °C	35 °C	45 °C
c_m	kJ/mol/*M*	4.19	5.00	5.37	5.58	5.28
m-value	kJ/mol/*M*	−9.5	−9.4	−9.1	−8.9	−8.5
$\Delta G°(0\ M)$	kJ/mol	39	47	49	50	45
$S_{D,ref}$	deg cm^2/dmol/res	−1.43	−1.80	−1.83	−1.95	−1.79
$S_{N,ref}$	deg cm^2/dmol/res	−10.16				
dS	deg cm^2/dmol/res/M	0.059				

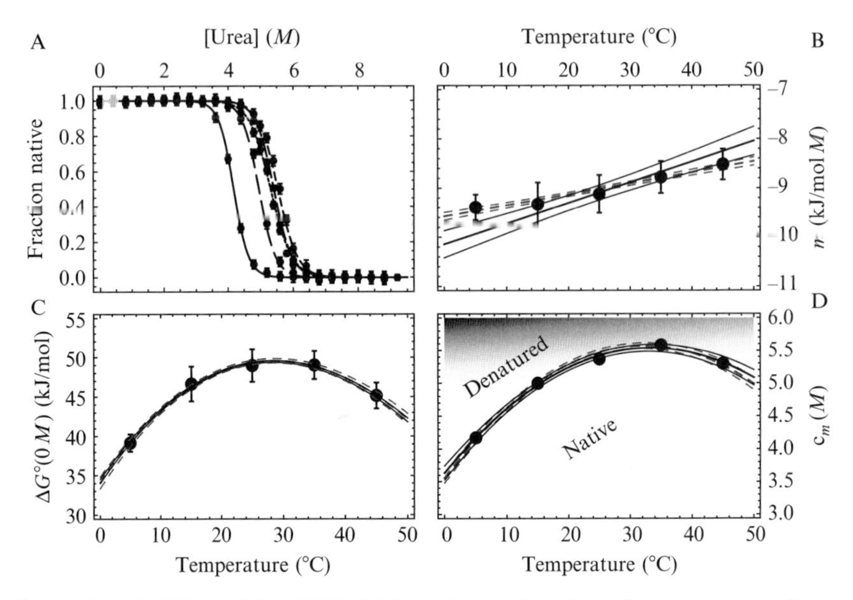

Figure 3.7 Stability of the VWF A3 domain as a function of temperature and urea concentration: (A) urea-induced unfolding at 5, 15, 25, 35, and 45 °C. The data (points with error bars) are normalized to fraction native. The lines represent the results from a global fit of the data to Eq. (3.22), using Eq. (3.53) for K. The shorter the dashes, the higher the temperature. (B) m-values derived from the individual curves (points with error bars), or from the global fit (continuous thick line with errors indicated by the thin continuous lines). A fit of the points to Eq. (3.57) is indicated as gray, thick, dashed curve. The fitting errors are indicated with the gray, thin, dashed curves. The meaning of the symbols and lines is the same in the following panels. (C) Stabilities at 0 M urea derived from the individual curves, global fit results, and direct fits of the points to Eq. (3.40). Most curves are tightly overlapping. (D) Phase diagram of $c_{\rm m}$ versus T derived from the individual curves (panel A), global fit results, and direct fit results of the points to Eq. (3.68).

Schellman, 1987). Also the denatured signal's dependence on temperature reflects the classical expectation of a decrease with temperature (Tiffany and Krimm, 1972). A less expected feature is the significant dependence of the m-value on temperature, since often constant or small dependencies have been observed (Auton *et al.*, 2009; Baskakov and Bolen, 1999; DeKoster and Robertson, 1997; Felitsky and Record, 2003; Ferreon *et al.*, 2007; Giletto and Pace, 1999; Henkels and Oas, 2005; Henkels *et al.*, 2001; Hu *et al.*, 1992; Nicholson and Scholtz, 1996; Pace and Laurents, 1989; Zweifel and Barrick, 2002).

The parameters derived from the individual fits shown in Table 3.5 can be fitted to obtain additional thermodynamic information. The first and most obvious fit is shown in Fig. 3.7B, where the m-values as a function of temperature are fitted to Eq. (3.57). Within the errors of each point, there is a significant temperature dependence of the m-value, but this dependence is not pronounced enough to justify the use of a quadratic function. The linear fit is shown as a thick, gray, dashed line, and the propagated fitting error as thin, gray, dashed lines, resulting in $m_{k,\mathrm{ref}} = 3.64/M$ and $(\partial m_k/\partial\beta)_{\mathrm{ref}} =$ 15.7 kJ/mol M at $T_{\mathrm{ref}} = 25$ °C. There is a slight deviation of these results from the global fit (continuous lines), which is discussed in Section 4.4.

The data in Fig. 3.7C are stability points as a function of temperature and are thus represented by $\Delta G^\circ(0\ M) = -RT \ln K$, where $\ln K$ is given by Eq. (3.40) (when applying Eq. (3.40), it is important to remember to use the absolute temperature in K). The function has a clear curvature, and thus we need to use terms at least up to second order (including the $\Delta C_\mathrm{p}^\circ$ term) for the fit, resulting in $\ln K(T_{\mathrm{ref}}) = -19.9$, $\Delta H^\circ(T_{\mathrm{ref}}) = 12$ kJ/mol, and $\Delta C_\mathrm{p}^\circ(T_{\mathrm{ref}}) = 10.6$ kJ/mol K. This fit (gray lines) is so well superimposed with the global fit results (continuous lines) that it is barely visible in Fig. 3.7C. Note that $\Delta H^\circ(T_{\mathrm{ref}})$ is very close to zero relative to normal unfolding enthalpies measured at high temperatures. Those values are normally many hundreds of kJ/mol. We find the small value, because our reference temperature is close to the maximum of the $-R \ln K$ curve, where the slope (i.e., the enthalpy of unfolding) is zero.

It is also possible to fit the phase diagram of c_m versus T in Fig. 3.7D with Eq. (3.68). This equation is given by the negative ratio of the equations used for panels C and B. Since there are as many fitting parameters as data points, we keep the two m-value parameters fixed at their values obtained from the linear fit in Fig. 3.7B. The resulting values for $\ln K(T_{\mathrm{ref}})$, $\Delta H^\circ(T_{\mathrm{ref}})$, and $\Delta C_\mathrm{p}^\circ(T_{\mathrm{ref}})$ are exactly the same as obtained from the fit in Fig. 3.7C.

4.4. Global analysis

A global analysis of the data was performed using the equation

$$f_\mathrm{N} = \frac{1}{1+K}, \tag{3.22}$$

where K is taken from Eq. (3.53), using terms up to second order without $(\partial m_K/\partial c)$. As initial fitting parameters, we use the parameters that were obtained from the various fits shown in Fig. 3.7B–D (see Section 4.3). The reference temperature is fixed to $T_{\mathrm{ref}} = 25$ °C (298.15 K). Table 3.6 shows the fitting results, and (in square brackets) further parameters that were derived from the fitting parameters. The results are also plotted in Fig. 3.7 as black lines.

It is instructive to compare the results of the fits in Sections 4.3 and 4.4. Overall, the global fit results are consistent with the data points in all panels of Fig. 3.7. Only in panel B, there is one m-value with error limits that are slightly outside the error limits of the global fit (thin black lines). However,

Table 3.6 Global fit results for Case 3: von Willebrand Factor A3 domain

Parameter	Unit	Estimate	SE	Confidence interval
T_{ref}	°C	25	N/A	N/A
$\ln K_{ref}$	N/A	−20.0	0.3	(−20.5, −19.2)
ΔH°_{ref}	kJ/mol	38	14	(11, 65)
$\Delta C^{\circ}_{p,ref}$	kJ/mol K	9.7	0.3	(9.2, 10.3)
m_K	1/M	3.67	0.06	(3.55, 3.79)
$\partial m_K/\partial\beta_{ref}$	kJ/mol *M*	22	3	(16, 27)
$[m]$	kJ/mol *M*	−9.1	0.15	
$[\partial m/\partial T]$	kJ/mol *M* K	0.042	0.010	

fitting errors are notoriously optimistic, and thus we do not consider that point to be significantly different from the global fit. That is, the global fit is good and valid.

However, there are differences between the global fit and the fits to the points in Fig. 3.7B–D. Specifically, ΔH°_{ref} is 38 kJ/mol in the global fit, but only 12 kJ/mol in the previous fits. Also, $(\partial m_K/\partial\beta)_{ref}$ is 22 versus 16 kJ/mol *M*. This difference is sufficiently large to be easily detectable by eye in Fig. 3.7B. However, while in contrast, the difference in the enthalpy may look large, the values from both fits are actually so close to zero that the curves in Fig. 3.7C look indistinguishable. The stability at 0 *M* urea in Fig. 3.7C is but a derived quantity that is not directly observed in the experiment and may not be as instructive as the following property. The most fundamental experimental observation in solvent-dependent denaturation is the midpoint concentration c_m, which is shown in Fig. 3.7D. Here it is seen that the global fit captures better the c_m value at 25 °C than the other fits do. Although the global fit is the best fit on the basis of the existing data, it may be worthwhile to collect an additional data set at −5 °C to check whether the *m*-values in Fig. 3.7B properly follow the solid or the dashed lines.

5. Thermal Unfolding in the Presence of Osmolytes

5.1. General outline how to use osmolytes in thermal scans

Thermal unfolding is widely used as a method for studying the stability of proteins. Because protecting osmolytes (such as sarcosine, TMAO, or betaine) and chaotropes (like urea and guanidinium chloride) affect the stability of proteins, their presence causes a shift in the melting temperature,

T_m, of the protein. Observing the dependence of T_m on osmolyte concentration is a particularly straightforward approach to sampling the effect of the additives on the stability of folded proteins for several reasons. (1) Both stabilizers and denaturants give a response, viz. increase or decrease in T_m, respectively; (2) even small effects can be detected—as long as T_m can be determined with good precision; (3) much fewer samples are needed in comparison to classical titration experiments; (4) every sample yields a full transition curve; and (5) the measurements are usually automated and run in an unattended manner. However, there are also disadvantages and limitations. (1) The thermal unfolding should be reasonably reversible, which is often not the case; (2) protein aggregation becomes more likely at higher temperatures; and (3) it may be difficult to extrapolate the results back to lower temperatures.

Methods: For reasons of baseline stability discussed above (Section 3.1), CD is normally preferred over fluorescence spectroscopy. However, use of fluorescence may allow for doing experiments in high-throughput mode on microplates (Ericsson *et al.*, 2006; Matulis *et al.*, 2005; Todd and Salemme, 2003). Depending on the severity of the baseline issues, it may or may not be possible to extract good quantitative information from such a high-throughput approach.

A good temperature equilibration is important in thermal scans. Whenever feasible, the solution should be well mixed, for example, using a stirring bar. Also, it should be kept in mind that temperature discrepancies may exist even after good thermal equilibration. For the data shown below, it was found that the temperature displayed in the controlling software deviated significantly from the actual temperature in the cuvette. It is therefore strongly recommended to perform a temperature calibration for each type of cuvette, and each scan rate, throughout the entire range of experimental temperatures.

Reversibility: One of the big issues in thermal unfolding studies is the degree of thermodynamic reversibility. Data analysis is only straightforward if the protein unfolding reaction is reversible. Reduction of reversibility may be due to two main reasons: (1) the denatured protein is not able to refold, or (2) the experiment is performed faster than the protein-folding equilibrates. Most researchers recognize point one and try to demonstrate reversibility by repeating an experiment with a sample that was already heated once. However, such experiment merely demonstrates repeatability at best. A repeatable experiment may or may not be a thermodynamically reversible one, depending on point two given above. Particularly when denaturant is added, the folding equilibrium may drastically slow down, making the experiment irreversible in a thermodynamic sense: it is performed faster than the system can equilibrate (Plaza del Pino *et al.*, 1992). These two aspects of thermodynamic reversibility can be illustrated with the following model of protein (un)folding that considers denaturation as a combination of a reversible process between native N and denatured D protein, with an irreversible step toward species I (Lumry and Eyring, 1954):

$$N \rightleftharpoons D \rightarrow I. \tag{3.23}$$

The experiment has to be performed sufficiently fast to prevent a significant population of I from accumulating. But the experiment also needs to be sufficiently slow to allow for proper equilibration between N and D. Depending on the velocity of the two steps in Eq. (3.23) relative to each other, such conditions could be possible or impossible to achieve.

How then can we find and demonstrate reversibility? The true thermodynamic criterion for reversibility is that the results are independent of the speed with which the experiment was performed. It is therefore necessary to repeat a thermal unfolding experiment with fresh solution at a different heating rate to be able to test for thermodynamic reversibility. When the resulting denaturation curve is independent of the heating rate, the reaction is thermodynamically reversible. More often than not, the irreversible step in Eq. (3.23) is fast. However, the irreversible species may not accumulate significantly until the denaturation transition is mostly over. During the subsequent sampling of the post-denaturation baseline, the irreversible species can become predominant. In that (typical) case, the unfolding curve may be thermodynamically reversible, without being repeatable.

The degree of reversibility does depend on various factors, including the solution conditions. Stabilizing osmolytes, in particular, generally decrease the solubility of proteins (Bolen, 2004) and thus pose a danger to the reversibility by precipitating the protein. Especially, the denatured state can be of much lower solubility than the native state (Pace *et al.*, 2004). The addition of solubilizing osmolytes, such as proline, may help in such cases. Other strategies for improving reversibility include change in pH and ionic strength.

Sample preparation: An important interfering factor in thermal scans is the formation of bubbles in the solution. Substantial amounts of air can dissolve in aqueous solution, particularly at low temperatures. As the solution is heated, the solubility decreases, and air bubbles can form when the solution becomes supersaturated with air in this process. The most basic precaution is to never take cold solution from the fridge for thermal scans. Figure 3.8 shows that the primary components of air, nitrogen, and oxygen have an especially steep decline in aqueous solubility between 0 °C and room temperature. Beyond 30 °C there is not much loss in solubility until close to the boiling point of water. Thus, as a minimal precaution, let the solution warm to room temperature, followed by centrifugation. The vibrations in the centrifuge tend to nucleate bubble formation, and the additional gravity helps the bubbles escape from the solution. Thus, centrifugation is a simple

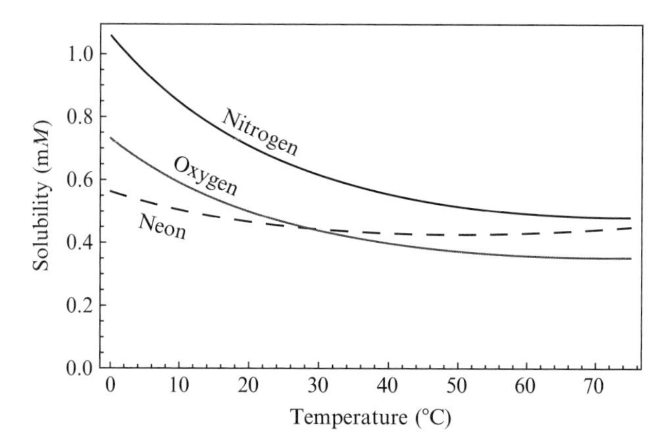

Figure 3.8 Solubility of gases in water according to Lide (2004). The solubility of neon depends very little on temperature. The decrease in solubilities of N_2 and O_2 occurs primarily below room temperature. Nitrogen and oxygen bubble formation in temperature scans is therefore most likely occurring from solutions that were stored in the cold.

degassing step. The degassing works even better, when a mild vacuum is applied during the centrifugation, for example, in a SpeedVac® instrument.

Such treatment is normally sufficient, but some methods may require particularly careful removal of air from the solution—for example, differential scanning densimetry (Hinz *et al.*, 1991). A simple method of efficient air removal can be done in small plastic syringes that have about twice the volume of the solution that has to be degassed. The syringe is loaded and the opening sealed with parafilm, firmly held in place with the index finger. Then the piston is retracted to produce a vacuum and held in place using other fingers of the same hand sealing the syringe opening. The strength of the vacuum can be easily fine-tuned by varying the amount of air that is in the syringe before pulling the vacuum, and by the degree to which the piston is retracted. Nucleation can be induced by tapping against the wall of the syringe. The degree of nucleation can be varied by the force of tapping and the hardness of the material used. While the vacuum is still intact, most of the bubbles float up quickly. Then the vacuum is released, excess gas pushed out of the top of the syringe, and the cycle repeated with stronger vacuum. After several cycles, the piston is adjusted so that all gas is removed from the syringe, and a strong vacuum carefully produced by pulling the piston slowly out. Residual bubbles will then float up.

Another strategy to avoid bubbles is to saturate all solutions with neon. This noble gas is peculiar in that its solubility hardly changes with temperature, and even increases beyond 50 °C (Lide, 2004), as seen in Fig. 3.8. Thus, the likelihood of bubble formation decreases during the thermal scan.

5.2. Case 4: Nank4-7* thermal stability in sarcosine solutions

As in Section 3.2, we use the Nank4-7★ system, a marginally stable ($T_m \approx$ 34 °C) 15 kDa truncated version of the ankyrin domain of the *Drosophila* notch receptor, for the thermal denaturations in the presence of osmolytes. A series of thermal denaturations was carried out in the presence of varying concentrations of the protecting/stabilizing osmolyte sarcosine, ranging from 0 to 1 *M* (Holthauzen *et al.*, 2010). The details of the sample preparation are the same as in Sections 2.2 and 3.2. The experiments were again carried out in a Jasco J-720 Polarimeter, equipped with a peltier controlled cell holder.

Optimization of experimental conditions: Most of the optimization procedure from Section 3.2 also applies here. There are, however, some additional aspects to consider.

Evaporation of the sample can be an issue at elevated temperatures. Therefore, it is important for the CD cuvette to be tightly capped. Completely filling the cuvette removes all free gases from the cuvette, which helps against evaporation at high temperatures, and against dissolution of gases at low temperatures (our first data point is at 5 °C). Such dissolved gases could lead to bubble formation at higher temperatures during the temperature scan.

As pointed out above, the reversibility of the reaction should be demonstrated by performing the experiments at different heating rates. One should note as well that the protecting osmolytes favor the aggregation of proteins—especially at high temperatures. Care must therefore be taken to find the best experimental conditions that minimize this type of problem. Ideally, the protein should spend the least amount of time as possible at high temperatures and high protecting osmolyte concentrations. We have performed thermal scans in two different ways: (1) continuous thermal scans at 1 °C/min and (2) data collection at a much smaller number of temperatures, with temperature jumps between. Because in (2) the resulting heating rate is higher, the protein spends less time at the high temperature. The transition region is identical with both strategies (demonstrated reversibility). In addition, aggregation problems beyond the transition region are minimized. Using 1 °C/min, the denatured state baseline starts drifting at high temperatures and 1 *M* sarcosine, which is a symptom of aggregation. However, the faster method (2) improves the postdenaturation baseline at high sarcosine.

5.3. Data treatment: Independent analysis of the individual curves

The thermal denaturation curves are shown in Fig. 3.9A. Results can be extracted from these data in several steps, involving first individual fits of each curve to obtain the midpoint temperature of the transition T_m, the

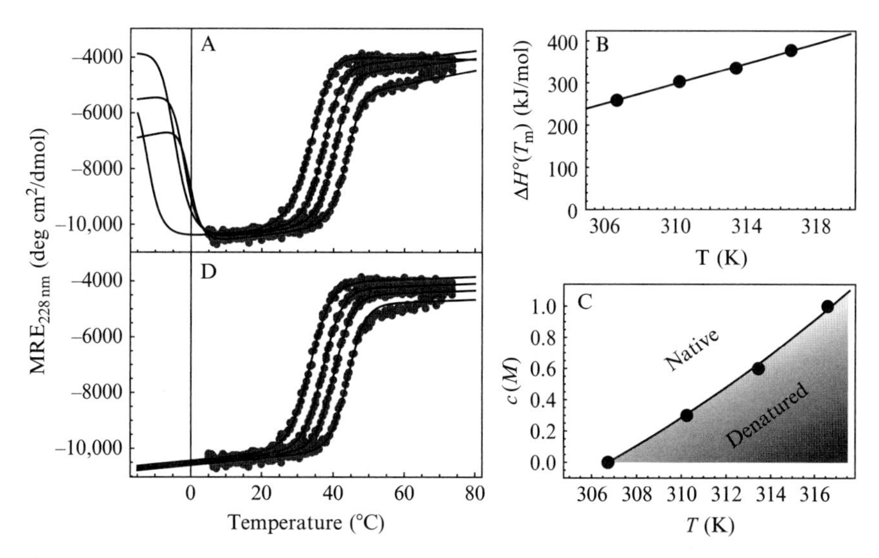

Figure 3.9 Thermal unfolding of Nank4-7* in sarcosine solutions: (A) individual fits to the data; (B) transition temperatures and enthalpies derived from A; and (C) phase diagram with 50% native/denatured line: above the line the proteins are predominantly native, and below mostly denatured. The *m*-value can be derived either from a fit (Eq. (3.26)) or from the slope (Eq. (3.27)); (D) global fit of the data (results given in Table 3.7).

enthalpy $\Delta H^\circ(T_\mathrm{m})$, and the heat capacity $\Delta C_\mathrm{p}(T_\mathrm{m})$. Then the *m*-value can be derived from the slope of the curve of osmolyte concentration versus midpoint temperature (Ferreon *et al.*, 2007). This process is explained in the following, using equations that are derived further below (Section 6.5).

Fitting each thermal scan: The individual fits in Fig. 3.9A were done with the equation

$$
\begin{aligned}
S = & S_{\mathrm{N,m}} + \mathrm{d}S_{\mathrm{N,T}}(T - T_m) \\
& + \left[S_{\mathrm{D,m}} - S_{\mathrm{N,m}} + \left(\mathrm{d}S_{\mathrm{D,T}} - \mathrm{d}S_{\mathrm{N,T}}\right)(T - T_\mathrm{m})\right]\frac{K}{1+K},
\end{aligned}
\tag{3.24}
$$

with K given by Eq. (3.41), temperature T, and all other parameters are constants. One of the first obvious features of the fits is that the curves match the data well but exhibit unreasonably looking extrapolations toward lower temperatures, where cold denaturation is (falsely) predicted to occur right below the experimental range of temperatures. Such behavior is typical for an overestimation of the curvature of the protein stability with temperature, $\Delta C_\mathrm{p}(T_\mathrm{m})$. This can be seen from Fig. 3.10, where an increase of the curvature (while T_m remains fixed) would increase the cold denaturation temperature, that is, the lower T where $\Delta G^\circ = 0$. In contrast to the gross

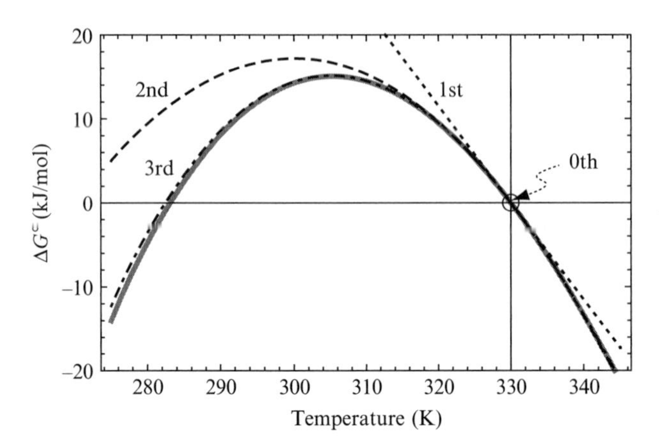

Figure 3.10 Effect of including more Taylor expansion terms: the expansion is anchored at $T_m = 330$ K, where the stability is zero. Including the first-order term (enthalpy, dotted line) gives a good approximation in a short temperature interval, approximately coinciding with the transition. Adding the second-order term (heat capacity, dashed line) and higher ones (dash-dotted line) progressively give a better representation of the true curve (gray continuous line).

overestimate of $\Delta C_p(T_m)$, the main parameters seem reasonably well fit, viz. the position of the transition T_m, and the width of the transition, $\Delta H^\circ(T_m)$. Figure 3.9B plots these parameters, along with a fit to Eq. (3.45), which is to first order

$$\Delta H^\circ(T) = \Delta H^\circ(T_{\text{ref}}) + \Delta C_p^\circ(T_{\text{ref}})(T - T_{\text{ref}}), \tag{3.25}$$

where we used the T_m at 0 M osmolyte as T_{ref}. The resulting $\Delta C_p^\circ(T_{\text{ref}})$ is 12 kJ/mol K, which is consistent with the heat capacities from each individual fit.

Deriving the m-value via curve-fitting: More information can be quickly derived from a phase diagram (Ferreon *et al.*, 2007; Rösgen and Hinz, 2003), which plots the 50% population line of protein states (here native and denatured). Such a diagram contains salient thermodynamic information that can be difficult to obtain otherwise. The phase diagram for our case is given in Fig. 3.9C, showing how the midpoint temperature of the transition depends on sarcosine concentration. The parameters obtained from the data in Fig. 3.9B define the temperature-dependent stability $\Delta G^\circ(T) = -RT \ln K$ (see Eq. (3.41)). A fit of the curve of osmolyte concentration versus T_m (Fig. 3.9C) can then be performed to yield the m-value. The needed equation is (Ferreon *et al.*, 2007)

$$c(T) = -\frac{\Delta G^\circ(T)}{m}, \tag{3.26}$$

as explained in Section 6.5. The m-value is 10.4 ± 0.18 kJ/mol M, with a confidence interval between 9.8 and 11.0 kJ/mol M. Thus, the result is but a little lower than the m-value derived from the global fit of the isothermal titrations (Table 3.4) and well within the range of the m-values derived by other methods (Fig. 3.5).

Deriving the m-value without additional curve-fitting: The m-value can be obtained even without a curve-fit of the phase diagram, namely from the slope of the phase separation line in Fig. 3.9C by the relation (Ferreon *et al.*, 2007)

$$m = \frac{\Delta H^\circ / T}{\partial c / \partial T}. \tag{3.27}$$

We consider the lowest concentrations of sarcosine (leftmost points in Figs. 3.9B and C). From panel B, we have $\Delta H^\circ \approx 260$ kJ/mol for the leftmost point, and from the first two points in panel C, the slope is about 0.3 M/3.5 K. Then, we get

$$m = \frac{260\,\mathrm{kJ/mol}}{307\,\mathrm{K}} \frac{3.5\,\mathrm{K}}{0.3\,\mathrm{M}} = 10\,\mathrm{kJ/mol\ M} \tag{3.28}$$

in good agreement with the fit using Eq. (3.26). Thus it is clear that a reasonable estimate of m may be possible with performing as little as two (reversible) thermal scans.

So far the results look nicely consistent, except for the apparent overestimation of $\Delta C_\mathrm{p}^\mathrm{o}(T_\mathrm{ref})$. This value is the slope of ΔH° with temperature. In our case, however, not only is the temperature varied as the various ΔH° (T_m) are sampled, but also the concentration of osmolyte is increased. Thus, the apparent $\Delta C_\mathrm{p}^\mathrm{o}(T_\mathrm{ref})$ contains contributions from the change of ΔH° with osmolyte concentration. In the individual fits, we could only neglect such contribution, but the global fit shown below allows to properly take it into account.

5.4. Global analysis

Analyzing all data simultaneously is the best use of the information, as long as the signal level is reproducible. Figure 3.9D shows that the native state baseline (gray data points below 20 °C) is reproducible within error. It is clear that the native state signal mainly depends on temperature and little on sarcosine concentration. However, the denatured state's signal clearly depends on sarcosine concentration and less on temperature. The overall signal can again be expressed as

$$S = S_{\mathrm{N,m}} + \mathrm{d}S_{\mathrm{N,S}}c + \mathrm{d}S_{\mathrm{N,T}}(T - T_m) + \left[S_{\mathrm{D,m}} - S_{\mathrm{N,m}} + \left(\mathrm{d}S_{\mathrm{D,S}} - \mathrm{d}S_{\mathrm{N,S}}\right)c + \left(\mathrm{d}S_{\mathrm{D,T}} - \mathrm{d}S_{\mathrm{N,T}}\right)(T - T_{\mathrm{m}})\right]\frac{K}{1+K}, \quad (3.29)$$

where the baseline parameters, $S_{\mathrm{N,m}}$, $S_{\mathrm{D,m}}$, etc., are constants, c is the sarcosine concentration, and K is given by Eq. (3.53), using terms up to second order without $(\partial m_K/\partial c)$.

Table 3.7 shows the results of a global fit of Eq. (3.29) to the data. Parameters that were not fitted, but afterwards derived from the fit are in square brackets. The m-value for sarcosine is slightly, but not significantly lower than in the previous fit (Table 3.4). We can also see that the m-value has a significant temperature dependence $(\partial m/\partial T)$ corresponding to a large slope of $\Delta H°$ with sarcosine concentration, $(\partial \Delta H°/\partial c) = -(\partial m_K/\partial \beta) = 51$ kJ/mol M.

This large effect, combined with the relatively moderate $\Delta C_{\mathrm{p}}^{\circ}$ of 2.8 kJ/mol K, results in the unreasonably large "heat capacity" obtained from the individual fits (Eq. (3.25)).

Small or negligible temperature dependences of urea m-values have been previously observed, typically between about 0 and 0.03 kJ/mol K (Baskakov and Bolen, 1999; DeKoster and Robertson, 1997; Felitsky and Record, 2003; Ferreon *et al.*, 2007; Giletto and Pace, 1999; Henkels and Oas, 2005; Henkels *et al.*, 2001; Hu *et al.*, 1992; Nicholson and Scholtz, 1996; Pace and Laurents, 1989), with extremes extending to 0.1 kJ/mol K (Auton *et al.*, 2009; Zweifel and Barrick, 2002). It is understandable that m normally does not depend much on temperature, based on the following considerations. According to Eq. (3.50), there is a factor of RT^2 between

Table 3.7 Global fit results for Case 3: Nank4-7* thermal unfolding

Parameter	Unit	Estimate	SE	Confidence interval
T_{m}	K	306.72	0.06	(306.59, 306.84)
$\Delta H°(T_{\mathrm{m}})$	kJ/mol	265	4	(257, 274)
$\Delta C_{\mathrm{p}}^{\circ}(T_{\mathrm{m}})$	kJ/mol K	2.8	1.3	(0.3, 5.4)
m_K	1/M	−4.26	0.1	(−4.46, −4.07)
$(\partial m_K/\partial \beta)$	kJ/mol M	−51	11	(−72, −29)
$S_{\mathrm{N,m}}$	deg cm^2/dmol	−10,020	30	(−10,080, −9960)
$S_{\mathrm{D,m}}$	deg cm^2/dmol	−4000	32	(−4070, −3940)
$\mathrm{d}S_{\mathrm{N,S}}$	deg cm^2/dmol M	−171	24	(−218, −124)
$\mathrm{d}S_{\mathrm{D,S}}$	deg cm^2/dmol M	−908	24	(−955, −862)
$\mathrm{d}S_{\mathrm{N,T}}$	deg cm^2/dmol K	12	1	(10, 14)
$\mathrm{d}S_{\mathrm{D,T}}$	deg cm^2/dmol K	2.7	1	(0.5, 4.9)
$[m]$	kJ/mol M	10.9	0.3	
$[\partial m/\partial T]$	kJ/mol M K	−0.130	0.036	

the temperature dependence of m_K, $(\partial m_K/\partial T)$, and the concentration dependence of the enthalpy, $(\partial \Delta H^\circ/\partial c)$. Because of $m = -RTm_K$ (Eq. (3.49)), there should then be a large, magnifying factor of about $-T$ between the concentration dependence of the enthalpy and the temperature dependence of the m-value. Only an exceedingly large $(\partial \Delta H^\circ/\partial c)$ would then lead to a substantial $(\partial m/\partial T)$. The temperature dependence of the m-value is comparably large in the current case, similar to previous measurements with a related protein (Zweifel and Barrick, 2002).

6. Where Do the Little Equations Come From?

In this section, we provide the background on the more complicated equations used throughout this chapter. Some of these are already moderately complex in 1D (e.g., temperature) and can become exceedingly difficult when an additional dimension is added. We therefore provide both the classical approaches, and alternatives that greatly facilitate the needed mathematics.

6.1. What is a Taylor expansion?

When a physical property has an entirely unknown mathematical form, it is often still possible to write down an equation that represents a good approximation of the unknown function. The principle is to know as much as possible about this function in one specific point, for example, the transition midpoint. This is illustrated in Fig. 3.10, which shows the stability of a protein ΔG° as a function of temperature. The coordinate system is set to highlight the midpoint temperature of thermal unfolding $T_m = 330$ K. The function value at this temperature $\Delta G^\circ(T_m)$ is the "0th derivative." Additional knowledge of the slope (first derivative) allows extrapolation of the stability to both higher and lower temperatures, as indicated by the short dashed line. Further taking into account the curvature (second derivative) and how the curvature changes with temperature (third derivative) results in progressively better approximation of the ΔG° (bold continuous line). As can be seen from Fig. 3.10, including more terms in the expansion extends the range in which the original function is well approximated. The derivatives are often directly related to physically meaningful parameters, such as the enthalpy and heat capacity (first and second temperature derivatives of the stability).

The Taylor expansion for a general function f (which, e.g., could be the protein stability ΔG°) is given by the following expression:

$$
\begin{aligned}
f(x) = \sum_{n=0}^{\infty} \frac{(x - x_0)^n}{n!} \left(\frac{\partial^n f(x)}{\partial x^n} \right)_{x_0} &= f(x_0) + (x - x_0) \left(\frac{\partial f(x)}{\partial x} \right)_{x_0} \\
+ \frac{(x - x_0)^2}{2} \left(\frac{\partial^2 f(x)}{\partial x^2} \right)_{x_0} &+ \frac{(x - x_0)^3}{6} \left(\frac{\partial^3 f(x)}{\partial x^3} \right)_{x_0} + \cdots
\end{aligned} \tag{3.30}
$$

Note that the value of the function $f(x_0)$ at some reference point x_0 is needed, as well as the derivatives of f with respect to x at the same point x_0. These are normally fitting parameters that are roughly known and will be optimized to their most likely values in a fitting routine that fits $f(x)$ to a given data set. Of course, it is impractical to use an infinitely large number of terms. In the example above, we list terms up to third order. How many terms should be used depends in general on the quality and extent of the data. Employing too few terms results in a curve that is not well represented by the Taylor expansion; employing too many terms results in large fitting errors that render meaningless the fitting parameters (x_0, and/or the various derivatives of $f(x)$ at x_0). For example, assume that data are available for the $\Delta G°$ displayed in Fig. 3.10, covering the temperature interval from 325 to 335 K. If the noise in the data conceals the degree of curvature, it would be sufficient to use terms up to first order. If, however, a curvature is detectable, the second-order term should be included. But it would be too much to use a third-order term, because in the given temperature interval there is no significant difference between the expansion up to second or third order.

When more than two independent variables are given (e.g., temperature and osmolyte concentration), the expansion of function $f(x,y)$ becomes longer. In the following equation, we represent the 2D expansion in three different ways:

$$
\begin{aligned}
f(x, y) &= \sum_{n=0}^{\infty} \sum_{k=0}^{\infty} \frac{(x - x_0)^n (y - y_0)^k}{n!k!} \left(\frac{\partial^{n+k} f(x, y)}{\partial x^n \partial y^k} \right)_{x_0, y_0} \\
&= \sum_{n=0}^{\infty} \left[\sum_{i=0}^{n} \frac{(x - x_0)^i (y - y_0)^{n-i}}{i!(n - i)!} \left(\frac{\partial^n f(x, y)}{\partial x^i \partial y^{n-i}} \right)_{x_0, y_0} \right]
\end{aligned} \tag{3.31}
$$

$$
\begin{aligned}
&= f(x_0, y_0) + \left[(x - x_0) \left(\frac{\partial f(x, y)}{\partial x} \right)_{x_0, y_0} + (y - y_0) \left(\frac{\partial f(x, y)}{\partial y} \right)_{x_0, y_0} \right] \\
&+ \sum_{n=2}^{\infty} \left[\sum_{i=0}^{n} \frac{(x - x_0)^i (y - y_0)^{n-i}}{i!(n - i)!} \left(\frac{\partial^n f(x, y)}{\partial x^i \partial y^{n-i}} \right)_{x_0, y_0} \right] + \cdots
\end{aligned} \tag{3.32}
$$

The first expression gives the general definition. In the second expression, the terms are sorted with respect to the number of derivatives, as illustrated in the third expression for the 0th and 1st derivative. Note that in the final double sum n starts from 2, rather than 0, because the first two terms are already written out. The second-order expression ($n = 2$) in Eq. (3.32) is:

$$\frac{(x-x_0)^2}{2}\left(\frac{\partial^2 f(x,y)}{\partial x^2}\right)_{x_0,y_0} + (x-x_0)(y-y_0)\left(\frac{\partial^2 f(x,y)}{\partial x \partial y}\right)_{x_0,y_0} + \frac{(y-y_0)^2}{2}\left(\frac{\partial^2 f(x,y)}{\partial y^2}\right)_{x_0,y_0} \tag{3.33}$$

As in the case of the 1D Taylor expansion (Eq. (3.30)), it depends on the data to be fitted, how many terms should be used. We indicated in Section 2.3 that ΔG° is normally a linear function in the osmolyte concentration c. That means terms up to first order in c are needed, and also the mixed terms should not exceed first order in c. The second-order terms from Eq. (3.33) would then become

$$\frac{(T-T_0)^2}{2}\left(\frac{\partial^2 \ln K(T,c)}{\partial T^2}\right)_{T_0,c_0} + (T-T_0)(c-c_0)\left(\frac{\partial^2 \ln K(T,c)}{\partial T \partial c}\right)_{T_0,c_0} + 0. \tag{3.34}$$

6.2. Protein stability as a function of temperature

6.2.1. Standard Taylor expansion

A regular power expansion in the temperature dimension is not as straightforward as we have seen at constant temperature (Sections 2 and 3). First of all, we should preferably work in the $1/RT$ dimension, rather than in the T dimension, because then the temperature dependence of the equilibrium constant K relates in a straightforward manner to the transition enthalpy at the reference temperature $\Delta H^\circ(T_{ref})$:

$$\left(\frac{\partial \ln K}{\partial 1/RT}\right)_{T_{ref}} = -\Delta H^\circ(T_{ref}). \tag{3.35}$$

For this reason, we also use ln K in place of $\Delta G^\circ = -RT \ln K$ and multiply the result later by $-RT$. However, nice as this works up to the first derivative, there is a problem resulting from a clash in definitions in the higher derivatives. Equation (3.35) shows that the first derivative of ln K is most conveniently done with respect to $1/RT$. But the second derivative of $\ln K$ (i.e., the derivative of Eq. (3.35)) is traditionally only available in the T dimension, rather than the $1/RT$ dimension:

$$\left(\frac{\partial \Delta H^{\circ}(T)}{\partial T}\right)_{T_{\mathrm{ref}}} = \Delta C_{\mathrm{p}}^{\circ}(T_{\mathrm{ref}}), \tag{3.36}$$

where $\Delta C_{\mathrm{p}}{}^{\circ}$ is the isobaric heat capacity change upon unfolding. For this reason, a nonstandard Taylor expansion is normally used for the temperature dependence of the protein stability. This will be explained further below in this section. But for now, we continue with a classical Taylor expansion for two reasons: First, to show in how far Eqs. (3.35) and (3.36) lead to problems; second, to point out that it is sometimes desirable to use the standard expansion, particularly when more variables are involved than just temperature (Sections 4 and 5).

Expanding $\ln K$ according to Eq. (3.30) up to second order in $1/RT$ results in

$$\begin{aligned} \ln K = \ln K_{\mathrm{ref}} &+ \left(\frac{\partial \ln K}{\partial 1/RT}\right)_{T_{\mathrm{ref}}} \left(\frac{1}{RT} - \frac{1}{RT_{\mathrm{ref}}}\right) \\ &+ \frac{1}{2}\left(\frac{\partial^2 \ln K}{\partial (1/RT)^2}\right)_{T_{\mathrm{ref}}} \left(\frac{1}{RT} - \frac{1}{RT_{\mathrm{ref}}}\right)^2 + \cdots. \end{aligned} \tag{3.37}$$

For simplicity, we write $\ln K_{\mathrm{ref}}$ in place of $\ln K(T_{\mathrm{ref}})$. If T_{ref} is chosen to be the midpoint temperature of the transition T_{m} we have $\ln K_{\mathrm{ref}} = \ln([\mathrm{D}]/[\mathrm{N}])_{\mathrm{ref}} = \ln 1 = 0$ for monomeric proteins, because the concentrations of native and denatured protein are equal at T_{m}. The first derivative is already given by the enthalpy (Eq. (3.35)), and the second one is

$$\left(\frac{\partial^2 \ln K}{\partial (1/RT)^2}\right)_{T_{\mathrm{ref}}} = \left(\frac{\partial - \Delta H^{\circ}}{\partial 1/RT}\right)_{T_{\mathrm{ref}}} = RT_{\mathrm{ref}}^2 \left(\frac{\partial \Delta H^{\circ}}{\partial T}\right)_{T_{\mathrm{ref}}} = RT_{\mathrm{ref}}^2 \Delta C_{\mathrm{p}}^{\circ}(T_{\mathrm{ref}}). \tag{3.38}$$

Upon calculation of the third and further derivatives, a problem becomes obvious. The heat capacity $\Delta C_{\mathrm{p}}^{\circ}(T_{\mathrm{ref}})$ is occurring in every term starting from the second-order one. We give just the third-order term for illustration purposes:

$$\begin{aligned} \left(\frac{\partial^3 \ln K}{\partial (1/RT)^3}\right) &= \left(\frac{\partial RT^2 \Delta C_{\mathrm{p}}^{\circ}}{\partial 1/RT}\right)_{T_{\mathrm{ref}}} = -RT_{\mathrm{ref}}^2 \left(\frac{\partial RT^2 \Delta C_{\mathrm{p}}^{\circ}}{\partial T}\right)_{T_{\mathrm{ref}}} \\ &= -RT_{\mathrm{ref}}^2 \left[RT_{\mathrm{ref}}^2 \left(\frac{\partial \Delta C_{\mathrm{p}}^{\circ}}{\partial T}\right)_{T_{\mathrm{ref}}} + 2RT_{\mathrm{ref}} \Delta C_{\mathrm{p}}^{\circ}(T_{\mathrm{ref}})\right] \end{aligned} \tag{3.39}$$

The expansion is typically done up to the second derivative. Consequently, this results in neglecting part of the contribution of a measurable quantity, $\Delta C_\mathrm{p}^\circ$, to the protein stability. Then, in face of confounding mathematical complexities that occur at higher dimensions, it is preferable to switch to the mathematically and physically more practical use of $(\partial\Delta H^\circ/\partial\beta)$ as the temperature derivative of the enthalpy instead of using the classical heat capacity. The latter one can still be calculated, and we will provide equations for this step with each experimental situation below.

Using Eq. (3.37) up to second order, we can insert Eqs. (3.35) and (3.38) to get

$$\begin{aligned}\ln K &= \ln K_\mathrm{ref} - \frac{\Delta H^\circ(T_\mathrm{ref})}{R}\left(\frac{1}{T}-\frac{1}{T_\mathrm{ref}}\right) + \frac{T_\mathrm{ref}^2\Delta C_\mathrm{p}^\circ(T_\mathrm{ref})}{2R}\left(\frac{1}{T}-\frac{1}{T_\mathrm{ref}}\right)^2 + \cdots \\ &= \ln K_\mathrm{ref} + \frac{\Delta H^\circ(T_\mathrm{ref})}{RTT_\mathrm{ref}}(T-T_\mathrm{ref}) + \frac{\Delta C_\mathrm{p}^\circ(T_\mathrm{ref})}{2RT^2}(T-T_\mathrm{ref})^2 + \cdots,\end{aligned} \tag{3.40}$$

where we brought the ratios in the brackets to their common denominator in the second step.

6.2.2. Classical protein stability curve: A nonstandard Taylor expansion

The classical protein stability curve up to second order has been discussed in detail by Becktel and Schellman (1987). A more general strategy is based on a double Taylor expansion (Franks *et al.*, 1988). It extends to higher-order terms and retains all occurrences of $\Delta C_\mathrm{p}^\circ(T_\mathrm{ref})$ and its derivatives within a single term each. In a nutshell, it is possible to write a Taylor expansion partly as a sum, and partly as an integral (the so-called Lagrange remainder). Used appropriately, we get a heat capacity term within this integral, and then do another Taylor expansion of the heat capacity:

$$\begin{aligned}\ln K = \ln K_\mathrm{ref} - \frac{\Delta H^\circ(T_\mathrm{ref})}{R}\left(\frac{1}{T}-\frac{1}{T_\mathrm{ref}}\right) - \int_{T_\mathrm{ref}}^{T}\left(\frac{1}{T}-\frac{1}{t}\right)\frac{\Delta C_\mathrm{p}^\circ(t)}{R}\mathrm{d}t = \ln K_\mathrm{ref} \\ + \frac{\Delta H^\circ(T_\mathrm{ref})}{RTT_\mathrm{ref}}(T-T_\mathrm{ref}) - \frac{\Delta C_\mathrm{p}^\circ(T_\mathrm{ref})}{RT}\left(T-T_\mathrm{ref}-T\ln\frac{T}{T_\mathrm{ref}}\right) + \cdots\end{aligned} \tag{3.41}$$

Here, we just show the results up to the zero-order term of $\Delta C_\mathrm{p}^\circ(T)$. Higher-order terms are (Rösgen and Hinz, 2000)

$$-\frac{\left(\partial_T\Delta C_\mathrm{p}^\circ\right)(T_\mathrm{ref})}{RT}\left(\frac{T_\mathrm{ref}^2-T^2}{2}+TT_\mathrm{ref}\ln\frac{T}{T_\mathrm{ref}}\right) \tag{3.42}$$

for the first temperature derivative of $\Delta C_\mathrm{p}^\circ$ at the reference temperature, $(\partial_T \Delta C_\mathrm{p}^\circ)(T_\mathrm{ref})$, and

$$-\frac{\left(\partial_T^2 \Delta C_\mathrm{p}^\circ\right)(T_\mathrm{ref})}{2RT}\left(\frac{T^3 - T_\mathrm{ref}^3}{3} + \frac{T(T - T_\mathrm{ref})^2}{2} - TT_\mathrm{ref}^2 \ln\frac{T}{T_\mathrm{ref}}\right) \quad (3.43)$$

for the second one, $(\partial_T{}^2 \Delta C_\mathrm{p}^\circ)(T_\mathrm{ref})$. For brevity, we used the symbols ∂_T and $\partial_T{}^2$ for first and second temperature derivatives.

It is possible to implement in two different ways the various fitting constants at T_ref, such as $\Delta H^\circ(T_\mathrm{ref})$, $\Delta C_\mathrm{p}^\circ(T_\mathrm{ref})$, $(\partial_T \Delta C_\mathrm{p}^\circ)(T_\mathrm{ref})$, $(\partial_T{}^2 \Delta C_\mathrm{p}^\circ)(T_\mathrm{ref})$. Either they can be directly used as constants, and then we could refer to them with subscripts as $\Delta HT_\mathrm{ref}^\circ$, $\Delta C_{\mathrm{p},T_\mathrm{ref}}^\circ$, $(\partial_T \Delta C_\mathrm{p}^\circ)_{T_\mathrm{ref}}$, $(\partial_T{}^2 \Delta C_\mathrm{p}^\circ) T_\mathrm{ref}$. Or, alternatively, the constants can be written as functions that are evaluated at T_ref. For example, the enthalpy equation can be defined

$$\Delta H^\circ(T) = \Delta H_{T_{\mathrm{ref}_2}}^\circ + \Delta C_{\mathrm{p},T_{\mathrm{ref}_2}}^\circ\left(T - T_{\mathrm{ref}_2}\right) + \frac{\left(\partial_T \Delta C_\mathrm{p}^\circ\right)_{T_{\mathrm{ref}_2}}}{2}\left(T - T_{\mathrm{ref}_2}\right)^2 + \frac{\left(\partial_T^2 \Delta C_\mathrm{p}^\circ\right)_{T_{\mathrm{ref}_2}}}{6}\left(T - T_{\mathrm{ref}_2}\right)^3 + \cdots, \quad (3.44)$$

the heat capacity,

$$\Delta C_\mathrm{p}^\circ(T) = \Delta C_{\mathrm{p},T_{\mathrm{ref}_2}}^\circ + (\partial_T \Delta C^\circ)_{\mathrm{p},T_{\mathrm{ref}_2}}\left(T - T_{\mathrm{ref}_2}\right) + \frac{\left(\partial_T^2 \Delta C_\mathrm{p}^\circ\right)_{T_{\mathrm{ref}_2}}}{2}\left(T - T_{\mathrm{ref}_2}\right)^2 + \cdots, \quad (3.45)$$

and, if necessary, also equations for the first (or even higher) derivative of the heat capacity

$$\left(\partial_T \Delta C_\mathrm{p}^\circ\right)(T) = \left(\partial_T \Delta C_\mathrm{p}^\circ\right)_{\mathrm{p},T_{\mathrm{ref}_2}} + \left(\partial_T^2 \Delta C_\mathrm{p}^\circ\right)_{\mathrm{p},T_{\mathrm{ref}_2}}\left(T - T_{\mathrm{ref}_2}\right) + \cdots. \quad (3.46)$$

Note that it is permissible to choose a value for T_{ref_2} in these equations that is different from the one used for T_ref in Eq. (3.41). This can be useful when parameters that were determined at one temperature (e.g., at T_m in the absence of osmolyte) are used for an expansion that is anchored at a different temperature (e.g., room temperature). Such situation may arise when more than one independent variable is used, for example, temperature and osmolyte concentration.

6.3. Protein stability as a function of temperature and osmolyte concentration

In this section, we assemble a Taylor expansion from the one in the temperature direction (Eq. (3.40)) and the one in the single concentration direction (either of Eqs. (3.14)–(3.16)). Because solvent-dependent protein stability is linear (first-order Taylor expansion), we normally do not need to go beyond $k = 1$ in Eq. (3.31) in the concentration dimension. In the temperature dimension, however, it is normal to find significant ΔC_p° contributions, which necessitates $n \geq 2$ in Eq. (3.31) in the temperature dimension. In some cases, where ligand binding is involved, higher-order terms may also be necessary in the concentration direction to properly capture the ligand binding as a function of osmolyte (Ferreon *et al.*, 2007; Gulotta *et al.*, 2007).

6.3.1. Issues with the classical stability equation

The ease or difficulty of formulating the 2D Taylor expansion greatly depends on the question whether cross correlations have to be taken into account. These correlations occur in the expansion as mixed derivatives, whose values should not depend on the order in which the derivatives are performed. With the mixed derivatives, we are again faced with an issue of a mismatch in definitions, as we saw in the case of the temperature derivatives in Section 6.2. The m-value is defined as the slope of the stability

$$m = \left(\frac{\partial \Delta G^\circ}{\partial c}\right)_T = -RT\left(\frac{\partial \ln K}{\partial c}\right)_T. \tag{3.47}$$

The problem is that this definition contains a factor of T. This leads to complications in the mixed temperature/concentration derivative

$$\left(\frac{\partial m}{\partial T}\right)_c = -R\left(\frac{\partial \ln K}{\partial c}\right)_T - RT\left(\frac{\partial^2 \ln K}{\partial c \partial T}\right) = \frac{m}{T} - \frac{1}{T}\left(\frac{\partial \Delta H^\circ}{\partial c}\right)_T, \tag{3.48}$$

which has on the right-hand side two terms that contain T explicitly. As a result, m reoccurs in every temperature derivative and is thus scattered over all remaining terms, just as ΔC_p° was earlier (see Eq. (3.39)). Even worse, if we assume there is no cross-correlation ($\partial m/\partial T = 0$), then we get the self-conflicting result that there is cross-correlation, because Eq. (3.48) tells us that the equivalent cross-correlation term $\partial \Delta H^\circ / \partial T$ is nonzero and equals m in that case. It is therefore highly reasonable to define an alternative m-value, based on the equilibrium constant, rather than on the stability

$$m_K = \left(\frac{\partial \ln K}{\partial c}\right)_T = -\frac{m}{RT}. \tag{3.49}$$

Then, the mismatch in Eq. (3.48) is resolved

$$\left(\frac{\partial m_K}{\partial 1/RT}\right)_c = -\left(\frac{\partial \Delta H^\circ}{\partial c}\right)_T. \tag{3.50}$$

This definition avoids running into much confusion in Taylor expansions that involve both temperature and concentration axes.

The implementation of the definition in Eq. (3.50) is possible in two ways. In the first possible implementation, the classical stability equation can be duplicated for each concentration derivative, that is,

$$\begin{aligned}\ln K = \ln K_{\text{ref}} + \frac{\Delta H^\circ_{\text{ref}}}{RTT_{\text{ref}}}(T - T_{\text{ref}}) - \frac{\Delta C^\circ_{\text{p,ref}}}{RT}\left(T - T_{\text{ref}} - T\ln\frac{T}{T_{\text{ref}}}\right) + m_{K,\text{ref}} + \\ \frac{(\partial_c \Delta H^\circ)_{\text{ref}}}{RTT_{\text{ref}}}(T - T_{\text{ref}}) - \frac{\left(\partial_c \Delta C^\circ_{\text{p}}\right)_{\text{ref}}}{RT}\left(T - T_{\text{ref}} - T\ln\frac{T}{T_{\text{ref}}}\right) + \cdots\end{aligned} \tag{3.51}$$

However, this approach leads to difficulties when one wants to convert mixed derivatives into each other. For example, $(\partial_c \Delta C^\circ_{\text{p}})_{\text{ref}}$ is representative of both the concentration dependence of the heat capacity and of the second temperature derivative of the m-value. But the current implementation makes such conversion cumbersome. The better implementation of the definition in Eq. (3.50) is based on a standard Taylor expansion, as discussed now.

6.3.2. A better approach: Standard Taylor expansion

Two- and higher-dimensional expansions can quickly get very bulky, and thus we use a notation from physics that defines the inverse temperature β and differences that often occur:

$$\beta = \frac{1}{RT}, \Delta\beta = \left(\frac{1}{RT} - \frac{1}{RT_{\text{ref}}}\right), \Delta\beta^n = \left(\frac{1}{RT} - \frac{1}{RT_{\text{ref}}}\right)^n. \tag{3.52}$$

The difference notation equally applies to other variables, for example, $\Delta c^n = (c - c_{\text{ref}})^n$. As pointed out in the previous section, it makes most

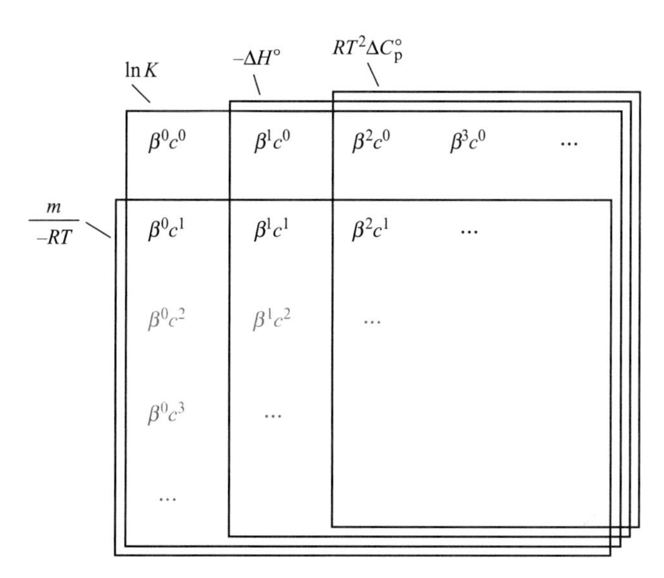

Figure 3.11 Illustration of a Taylor expansion of several thermodynamic properties with respect to inverse temperature ($\beta = 1/RT$) and concentration c. The superscripts of the variables β and c indicate the order of the term (corresponding to i and $n - i$ in Eq. (3.53)) in the expansion of ln K. Terms up to third order in both dimensions are represented, and their specific form is given in Eq. (3.53). For protein folding, the terms of second and higher order in c are normally insignificant and are grayed out here. The boxes indicate which terms are needed for calculating various thermodynamic properties (Eqs. (3.55)–(3.57)).

sense to do a standard expansion in β and c when we want to express the energetics of the equilibrium as a function of temperature and osmolyte concentration. Figure 3.11 illustrates this expansion, and how it has overlapping terms with other expansions given further below. For the respective expansions, the top left corner of each rectangle corresponds to the zeroth derivative with respect to both dimensions. The number of inverse-temperature (β) derivatives increases toward the right, and concentration derivatives toward the bottom. The entries in Fig. 3.11 merely give the order of the terms for ln K. The full equation (after the first equal sign) and the explicit terms up to second order (after the second equal sign) are

$$\begin{aligned}\ln K(T,c) &= \sum_{n=0}^{\infty}\sum_{i=0}^{n}\frac{\Delta\beta^i\Delta c^{n-i}}{i!(n-i)!}\left(\frac{\partial^n \ln K}{\partial\beta^i\partial c^{n-i}}\right)_{\beta_{\text{ref}},c_{\text{ref}}}\\ &= \ln K(T_{\text{ref}},c_{\text{ref}}) - \Delta\beta\Delta H^\circ(T_{\text{ref}},c_{\text{ref}}) + \Delta c\cdot m_K(T_{\text{ref}},c_{\text{ref}})\\ &\quad+\frac{\Delta\beta^2}{2}RT_{\text{ref}}^2\Delta C_{\text{p}}^\circ(T_{\text{ref}},c_{\text{ref}}) + \Delta\beta\Delta c\left(\frac{\partial m_K}{\partial\beta}\right)_{T_{\text{ref}},c_{\text{ref}}} + \frac{\Delta c^2}{2}\left(\frac{\partial m_K}{\partial c}\right)_{T_{\text{ref}},c_{\text{ref}}} + \cdots.\end{aligned} \tag{3.53}$$

Note that on the right-hand side of the equation there is one zero-order term, followed by two first-order terms and three second-order terms. As mentioned earlier, the Δc^2 term is usually negligible for protein folding, but may be significant in the case of ligand binding (Ferreon *et al.*, 2007; Gulotta *et al.*, 2007). The second temperature derivative in Eq. (3.53) was rewritten in terms of heat capacity according to Eq. (3.38). This step may or may not be desirable to do, depending on whether higher temperature derivatives are significant. If they are, we recommend to use $(\partial^2 \ln K/\partial\beta^2)_{\mathrm{ref}} = -(\partial\Delta H^\circ/\partial\beta)_{\mathrm{ref}}$ as parameter rather than substituting it with a heat capacity expression as done above. Among the third-order terms,

$$-\frac{\Delta\beta^2}{6}\left(\frac{\partial^2\Delta H^\circ}{\partial\beta^2}\right)_{\mathrm{ref}}+\frac{\Delta\beta^2\Delta c}{2}\left(\frac{\partial^2 m_K}{\partial\beta^2}\right)_{\mathrm{ref}}+\frac{\Delta\beta\Delta c^2}{2}\left(\frac{\partial^2 m_K}{\partial\beta\partial c}\right)_{\mathrm{ref}}+\frac{\Delta c^3}{6}\left(\frac{\partial^2 m_K}{\partial c^2}\right)_{\mathrm{ref}}, \tag{3.54}$$

we expect to rarely ever need to invoke the last two terms, which reflect concentration-dependent m-values, for protein conformational transitions. By the same token, all terms beyond first order ($n > 1$) should be restricted to derivatives of the type $\partial^{n-1}\Delta H^\circ/\partial\beta^{n-1}$ and $\partial^{n-1}m_K/\partial\beta^{n-1}$ (cf. the first two terms in Eq. (3.54)) whenever only proteins are involved in equilibrium. These terms correspond to the first two rows in Fig. 3.11.

The traditional thermodynamic parameters can be directly calculated (if desired) using the same parameters as in Eq. (3.53). Specifically, the enthalpy

$$\begin{aligned}\Delta H^\circ(T,c) &= -\left(\frac{\partial \ln K}{\partial\beta}\right)_c = -\sum_{n=0}^{\infty}\sum_{i=1}^{n}\frac{\Delta\beta^{i-1}\Delta c^{n-i}}{(i-1)!(n-i)!}\left(\frac{\partial^n \ln K}{\partial\beta^i c^{n-i}}\right)_{\beta_{\mathrm{ref}},c_{\mathrm{ref}}}\\ &= \Delta H^\circ(T_{\mathrm{ref}},c_{\mathrm{ref}}) - \Delta\beta R T_{\mathrm{ref}}^2\Delta C_{\mathrm{p}}^\circ(T_{\mathrm{ref}},c_{\mathrm{ref}}) - \Delta c\left(\frac{\partial m_K}{\partial\beta}\right)_{T_{\mathrm{ref}},c_{\mathrm{ref}}} + \cdots,\end{aligned} \tag{3.55}$$

heat capacity

$$\begin{aligned}\Delta C_{\mathrm{p}}^\circ(T,c) &= \frac{-1}{RT^2}\left(\frac{\partial\Delta H^\circ}{\partial\beta}\right)_c = \frac{1}{RT^2}\sum_{n=0}^{\infty}\sum_{i=2}^{n}\frac{\Delta\beta^{i-2}\Delta c^{n-i}}{(i-2)!(n-i)!}\\ &\quad\left(\frac{\partial^n \ln K}{\partial\beta^i c^{n-i}}\right)_{\beta_{\mathrm{ref}},c_{\mathrm{ref}}} = \frac{T_{\mathrm{ref}}^2}{T^2}\Delta C_{\mathrm{p}}^\circ(T_{\mathrm{ref}},c_{\mathrm{ref}}) + \cdots,\end{aligned} \tag{3.56}$$

and m-value

$$m(T,c) = -RT\left(\frac{\partial \ln K}{\partial c}\right)_{\beta} = -RT\sum_{n=1}^{\infty}\sum_{i=0}^{n-1}\frac{\Delta\beta\Delta c^{n-i-1}}{i!(n-i-1)!}\left(\frac{\partial^n \ln K}{\partial\beta^i c^{n-i}}\right)_{\beta_{\text{ref}},c_{\text{ref}}}$$
$$= -RT\cdot m_K(T_{\text{ref}},c_{\text{ref}}) - RT\Delta\beta\left(\frac{\partial m_K}{\partial\beta}\right)_{T_{\text{ref}},c_{\text{ref}}} - RT\Delta c\left(\frac{\partial m_K}{\partial c}\right)_{T_{\text{ref}},c_{\text{ref}}} + \cdots \quad (3.57)$$

are directly obtained, as shown here up to second order. Note again that the concentration dependence of the m-value, $(\partial m_K/\partial c)$, is normally zero for protein transitions and is here just given for the sake of completeness. These various expansions (Eqs. (3.55)–(3.57)) use some, but not all terms from the expansion of ln K (Eq. (3.53)), as indicated in Fig. 3.11.

6.3.3. Implementing parameters that are already known

When we are dealing with a 2D expansion in temperature and concentration direction, we obtain from the experiments in each dimension predictions for the stability at one common reference point. This point is at zero molar additive and the experimental temperature from the osmolyte titration (often room temperature). Therefore, the following equation holds (in terms of the classical stability equation):

$$m\cdot(0-c_{\text{m}}) = \Delta H^{\circ}(T_{\text{m}})\left(1-\frac{T_{\text{c}}}{T_{\text{m}}}\right) + \Delta C_{\text{p}}^{\circ}(T_{\text{m}})\left(T_{\text{c}} - T_{\text{m}} - T_{\text{c}}\ln\frac{T_{\text{c}}}{T_{\text{m}}}\right) + \cdots, \quad (3.58)$$

where T_c is the experimental temperature in the isothermal titration with the osmolyte. This equation can be solved for any of the parameters to eliminate the need to fit it (Ferreon *et al.*, 2007). This is preferably done with a parameter that is suspected to be of significance but cannot be determined with certainty otherwise. For example, Eq. (3.58) could be expanded up to the third-order term (first temperature derivative of the heat capacity, Eq. (3.42)) and then solved for $\partial_T \Delta C^{\circ}_{\text{p, Tm}}$:

$$\partial_T \Delta C^{\circ}_{\text{p},T_{\text{m}}} = \frac{-mc_{\text{m}} - \Delta H^{\circ}(T_{\text{m}})(1-(T_{\text{c}}/T_{\text{m}})) - \Delta C_{\text{p}}^{\circ}(T_{\text{m}})(T_{\text{c}} - T_{\text{m}} - T_{\text{c}}\ln(T_{\text{c}}/T_{\text{m}}))}{(T_{\text{ref}}^2 - T_{\text{c}}^2)/2 + T_{\text{c}}T_{\text{ref}}\ln(T_{\text{c}}/T_{\text{ref}})}. \quad (3.59)$$

Every occurrence of $\partial_T \Delta C^{\circ}_{\mathrm{p},T_{\mathrm{m}}}$ can then be replaced with Eq. (3.59) in the 2D curve fit.

When the new alternative approach is used instead of the classical stability equation, we have

$$\sum_{n=1}^{n_c} \frac{(0-c_{\mathrm{m}})^n}{n!} \left(\frac{\partial^n \ln K}{\partial c^n}\right)_{T_{\mathrm{c}},c_{\mathrm{m}}} = \sum_{n=1}^{n_\beta} \frac{(1/T_{\mathrm{c}} - 1/T_{\mathrm{m}})^n}{n!R} \left(\frac{\partial^n \ln K}{\partial \beta^n}\right)_{T_{\mathrm{m}},c=0} \tag{3.60}$$

instead of Eq. (3.58), and we have $n_c = 1$ and $n_\beta = 3$ for the situation given in Eq. (3.59). We can also here solve for any desired parameter, for example, to match the situation in Eq. (3.59)

$$\left(\frac{\partial^3 \ln K}{\partial \beta}\right)_{T_{\mathrm{m}},c=0} = \frac{-c_{\mathrm{m}} m_{K,0} + \Delta^{\mathrm{c}}_{\mathrm{m}}\beta \Delta H^{\circ}_0 - \Delta^{\mathrm{c}}_{\mathrm{m}}\beta^2 (R/2)\left(\partial^2 \ln K/\partial \beta^2\right)_{T_{\mathrm{m}},c=0}}{(R^2/6)\Delta^{\mathrm{c}}_{\mathrm{m}}\beta^3}, \tag{3.61}$$

where $m_{K,0}$ is the parameter obtained at T_{c} and c_{m}, and ΔH°_0 is the one obtained at T_{m} and $c = 0$. We defined $\Delta_{\mathrm{m}}{}^{\mathrm{c}}\beta = \beta_{\mathrm{c}} - \beta_{\mathrm{m}}$, and $\Delta_{\mathrm{m}}{}^{\mathrm{c}}\beta^n = (\beta_{\mathrm{c}} - \beta_{\mathrm{m}})^n$ to get a more compact, less lengthy expression.

6.4. Protein stability as a function of temperature and the concentrations of two osmolytes

The equations in 3D (concentrations of two osmolytes and temperature) follow the same principles as those in two. The general equation and terms up to second order are

$$\begin{aligned}
\ln K(T, c_1, c_2) &= \sum_{n=0}^{\infty} \sum_{i+j+k=n} \frac{\Delta\beta^i \Delta c_1^j \Delta c_2^k}{i!j!k!} \left(\frac{\partial^n \ln K}{\partial \beta^i \partial c_1^j \partial c_2^k}\right)_{\mathrm{ref}} \\
&= \ln K_{\mathrm{ref}} - \Delta\beta \Delta H^{\circ}_{\mathrm{ref}} + \Delta c_1 \cdot m_{K,1,\mathrm{ref}} + \Delta c_2 \cdot m_{K,2,\mathrm{ref}} \\
&\quad + \frac{\Delta\beta^2}{2} R T^2_{\mathrm{ref}} \Delta C^{\circ}_{\mathrm{p,ref}} + \Delta c_1 \Delta c_2 \left(\frac{\partial m_{K,1}}{\partial c_2}\right)_{\mathrm{ref}} \\
&\quad + \Delta\beta \Delta c_1 \left(\frac{\partial m_{K,1}}{\partial \beta}\right)_{\mathrm{ref}} + \Delta\beta \Delta c_2 \left(\frac{\partial m_{K,2}}{\partial \beta}\right)_{\mathrm{ref}} \\
&\quad + \frac{\Delta c_1^2}{2} \left(\frac{\partial m_{K,1}}{\partial c_1}\right)_{\mathrm{ref}} + \frac{\Delta c_2^2}{2} \left(\frac{\partial m_{K,2}}{\partial c_2}\right)_{\mathrm{ref}} + \cdots
\end{aligned} \tag{3.62}$$

There is one zero-order term, ln K_{ref}, followed by three first-order terms and six second-order terms. Plotting this equation directly is not possible because it has 4D, T, c_1, c_2, and ln K. But it is possible to show a phase diagram (see section 6.5), utilizing some of the results shown earlier in this chapter.

For the protein Nank4-7*, we already have most of the needed parameters in Tables 3.4 and 3.7 and use them as follows. The temperature dependence of the urea m_K-value, $(\partial m_{K,\text{U}} \partial \beta)_{\text{ref}}$, is not known and therefore set to zero. Also, the Δc_i^2 terms are set to zero, as usual. We choose T_{m} as reference temperature, and thus ln $K_{\text{ref}} = 0$. The other reference conditions are $c_{\text{ref,U}} = 0\ M$ and $c_{\text{ref,S}} = 0\ M$, and thus $\Delta c_{\text{u}} = c_{\text{u}}$ and $\Delta c_{\text{S}} = c_{\text{S}}$. The m_K values, as well as the mixed concentration derivative, $(\partial m_{K,\text{S}} / \partial c_{\text{U}}) = (\partial m_{K,\text{U}} / \partial c_{\text{S}})$, are calculated from the corresponding m_{U}, m_{S}, and $m_{\text{U,S}}$ values in Table 3.4 through division with $-RT$ (Eq. (3.49)). The residual parameters are taken from Table 3.7.

Figure 3.12 shows phase separation lines of ln $K = 0$, that is, 50% native, in the two concentration dimensions at different temperatures. Each phase separation line in the figure may be either interpreted as the midpoint concentration of Nank4-7* denaturation by urea in the presence of various amounts of sarcosine; or it could conversely be the midpoint concentration of forced folding with sarcosine at various amounts of urea in the

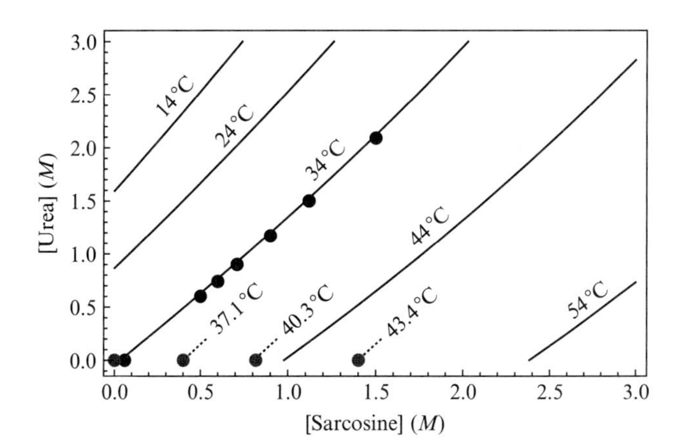

Figure 3.12 Phase diagram of 50% native Nank4-7* at different temperatures: the solid lines are calculated from Eq. (3.71). The protein is predominantly denatured under conditions above the lines, and native below. The black points are the midpoints of the urea unfolding curves at various background sarcosine concentrations and of the sarcosine folding curves at various urea background concentrations at 34 °C (Fig. 3.5). The gray points are calculated from the T_m values of Nank4-7* denaturation in the presence of 0, 0.3, 0.6, and 1 M sarcosine, and 0 M urea (Fig. 3.9).

background. These lines are slightly curved, because of the small degree of synergy, $m_{U,S}$, between urea and sarcosine: from the principles given by Ferreon *et al.* (2007), it can be deduced that the slope of the phase separation lines in Fig. 3.12 is

$$\left(\frac{\partial c_U}{\partial c_S}\right) = -\frac{m_S}{m_U}, \tag{3.63}$$

which would be a constant ratio of constants in the absence of synergy (see also Section 6.5 below).

Similar to Section 6.3.2, the traditional thermodynamic parameters can be directly calculated using the same parameters as in Eq. (3.62). Specifically, the enthalpy

$$\begin{aligned}\Delta H^\circ(T, c_1, c_2) &= -\left(\frac{\partial \ln K}{\partial \beta}\right)_c \\ &= -\sum_{n=1}^{\infty} \sum_{\substack{i+j+k=n,\\ i>0}} \frac{\Delta\beta^{i-1}\Delta c_1^j \Delta c_2^k}{(i-1)!j!k!}\left(\frac{\partial^n \ln K}{\partial \beta^i \partial c_1^j \partial c_2^k}\right)_{\text{ref}} \\ &= \Delta H^\circ_{\text{ref}} - \Delta\beta R T^2_{\text{ref}} \Delta C^\circ_{\text{p,ref}} - \Delta c_1 \left(\frac{\partial m_{K,1}}{\partial \beta}\right)_{\text{ref}} - \Delta c_2 \left(\frac{\partial m_{K,2}}{\partial \beta}\right)_{\text{ref}} + \cdots,\end{aligned} \tag{3.64}$$

heat capacity

$$\begin{aligned}\Delta C^\circ_{\text{p}}(T, c_1, c_2) &= \frac{-1}{RT^2}\left(\frac{\partial \Delta H^\circ}{\partial \beta}\right)_c = \frac{1}{RT^2}\sum_{n=2}^{\infty} \sum_{\substack{i+j+k=n,\\ i>1}} \\ &\frac{\Delta\beta^{i-2}\Delta c_1^j \Delta c_2^k}{(i-2)!j!k!}\left(\frac{\partial^n \ln K}{\partial \beta^i \partial c_1^j \partial c_2^k}\right)_{\text{ref}} = \frac{T^2_{\text{ref}}}{T^2}\Delta C^\circ_{\text{p,ref}} + \cdots,\end{aligned} \tag{3.65}$$

and m_1-value

$$
\begin{aligned}
m_1(T, c_1, c_2) &= -RT\left(\frac{\partial \ln K}{\partial c_1}\right)_\beta \\
&= -RT\sum_{n=1}^{\infty}\sum_{\substack{i+j+k=n,\\ j>0}} \frac{\Delta\beta^i \Delta c_1^{j-1} \Delta c_2^k}{i!(j-1)!k!}\left(\frac{\partial^n \ln K}{\partial\beta^i \partial c_1^j \partial c_2^k}\right)_{\text{ref}} \\
&= -RTm_{K,1,\text{ref}} - RT\Delta\beta\left(\frac{\partial m_{K,1}}{\partial\beta}\right)_{\text{ref}} \\
&\quad -RT\Delta c_2\left(\frac{\partial m_{K,1}}{\partial c_2}\right)_{\text{ref}} - RT\Delta c_1\left(\frac{\partial m_{K,1}}{\partial c_1}\right)_{\text{ref}} + \cdots
\end{aligned}
\tag{3.66}
$$

are directly obtained, as shown here up to second order. The m_2-value can be written analogously to m_1. Again, the concentration dependence of the m_1-value, $(\partial m_{K,1}/\partial c_1)$, is normally zero for protein transitions and is here just given for the sake of completeness.

6.5. Phase diagrams

The phase diagram method is a handy way to derive results from data that are difficult to get otherwise (Ferreon *et al.*, 2007; Rajagopalan *et al.*, 2005; Rösgen and Hinz, 2003). A protein phase diagram for our purposes is characterized by lines of 50% population of one or more protein states, such as folded, unfolded, ligand bound, ligand free, etc. These lines are also called phase separation lines. Throughout this chapter, there are several examples for phase diagrams. Figs. 3.6 and 3.12 show at which urea and sarcosine concentrations Nank4-7⋆ has a 50% native population, Figure 3.7D shows the 50% native line for the A3 domain of von Willebrand factor in the urea-temperature plane, and Fig. 3.9C shows a phase separation line (50% population) of the unfolding equilibrium of Nank4-7⋆. In the following, we explain how to derive the corresponding equations, focusing on the one used for Fig. 3.9C.

On the line of 50% native and unfolded population, we have $K = [\mathrm{D}]/[\mathrm{N}] = 1$, and thus $\ln K = 0$. Then, Eq. (3.53) can be rewritten, so that all concentration independent terms are on the left-hand side, and all other terms on the right-hand side. Neglecting second and higher-order concentration terms is permissible for protein conformational changes, resulting in

$$-\sum_{n=0}^{\infty}\frac{\Delta\beta^n}{n!}\left(\frac{\partial^n \ln K}{\partial\beta^n}\right)_{\beta_{ref},c_{ref}}=\sum_{n=0}^{\infty}\frac{\Delta\beta^n\Delta c}{n!}\left(\frac{\partial^{n+1}\ln K}{\partial\beta^n\partial_c}\right)_{\beta_{ref},c_{ref}}. \tag{3.67}$$

This equation can directly be solved for Δc to yield

$$\Delta c=-\frac{\sum_{n=0}^{\infty}\frac{\Delta\beta^n}{n!}\left(\frac{\partial^n \ln K}{\partial\beta^n}\right)_{\beta_{ref},c_{ref}}}{\sum_{n=0}^{\infty}\frac{\Delta\beta^n}{n!}\left(\frac{\partial^{n+1}\ln K}{\partial\beta^n\partial c}\right)_{\beta_{ref},c_{ref}}}. \tag{3.68}$$

The numerator in this equation is the temperature dependence of the equilibrium constant at 0 M osmolyte, ln $K(T, 0\ M)$, and the denominator represents m_K (Eq. (3.49)) under the assumption that it does not depend on c. Equation (3.68) can thus be rewritten as

$$\Delta c=-\frac{\Delta G^{\circ}(T,0M)}{m_2(T)} \tag{3.69}$$

using Eq. (3.57) for m, and $\Delta G^{\circ}(T,\ 0\ M)=-RT$ ln $K(T,\ 0\ M)$. In Section 5.3, Eq. (3.26), the m-value was assumed to be independent of temperature, because this resulted in a good fit, without the need of employing further parameters. Only when either $\Delta C_{\mathrm{p}}^{\circ}$ can be determined with great confidence (which is rarely the case) or $\partial m/\partial T$ (or $\partial m_K/\partial\beta$) is it possible to distinguish between the effects of $\Delta C_{\mathrm{p}}^{\circ}$ and $\partial m/\partial T$ with this quick and simple approach.

The situation is similar in 3D (T, c_1, and c_2). Under the assumption that there are no significant quadratic or higher-order terms in Δc_2, we get from Eq. (3.62)

$$\Delta c_2=-\frac{\sum_{n=0}^{\infty}\sum_{i+j=n}\frac{\Delta\beta^i\Delta c_1^j}{i!j!}\left(\frac{\partial^n \ln K}{\partial\beta^i\partial c_1^j}\right)_{ref}}{\sum_{n=0}^{\infty}\sum_{i+j=n}\frac{\Delta\beta^i\Delta c_1^j}{i!j!}\left(\frac{\partial^n \ln K}{\partial\beta^i\partial c_1^j\partial c_2}\right)_{ref}}. \tag{3.70}$$

Again, this equation can be written shorter as

$$\Delta c_2=-\frac{\Delta G^{\circ}(T,c_1,0M)}{m_2(T,c_1)}, \tag{3.71}$$

where

$$-\frac{\Delta G^{\circ}(T,c_1,0M)}{RT}=\ln K_{\mathrm{ref}}-\Delta\beta\Delta H^{\circ}_{\mathrm{ref}}+\frac{\Delta\beta^2}{2}RT^2_{\mathrm{ref}}\Delta C^{\circ}_{\mathrm{p,ref}}+\Delta c_1\cdot m_{K,1,\mathrm{ref}}+\Delta\beta\Delta c_1\left(\frac{\partial m_{K,1}}{\partial\beta}\right)_{\mathrm{ref}}+\cdots \tag{3.72}$$

and

$$-\frac{m_2(T,c_1)}{RT}=m_{K,2,\mathrm{ref}}+\Delta c_1\left(\frac{\partial m_{K,1}}{\partial c_2}\right)_{\mathrm{ref}}+\Delta\beta\left(\frac{\partial m_{K,2}}{\partial\beta}\right)_{\mathrm{ref}}+\cdots, \tag{3.73}$$

showing terms up to second order.

For the case of isothermal folding/unfolding in the presence of osmolyte mixtures as discussed in Section 3, Eq. (3.71) becomes

$$\Delta c_2=\frac{\Delta G^{\circ}_{\mathrm{ref}}+\Delta c_1 m_1}{m_2+\Delta c_1 m_{12}}, \tag{3.74}$$

where we used $m_i = -RTm_{K,\,i}$, $m_{12} = -RT(\partial\, m_{K,\,1}/\partial\, c_2)_{\mathrm{ref}}$, and $\Delta G_{\mathrm{ref}}^{\circ} = -RT\ln K_{\mathrm{ref}}$.

6.6. Three-state versus two-state transitions

Many proteins exhibit two-state transitions between their native and denatured states. But especially in large proteins, there are often additional transitions. These can be taken into account according to the same principles outlined for two-state systems. Consider the equilibrium

$$N \rightleftharpoons I \rightleftharpoons D, \tag{3.75}$$

which contains an intermediate *I*. It is possible to define three equilibrium constants

$$K_{\mathrm{IN}}=\frac{[\mathrm{I}]}{[\mathrm{N}]},\ K_{\mathrm{DN}}=\frac{[\mathrm{D}]}{[\mathrm{N}]},\ K_{\mathrm{DI}}=\frac{[\mathrm{D}]}{[\mathrm{I}]} \tag{3.76}$$

for the equilibria between *N* and *I*, between *N* and *D*, and between *I* and *D*. Since $K_{\mathrm{DN}} = K_{\mathrm{DI}}K_{\mathrm{IN}}$, we can express the fraction of denatured molecules in two alternative ways:

$$f_{\mathrm{D}} = \frac{[\mathrm{D}]}{[\mathrm{N}] + [\mathrm{I}] + [\mathrm{D}]} = \frac{K_{\mathrm{DN}}}{1 + K_{\mathrm{IN}} + K_{\mathrm{DN}}} = \frac{K_{\mathrm{DI}}K_{\mathrm{IN}}}{1 + K_{\mathrm{IN}} + K_{\mathrm{DI}}K_{\mathrm{IN}}}. \tag{3.77}$$

Similarly, the fraction of intermediates can be written as:

$$f_{\mathrm{I}} = \frac{[\mathrm{I}]}{[\mathrm{N}] + [\mathrm{I}] + [\mathrm{D}]} = \frac{K_{\mathrm{IN}}}{1 + K_{\mathrm{IN}} + K_{\mathrm{DN}}} = \frac{K_{\mathrm{IN}}}{1 + K_{\mathrm{IN}} + K_{\mathrm{DI}}K_{\mathrm{IN}}}. \tag{3.78}$$

Whenever the intermediate is not much populated, it is convenient to choose the variants with K_{DN}. However, the variants containing $K_{\mathrm{DI}}K_{\mathrm{IN}}$ may be preferable if the transition between I and D is well defined and visible. Each of the two chosen K is expressed according to the experimental setup, as discussed in Sections 2.3.2 (as a function of osmolyte concentration), 3.4 (as a function of two osmolytes), 4.4, 5.3, 5.4, and 6.3.2 (as a function of temperature, or temperature and osmolyte), and 6.4 (as a function of temperature and several osmolytes).

The equation for the experimental signal

$$S = S_{\mathrm{N}} + (S_{\mathrm{I}} - S_{\mathrm{N}})f_{\mathrm{I}} + (S_{\mathrm{D}} - S_{\mathrm{N}})f_{\mathrm{D}} \tag{3.79}$$

is very similar to the ones shown for the two-state transitions. Note that the signals of the three states, S_{N}, S_{I}, and S_{D}, can be dependent on concentrations and temperature, for example,

$$S_{\mathrm{N}} = S_{\mathrm{N,ref}} + \Delta c_1 \left(\frac{\partial S_{\mathrm{N}}}{\partial c_1}\right)_{ref} + \Delta c_2 \left(\frac{\partial S_{\mathrm{N}}}{\partial c_2}\right)_{ref} + \Delta T \left(\frac{\partial S_{\mathrm{N}}}{\partial T}\right)_{ref}. \tag{3.80}$$

6.7. Oligomeric versus monomeric proteins

So far, we have only dealt with the unfolding of monomeric proteins

$$N \rightleftharpoons D, \quad K = \frac{[\mathrm{D}]}{[\mathrm{N}]}. \tag{3.81}$$

With oligomeric proteins, the same principles do apply as in the case of monomeric proteins. The general transition for homo-oligomeric proteins is:

$$N_n \rightleftharpoons nD, \quad K = \frac{[\mathrm{D}]^n}{[\mathrm{N}_n]}. \tag{3.82}$$

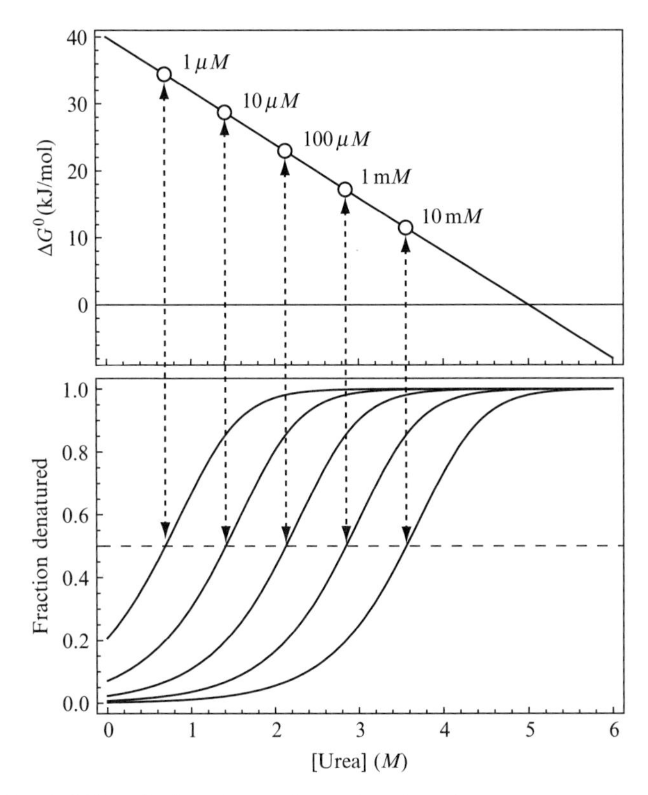

Figure 3.13 Stability of a dimeric protein as a function of urea concentration: the upper panel shows the stability curve, calculated according to Eq. (3.15), with $m = -8$ kJ/mol M, and $c_g = 5$ M. The circles and arrows indicate the stabilities, ΔG°, and midpoint concentrations, c_m, of the transition at the indicated protein concentrations, c_P. Equation (3.86) was used for calculating c_m (using $n = 2$ for dimers). The fraction of denatured proteins is shown in the lower panel. The curves were calculated from Eq. (3.88).

The peculiarity of oligomeric proteins is that the transition midpoint depends on their concentration as shown in Fig. 3.13. Accordingly, the midpoint temperature of the transition, T_m, is different from the temperature, T_g, at which the stability equals zero. Similarly, also the midpoint concentration, c_m, and the concentration of zero stability, c_g, are different. If T_g or c_g is chosen as a reference point, then the stability equations listed above are all valid as they are shown. In case T_m or c_m is desired as the reference conditions, a term of $-RT \ln K_m$ has to be added. The concentration of [D] equals that of $[N_n]$ in terms of monomeric units at the midpoint. To formulate K_m, we first define the total protein concentration c_P in terms of monomer units

$$c_P = n[N_n] + [D]. \tag{3.83}$$

Then, we have $[D]_m = c_P/2$ and $[N_n]_m = c_P/(2n)$ as the concentrations of native and denatured protein units in the middle of the transition, and the equilibrium constant is

$$K_m = \frac{(c_P/2)^n}{c_p/(2n)} = n\left(\frac{c_P}{2}\right)^{n-1}. \tag{3.84}$$

Accordingly, the stability ΔG_m° at the midpoint of the transition is

$$\Delta G_m^\circ = -RT\ln\left[n\left(\frac{c_P}{2}\right)^{n-1}\right], \tag{3.85}$$

where the midpoint could be, for example, at the midpoint temperature T_m or at the midpoint concentration c_m. In an isothermal solvent titration, we have then as a midpoint

$$m\cdot(c_m - c_g) = -RT\ln\left[n\left(\frac{c_P}{2}\right)^{n-1}\right] \Rightarrow c_m = c_g - \frac{RT\ln\left[n(c_P/2)^{n-1}\right]}{m}. \tag{3.86}$$

Another complication with oligomeric proteins arises from the way the fraction of native and denatured proteins is written, compared to monomeric proteins (Eq. (3.3)). Since this equation is only straightforward for n-mers with $n \leq 2$, we restrict the following calculation to dimers.

To obtain the fraction of denatured dimeric proteins, we combine Eq. (3.83) with Eq. (3.82) to get

$$c_P = 2[N_2] + [D] = 2\frac{[D]^2}{K} + [D] \Rightarrow [D] = \frac{-1+\sqrt{1+8c_P/K}}{4/K}. \tag{3.87}$$

The fraction of denatured molecules is then

$$f_D = \frac{[D]}{c_P} = \frac{-1+\sqrt{1+8c_P/K}}{4c_P/K}. \tag{3.88}$$

This expression replaces, as applicable, either the term $K/(1 + K)$ (in Eqs. (3.4), (3.24), and (3.29)), the corresponding exponential ratios, (in Eqs. (3.13) and (3.20)), or f_D in $f_N = 1 - f_D$ (in Eq. (3.22)).

ACKNOWLEDGMENT

The authors gratefully acknowledge support by NIH through grant GM049760 to J. R.

REFERENCES

Auton, M., and Bolen, D. W. (2005). Predicting the energetics of osmolyte-induced protein folding/unfolding. *Proc. Natl. Acad. Sci. USA* **102**(42), 15065–15068.

Auton, M., Cruz, M. A., and Moake, J. (2007a). Conformational stability and domain unfolding of the von willebrand factor a domains. *J. Mol. Biol.* **366**(3), 986–1000.

Auton, M., Holthauzen, L. M., and Bolen, D. W. (2007b). Anatomy of energetic changes accompanying urea-induced protein denaturation. *Proc. Natl. Acad. Sci. USA* **104**(39), 15317–15322.

Auton, M., Sedlak, E., Marek, J., Wu, T., Zhu, C., and Cruz, M. A. (2009). Changes in thermodynamic stability of von willebrand factor differentially affect the force-dependent binding to platelet gpibalpha. *Biophys. J.* **97**(2), 618–627.

Auton, M., Sowa, K. E., Smith, S. M., Sedlak, E., Vijayan, K. V., and Cruz, M. A. (2010). Destabilization of the a1 domain in von willebrand factor dissociates the a1a2a3 tri-domain and provokes spontaneous binding to glycoprotein ibalpha and platelet activation under shear stress. *J. Biol. Chem.* **285**(30), 22831–22839.

Bai, Y., Milne, J. S., Mayne, L., and Englander, S. W. (1994). Protein stability parameters measured by hydrogen exchange. *Proteins* **20**(1), 4–14.

Baskakov, I., and Bolen, D. W. (1998). Forcing thermodynamically unfolded proteins to fold. *J. Biol. Chem.* **273**(9), 4831–4834.

Baskakov, I. V., and Bolen, D. W. (1999). The paradox between m values and deltacp's for denaturation of ribonuclease t1 with disulfide bonds intact and broken. *Protein Sci.* **8**(6), 1314–1319.

Becktel, W. J., and Schellman, J. A. (1987). Protein stability curves. *Biopolymers* **26**(11), 1859–1877.

Bedard, S., Krishna, M. M., Mayne, L., and Englander, S. W. (2008). Protein folding: Independent unrelated pathways or predetermined pathway with optional errors. *Proc. Natl. Acad. Sci. USA* **105**(20), 7182–7187.

Bienkowska, J., Cruz, M., Atiemo, A., Handin, R., and Liddington, R. (1997). The von willebrand factor a3 domain does not contain a metal ion-dependent adhesion site motif. *J. Biol. Chem.* **272**(40), 25162–25167.

Bolen, D. W. (2004). Effects of naturally occurring osmolytes on protein stability and solubility: Issues important in protein crystallization. *Methods* **34**(3), 312–322.

Chang, Y. C., and Oas, T. G. (2010). Osmolyte-induced folding of an intrinsically disordered protein: Folding mechanism in the absence of ligand. *Biochemistry* **49**(25), 5086–5096.

Chen, B. L., Baase, W. A., and Schellman, J. A. (1989). Low-temperature unfolding of a mutant of phage-t4 lysozyme. 2. Kinetic investigations. *Biochemistry* **28**(2), 691–699.

Courtenay, E. S., Capp, M. W., Saecker, R. M., and Record, M. T. J. (2000). Thermodynamic analysis of interactions between denaturants and protein surface exposed on unfolding: Interpretation of urea and guanidinium chloride m-values and their correlation with changes in accessible surface area (asa) using preferential interaction coefficients and the local-bulk domain model. *Proteins* **4**(Suppl), 72–85.

Crouch, T. H., and Kupke, D. W. (1977). Volume change by density—Ribonuclease in 0–8-m guanidinium chloride. *Biochemistry* **16**(12), 2586–2593.

DeKoster, G. T., and Robertson, A. D. (1997). Calorimetrically-derived parameters for protein interactions with urea and guanidine-HCl are not consistent with denaturant *m* values. *Biophys. Chem.* **64**(1–3), 59–68.

Dyson, H. J., and Wright, P. E. (2005). Intrinsically unstructured proteins and their functions. *Nat. Rev. Mol. Cell Biol.* **6**(3), 197–208.

Eftink, M. R. (1994). The use of fluorescence methods to monitor unfolding transitions in proteins. *Biophys. J.* **66**(2 Pt. 1), 482–501.

Ericsson, U. B., Hallberg, B. M., Detitta, G. T., Dekker, N., and Nordlund, P. (2006). Thermofluor-based high-throughput stability optimization of proteins for structural studies. *Anal. Biochem.* **357**(2), 289–298.

Erilov, D., Puorger, C., and Glockshuber, R. (2007). Quantitative analysis of nonequilibrium, denaturant-dependent protein folding transitions. *J. Am. Chem. Soc.* **129**(29), 8938–8939.

Felitsky, D. J., and Record, M. T. (2003). Thermal and urea-induced unfolding of the marginally stable lac repressor DNA-binding domain: A model system for analysis of solute effects on protein processes. *Biochemistry* **42**(7), 2202–2217.

Ferreon, A. C., and Bolen, D. W. (2004). Thermodynamics of denaturant-induced unfolding of a protein that exhibits variable two-state denaturation. *Biochemistry* **43**(42), 13357–13369.

Ferreon, A. C., Ferreon, J. C., Bolen, D. W., and Rösgen, J. (2007). Protein phase diagrams ii: Nonideal behavior of biochemical reactions in the presence of osmolytes. *Biophys. J.* **92**(1), 245–256.

Franks, F., Hatley, R. H., and Friedman, H. L. (1988). The thermodynamics of protein stability. Cold destabilization as a general phenomenon. *Biophys. Chem.* **31**(3), 307–315.

Garcia-Mira, M. M., and Sanchez-Ruiz, J. M. (2001). ph corrections and protein ionization in water/guanidinium chloride. *Biophys. J.* **81**(6), 3489–3502.

Garcia-Perez, A., and Burg, M. B. (1991). Renal medullary organic osmolytes. *Physiol. Rev.* **71**(4), 1081–1115.

Gerding, J. J., Koppers, A., Hagel, P., and Bloemendal, H. (1971). Cyanate formation in solutions of urea. ii. Effect of urea on the eye lens protein-crystallin. *Biochim. Biophys. Acta* **243**(3), 375–379.

Giletto, A., and Pace, C. N. (1999). Buried, charged, non-ion-paired aspartic acid 76 contributes favorably to the conformational stability of ribonuclease t1. *Biochemistry* **38** (40), 13379–13384.

Greene, R. F. J., and Pace, C. N. (1974). Urea and guanidine hydrochloride denaturation of ribonuclease, lysozyme, alpha-chymotrypsin, and beta-lactoglobulin. *J. Biol. Chem.* **249** (17), 5388–5393.

Gulotta, M., Qiu, L., Desamero, R., Rösgen, J., Bolen, D. W., and Callender, R. (2007). Effects of cell volume regulating osmolytes on glycerol 3-phosphate binding to triose-phosphate isomerase. *Biochemistry* **46**(35), 10055–10062.

Hagel, P., Gerding, J. J., Fieggen, W., and Bloemendal, H. (1971). Cyanate formation in solutions of urea. i. Calculation of cyanate concentrations at different temperature and ph. *Biochim. Biophys. Acta* **243**(3), 366–373.

Harms, M. J., Schlessman, J. L., Chimenti, M. S., Sue, G. R., Damjanovic, A., and Garcia-Moreno, B. (2008). A buried lysine that titrates with a normal pka: Role of conformational flexibility at the protein-water interface as a determinant of pka values. *Protein Sci.* **17**(5), 833–845.

Harries, D., and Rösgen, J. (2008). A practical guide on how osmolytes modulate macromolecular properties. *Methods Cell Biol.* **84,** 679–735.

Henkels, C. H., and Oas, T. G. (2005). Thermodynamic characterization of the osmolyte- and ligand-folded states of bacillus subtilis ribonuclease p protein. *Biochemistry* **44**(39), 13014–13026.

Henkels, C. H., Kurz, J. C., Fierke, C. A., and Oas, T. G. (2001). Linked folding and anion binding of the bacillus subtilis ribonuclease p protein. *Biochemistry* **40**(9), 2777–2789.
Hinz, H. J., Kuttenreich, H., Meyer, R., Renner, M., Frund, R., Koynova, R., Boyanov, A. I., and Tenchov, B. G. (1991). Stereochemistry and size of sugar head groups determine structure and phase behavior of glycolipid membranes: Densitometric, calorimetric, and x-ray studies. *Biochemistry* **30**(21), 5125–5138.
Holthauzen, L. M., and Bolen, D. W. (2007). Mixed osmolytes: The degree to which one osmolyte affects the protein stabilizing ability of another. *Protein Sci.* **16**(2), 293–298.
Holthauzen, L. M., Rösgen, J., and Bolen, D. W. (2010). Hydrogen bonding progressively strengthens upon transfer of the protein urea-denatured state to water and protecting osmolytes. *Biochemistry* **49**(6), 1310–1318.
Hopkins, F. G. (1930). Denaturation of proteins by urea and related substances. *Nature* **3175**(126), 328–330, 383–384.
Hu, C. Q., Sturtevant, J. M., Thomson, J. A., Erickson, R. E., and Pace, C. N. (1992). Thermodynamics of ribonuclease t1 denaturation. *Biochemistry* **31**(20), 4876–4882.
Kim, M. S., Song, J., and Park, C. (2009). Determining protein stability in cell lysates by pulse proteolysis and western blotting. *Protein Sci.* **18**(5), 1051–1059.
Kumar, R., Lee, J. C., Bolen, D. W., and Thompson, E. B. (2001). The conformation of the glucocorticoid receptor af1/tau1 domain induced by osmolyte binds co-regulatory proteins. *J. Biol. Chem.* **276**(21), 18146–18152.
Lide, D. (2004). CRC Handbook of Chemistry and Physics. CRC Press, Boca Raton, FL.
Limbourg, P. (1887). Beiträge zur chemischen nervenreizung und zur wirkung der salze. *Arch. Physiol.* **41,** 303–325.
Lin, T. Y., and Timasheff, S. N. (1994). Why do some organisms use a urea-methylamine mixture as osmolyte? Thermodynamic compensation of urea and trimethylamine n-oxide interactions with protein. *Biochemistry* **33**(42), 12695–12701.
Lin, M. F., Williams, C., Murray, M. V., Conn, G., and Ropp, P. A. (2004). Ion chromatographic quantification of cyanate in urea solutions: estimation of the efficiency of cyanate scavengers for use in recombinant protein manufacturing. *J. Chromatogr. B Analyt. Technol. Biomed. Life Sci.* **803**(2), 353–362.
Lumry, R., and Eyring, H. (1954). Conformation changes of proteins. *J. Phys. Chem.* **58**(2), 110–120.
Lumry, R., Biltonen, R., and Brandts, J. F. (1966). Validity of the "two-state" hypothesis for conformational transitions of proteins. *Biopolymers* **4,** 917–944.
Makhatadze, G. I. (1999). Thermodynamics of protein interactions with urea and guanidinium hydrochloride. *J. Phys. Chem. B* **103**(23), 4781–4785.
Manning, M. C., Chou, D. K., Murphy, B. M., Payne, R. W., and Katayama, D. S. (2010). Stability of protein pharmaceuticals: An update. *Pharm. Res.* **27**(4), 544–575.
Matulis, D., Kranz, J. K., Salemme, F. R., and Todd, M. J. (2005). Thermodynamic stability of carbonic anhydrase: Measurements of binding affinity and stoichiometry using thermofluor. *Biochemistry* **44**(13), 5258–5266.
Mello, C. C., and Barrick, D. (2003). Measuring the stability of partly folded proteins using tmao. *Protein Sci.* **12**(7), 1522–1529.
Mello, C. C., and Barrick, D. (2004). An experimentally determined protein folding energy landscape. *Proc. Natl. Acad. Sci. USA* **101**(39), 14102–14107.
Müller, C. W., Schlauderer, G. J., Reinstein, J., and Schulz, G. E. (1996). Adenylate kinase motions during catalysis: An energetic counterweight balancing substrate binding. *Structure* **4**(2), 147–156.
Neurath, H., and Saum, A. M. (1939). The denaturation of serum albumin—Diffusion and viscosity measurements of serum albumin in the presence of urea. *J. Biol. Chem.* **128**(1), 347–362.

Nicholson, E. M., and Scholtz, J. M. (1996). Conformational stability of the *Escherichia coli* hpr protein: Test of the linear extrapolation method and a thermodynamic characterization of cold denaturation. *Biochemistry* **35**(35), 11369–11378.

Nozaki, Y., and Tanford, C. (1967). Acid-base titrations in concentrated guanidine hydrochloride. Dissociation constants of the guanidinium ion and of some amino acids. *J. Am. Chem. Soc.* **89**(4), 736–742.

Pace, C. N. (1986). Determination and analysis of urea and guanidine hydrochloride denaturation curves. *Methods Enzymol.* **131,** 266–280.

Pace, C. N., and Laurents, D. V. (1989). A new method for determining the heat capacity change for protein folding. *Biochemistry* **28**(6), 2520–2525.

Pace, C. N., Vajdos, F., Fee, L., Grimsley, G., and Gray, T. (1995). How to measure and predict the molar absorption coefficient of a protein. *Protein Sci.* **4**(11), 2411–2423.

Pace, C. N., Trevino, S., Prabhakaran, E., and Scholtz, J. M. (2004). Protein structure, stability and solubility in water and other solvents. *Philos. Trans. R Soc. Lond. B Biol. Sci.* **359**(1448), 1225–1234, discussion 1234–1235.

Park, C., and Marqusee, S. (2004). Probing the high energy states in proteins by proteolysis. *J. Mol. Biol.* **343**(5), 1467–1476.

Pfeil, W., and Privalov, P. L. (1976). Thermodynamic investigations of proteins. ii. Calorimetric study of lysozyme denaturation by guanidine hydrochloride. *Biophys. Chem.* **4**(1), 33–40.

Plaza del Pino, I. M., Pace, C. N., and Freire, E. (1992). Temperature and guanidine hydrochloride dependence of the structural stability of ribonuclease t1. *Biochemistry* **31** (45), 11196–11202.

Qu, Y., and Bolen, D. W. (2003). Hydrogen exchange kinetics of rnase a and the urea:tmao paradigm. *Biochemistry* **42**(19), 5837–5849.

Ragone, R., Colonna, G., Balestrieri, C., Servillo, L., and Irace, G. (1984). Determination of tyrosine exposure in proteins by second-derivative spectroscopy. *Biochemistry* **23**(8), 1871–1875.

Rajagopalan, L., Rösgen, J., Bolen, D. W., and Rajarathnam, K. (2005). Novel use of an osmolyte to dissect multiple thermodynamic linkages in a chemokine ligand-receptor system. *Biochemistry* **44**(39), 12932–12939.

Rosengarth, A., Rösgen, J., and Hinz, H. J. (1999). Slow unfolding and refolding kinetics of the mesophilic rop wild-type protein in the transition range. *Eur. J. Biochem.* **264**(3), 989–995.

Rösgen, J. (2007). Molecular basis of osmolyte effects on protein and metabolites. *Methods Enzymol.* **428,** 459–486.

Rösgen, J., and Hinz, H. J. (2000). Response functions of proteins. *Biophys. Chem.* **83**(1), 61–71.

Rösgen, J., and Hinz, H. J. (2003). Phase diagrams: A graphical representation of linkage relations. *J. Mol. Biol.* **328**(1), 255–271.

Rösgen, J., Pettitt, B. M., and Bolen, D. W. (2004). Uncovering the basis for nonideal behavior of biological molecules. *Biochemistry* **43**(45), 14472–14484.

Russo, A. T., Rösgen, J., and Bolen, D. W. (2003). Osmolyte effects on kinetics of fkbp12 c22a folding coupled with prolyl isomerization. *J. Mol. Biol.* **330**(4), 851–866.

Sanchez-Ruiz, J. M., Lopez-Lacomba, J. L., Cortijo, M., and Mateo, P. L. (1988). Differential scanning calorimetry of the irreversible thermal denaturation of thermolysin. *Biochemistry* **27**(5), 1648–1652.

Santoro, M. M., and Bolen, D. W. (1988). Unfolding free energy changes determined by the linear extrapolation method. 1. Unfolding of phenylmethanesulfonyl alpha-chymotrypsin using different denaturants. *Biochemistry* **27**(21), 8063–8068.

Singh, R., Haque, I., and Ahmad, F. (2005). Counteracting osmolyte trimethylamine n-oxide destabilizes proteins at ph below its pka. Measurements of thermodynamic parameters of proteins in the presence and absence of trimethylamine n-oxide. *J. Biol. Chem.* **280**(12), 11035–11042.

Skerjanc, J., Dolecek, V., and Lapanje, S. (1970). The partial specific volume of chymotrypsinogen a in aqueous urea solutions. *Eur. J. Biochem.* **17**(1), 160–164.

Stark, G. R., Stein, W. H., and Moore, S. (1960). Reactions of the cyanate present in aqueous urea with amino acids and proteins. *J. Biol. Chem.* **235**(11), 3177–3181.

Stites, W. E., Byrne, M. P., Aviv, J., Kaplan, M., and Curtis, P. M. (1995). Instrumentation for automated determination of protein stability. *Anal. Biochem.* **227**(1), 112–122.

Tiffany, M. L., and Krimm, S. (1972). Effect of temperature on the circular dichroism spectra of polypeptides in the extended state. *Biopolymers* **11**(11), 2309–2316.

Timasheff, S. N., and Xie, G. (2003). Preferential interactions of urea with lysozyme and their linkage to protein denaturation. *Biophys. Chem.* **105**(2–3), 421–448.

Todd, M. J., and Salemme, F. R. (2003). Direct binding assays for pharma screening-assay tutorial: Thermofluor miniaturized direct-binding assay for hts & secondary screening. *Genet. Eng. News* **23**(3), 28–29.

Uversky, V. N., Gillespie, J. R., and Fink, A. L. (2000). Why are "natively unfolded" proteins unstructured under physiologic conditions? *Proteins* **41**(3), 415–427.

Wang, A., and Bolen, D. W. (1997). A naturally occurring protective system in urea-rich cells: Mechanism of osmolyte protection of proteins against urea denaturation. *Biochemistry* **36**(30), 9101–9108.

Wen, N. P., and Brooker, M. H. (1993). Urea protonation—Raman and theoretical-study. *J. Phys. Chem.* **97**(33), 8608–8616.

Wu, P., and Bolen, D. W. (2006). Osmolyte-induced protein folding free energy changes. *Proteins* **63**(2), 290–296.

Yan, H., and Tsai, M. D. (1999). Nucleoside monophosphate kinases: Structure, mechanism, and substrate specificity. *Adv. Enzymol. Relat. Areas Mol. Biol.* **73,** 103–134x.

Zweifel, M. E., and Barrick, D. (2002). Relationships between the temperature dependence of solvent denaturation and the denaturant dependence of protein stability curves. *Biophys. Chem.* **101–102,** 221–237.

CHAPTER FOUR

Small-Angle X-ray Scattering Studies of Peptide–Lipid Interactions Using the Mouse Paneth Cell α-Defensin Cryptdin-4

Abhijit Mishra,[*] Kenneth P. Tai,[†] Nathan W. Schmidt,[*,‡,§] André J. Ouellette,[†] *and* Gerard C. L. Wong[*,‡,§]

Contents

[*] Department of Bioengineering, University of California, Los Angeles, California, USA
[†] Department of Pathology and Laboratory Medicine, USC Norris Cancer Center, Keck School of Medicine of the University of Southern California, Los Angeles, California, USA
[‡] Department of Physics, University of Illinois, Urbana-Champaign, Illinois, USA
[§] Department of Materials Science, University of Illinois, Urbana-Champaign, Illinois, USA

Methods in Enzymology, Volume 492
ISSN 0076-6879, DOI: 10.1016/B978-0-12-381268-1.00016-1
© 2011 Elsevier Inc.
All rights reserved.

Abstract

In the presence of specialized proteins or peptides, a biological membrane can spontaneously restructure itself to allow communication between the intracellular and the extracellular sides. Examples of these proteins include cell-penetrating peptides and antimicrobial peptides (AMPs), which interact with cell membranes in complex ways. We briefly review cell-penetrating peptides and AMPs, and describe in detail how recombinant AMPs are made and their activity evaluated, using α-defensins as a specific example. We also review X-ray scattering methods used in studying peptide–membrane interactions, focusing on the procedures for small-angle X-ray scattering experiments on peptide–membrane interactions at realistic solution conditions, using both laboratory and synchrotron sources.

1. Introduction

With the exception of solute uptake by selective transmembrane pumps and transporters or endocytosis of receptor-bound ligands, biological membranes remain impervious to protein molecules. However, certain cell-penetrating peptides (CPPs), including the HIV TAT protein transduction domain (Frankel and Pabo, 1988; Green and Loewenstein, 1988), a short domain in the Drosophila *antennapaedia* homeotic transcription factor (Antp) (Joliot *et al.*, 1991), and the Herpes-Simplex-Virus-1 DNA binding protein VP22 (Elliott and O'Hare, 1997), enable spontaneous membrane restructuring, to allow molecules to traverse the boundary between the intracellular and extracellular sides. Antimicrobial peptides (AMPs) comprise a different set of host defense molecules with membrane disruptive activities, highly diverse primary and secondary structures. Most CPPs and AMPs are cationic and associate with electronegative microbial cell membranes via electrostatic interactions. They also have varying degrees of hydrophobicity and are therefore amphiphilic. The hydrophobic side chains perturb membrane self-assembly and stability, although the detailed molecular mechanisms of action have not been fully understood. Here, we describe X-ray scattering methods and illustrate their application in characterizing membrane interactions with the bactericidal α-defensin, cryptdin-4 (Crp4), in small-angle X-ray scattering (SAXS) experiments performed in solution using both laboratory and synchrotron sources.

1.1. Cell-penetrating peptides

CPPs are short (<20 amino acid) cationic peptides that can traverse cell membranes of various mammalian cells. A wide variety of macromolecules can be internalized while retaining their biological activity when attached to these peptides. This ability of CPPs to transport biologically active molecules across cell membranes makes them promising candidates for a broad range of drug-delivery applications. The CPPs can be classified into arginine-rich peptides and amphipathic peptides. The arginine-rich CPPs have been the most widely studied (El-Sayed *et al.*, 2009; Wender *et al.*, 2008). The exact molecular mechanism of cellular entry of arginine-rich CPPs is currently an active area of research. The cationic nature of the peptides is a necessary but not sufficient condition for translocation activity. It has been observed that arginine-rich oligomers can enter cells, but similar length polymers composed of other basic amino acids such as lysine, ornithine, or histidine cannot (Mitchell *et al.*, 2000). The guanidinium headgroup of arginine with its ability to form bidentate hydrogen bonds is the central structural feature required for peptide uptake (Rothbard *et al.*, 2005), and recent work suggests that such hydrogen bond patterns is related to the generation of specific types of membrane curvature topologically required for pore formation (Mishra *et al.*, 2008; Schmidt *et al.*, 2010b).

Experimental studies have shown evidence for many different entry mechanisms, including direct translocation and various endocytotic mechanisms. It is believed that more than one mechanism may be involved in translocation activity, with the dominant mechanism influenced by a variety of factors, including temperature, incubation time, cell type, cargo type and size, and linkage type and size (Wender *et al.*, 2008).

1.2. Antimicrobial peptides

AMPs are important mediators of an innate host defense system, with antimicrobial activities against a broad spectrum of microorganisms (Brogden, 2005; Hancock and Sahl, 2006; Shai, 1999; Zasloff, 2002). Most AMPs share two general structural features; they are amphipathic and cationic (Brogden, 2005; Zasloff, 2002). It is believed that AMPs disrupt membranes through a combination of electrostatic interactions between cationic amino acid side chains and electronegative components of the microbial cell envelope, followed by the insertion of hydrophobic patches into the nonpolar interior of the membrane bilayer (Brogden, 2005; Huang, 2000; Matsuzaki, 1999; Matsuzaki *et al.*, 1998; Shai, 1999; Zasloff, 2002).

Defensins constitute one of two major AMP families in mammals (Ganz, 2003; Lehrer, 2004; Selsted and Ouellette, 2005), the other being the cathelicidins (Zanetti, 2004). Defensins were among the first AMPs to be described (Lehrer *et al.*, 1983; Selsted *et al.*, 1983), consisting of three subfamilies of cationic, Cys-rich AMPs, the α-, β-, and θ-defensins, all of which have broad-spectrum antimicrobial activities and are defined by the disulfide connectivities (Selsted and Ouellette, 2005). The α-defensins are major granule constituents of mammalian phagocytic leukocytes and of small intestinal Paneth cells (Ganz, 2003). The β-defensins, discovered in cattle as AMPs of airway and lingual epithelial cells, and in bovine neutrophil granules (Diamond *et al.*, 1991; Schonwetter *et al.*, 1995; Selsted *et al.*, 1993), exist in diverse species and are expressed by many epithelial cell types and more widely than the α-defensins (Schutte *et al.*, 2002). The θ-defensins are unusual ~2 kDa peptides from rhesus macaque neutrophils and monocytes and are found only in Old World monkeys. θ-Defensins are the only macrocyclic peptides known in animals, and like all defensins, they are stabilized by three disulfide bonds. θ-Defensins assemble from two hemiprecursors that derive from α-defensin genes that have stop codons that terminate the peptide at residue position 12. The ligation mechanisms that circularize the closed θ-defensin polypeptide chain remain unknown.

1.3. α-Defensins

α-Defensins are ~4 kDa, cationic, and amphipathic peptides with broad-spectrum bactericidal activities. Structurally, they consist of a triple-stranded β-sheet structure that is established by three invariantly paired disulfide bonds (Fig. 4.1) (Lehrer, 2007). Despite having highly diverse primary structures (Ouellette, 2006; Ouellette and Bevins, 2001), α-defensins retain conserved biochemical features that include an invariant disulfide array (Selsted and Harwig, 1989), a canonical Arg–Glu salt bridge, a conserved

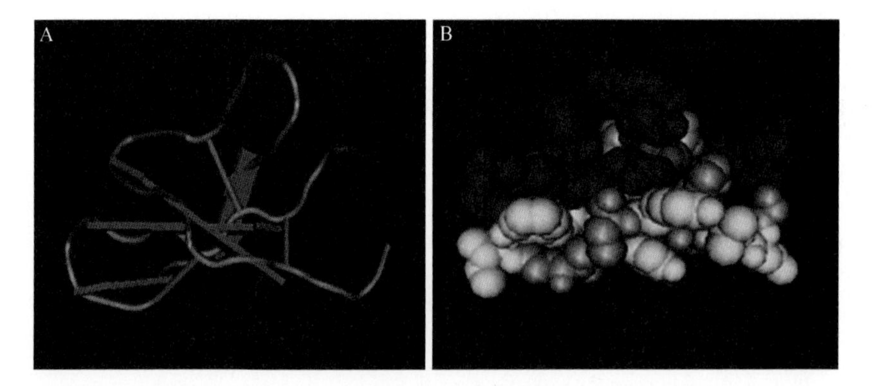

Figure 4.1 Solution structure of mouse Paneth cell alpha-defensin Crp-4 (protein data bank ID 2GW9) obtained by NMR. (A) Structure shown in worm rendering. The three disulfide bonds from six cysteines are displayed in orange and Beta-sheets are represented by purple arrows. Blue regions denote cationic amino acids (arginine, lysine, and histidine), while the anionic glutamic acid is red. (B) Space-filled structure illustrates the cationic (blue) and hydrophobic (yellow) patches of amphipathic Crp-4. Here, hydrophobic amino acids include leucine, isoleucine, valine, phenylalanine, and tyrosine. Neutral residues are colored gray. (See Color Insert.)

Gly residue at CysIII+8, and high Arg content relative to Lys (Lehrer, 2007). Studies have shown consistently that bactericidal activity is independent of these highly conserved features of the peptide family, with exception to the relatively high Arg content (Lehrer *et al.*, 1988; Maemoto *et al.*, 2004; Rajabi *et al.*, 2008; Rosengren *et al.*, 2006; Wu *et al.*, 2005).

In vitro, α-defensins are microbicidal against gram-positive and gram-negative bacteria, fungi, spirochetes, protozoa, and enveloped viruses (Aley *et al.*, 1994; Borenstein *et al.*, 1991; Ganz *et al.*, 1985; Lehrer, 2007; Zhu, 2008). Most α-defensins exert antibacterial effects by membrane disruption, inducing permeabilization of target cell membranes as inferred from the formation of transient defects or stable pores in model phospholipid bilayers (Hristova *et al.*, 1996; White *et al.*, 1995). For example, the bactericidal activity of the mouse α-defensin cryptdin-4 (Crp4) is directly related to peptide binding and disruption of phospholipid bilayers (Satchell *et al.*, 2003a). Crp4 exhibits strong interfacial binding to model membranes, inducing "graded" fluorophore leakage from model membrane vesicles (Cummings and Vanderlick, 2007; Cummings *et al.*, 2003; Satchell *et al.*, 2003b). Mammalian α-defensins secreted by Paneth cells determine the composition of the mouse small intestinal microbiome, apparently by selecting for peptide-tolerant microbial species as residents in that microbial ecosystem.

1.4. Peptide-induced membrane restructuring

The HIV TAT cell-penetrating peptide generates negative Gaussian membrane curvature in model membrane systems manifested in the generation of a Pn3m cubic phase. This type of curvature, also known as "saddle-splay" curvature, is characterized by saddle-shaped deformations. The ability of the guanidinium group of arginine to crosslink multiple lipid headgroups through bidentate hydrogen bonding facilitates the generation of negative Gaussian curvature (Mishra *et al.*, 2008). The negative Gaussian curvature, necessary to form the observed cubic phases, is topologically required for pore formation, and induction of that curvature can lower the free energy barriers, providing a range of entry mechanisms, including direct translocation as well as endocytotic pathways (Schmidt *et al.*, 2010b).

The generation of negative Gaussian curvature correlates with the permeation capability of a peptide. The arginine-rich cell-penetrating peptides, Antp and polyarginine, also induce negative Gaussian curvature, but polylysine (K_8), which has the same charge as TAT but cannot form bidentate hydrogen bonds, generates negative mean curvature but zero Gaussian curvature, resulting in inverted hexagonal H_{II} phases. In general, the interaction between charged polymers and charged membranes yields a rich polymorphism of phases with a broad range of applications (Liang *et al.*, 2005; Purdy Drew *et al.*, 2008; Rädler *et al.*, 1997; Wong *et al.*, 2000; Yang *et al.*, 2004).

The structural tendency to form negative Gaussian membrane curvature has also been observed in pore-forming AMPs and synthetic peptide mimics. Under specific lipid compositions and solution conditions, alamethicin (Keller *et al.*, 1996), gramicidin S (Prenner *et al.*, 1997; Staudegger *et al.*, 2000), lactoferricin (LF11)-derived peptides, VS1-13 and VS1-24 (Zweytick *et al.*, 2008), as well as protegrin-1 and peptidylglycylleucine-carboxyamide (Hickel *et al.*, 2008), also induce cubic phases. Defensins also restructure vesicles by inducing negative Gaussian curvature when the lipid composition of model membranes mimics that of bacterial membranes but not the composition of mammalian membrane bilayers (Schmidt *et al.*, 2010a). Likewise, curvature generation drives the formation of a sequence of phases, including cubic and hexagonal phases for synthetic molecules that mimic AMP action (Yang *et al.*, 2007, 2008). The preferential formation of high curvatures necessary for pore formation is favored in membranes rich in negative curvature lipids, such as those found in high concentrations in bacterial membranes (Som *et al.*, 2009; Yang *et al.*, 2008). The generation of negative Gaussian curvature requires both anionic and negative curvature lipids. Model membranes in water form bilayers in the absence of AMPs. The peptides interact differently with membranes of pure lipid species. For example, membranes composed of pure anionic lipids are "glued" together into a lamellar phase; membranes of pure negative curvature lipids interact weakly with peptides with no major reorganization of lipids. The induced phases have zero Gaussian curvature, in contrast to the behavior of composite membranes.

The methods outlined in this chapter are general, applicable to a broad range of AMPs and CPPs. Here, we focus on Crp4 as a model of peptide–membrane interactions. In order to elucidate the molecular mechanisms responsible for the bacterial killing activity, we investigate peptide–membrane interactions using SAXS. Below, we examine why and how X-rays are used for this purpose.

2. X-Rays as Structural Probes of Biological Systems Under Biomimetic Conditions

Historically, diffraction (or equivalently, "scattering")-based methods have contributed immensely to our understanding of structures at the nanoscale. Ideally, the scattering particle should interact only weakly with the system under study and its wavelength must be comparable to the length scale of the system (Chaikin and Lubensky, 1995). Electrons are scattered by the electrostatic forces between the electrons and the atoms within the system and require energies of $\sim$100 eV to probe nanoscale structures. However, typically, thin ($\sim$1 μm thick) samples are needed to prevent problems with multiple scattering. Moreover, electron-based probes typically require a vacuum, which is quite different from physiological conditions. However, this problem is sometimes partially circumvented through differential pumping. Neutrons have a much higher mass than electrons and hence require much lower energies ($\sim$0.1 eV) to probe nanostructures. Neutrons are scattered by nuclear forces or by the electron spins. They interact weakly with matter and have low absorption. Therefore, neutrons can penetrate samples several millimeters thick. However, because of this, large amounts of samples are needed for neutron scattering which is not always feasible for biological samples, such as the samples considered here. Also, neutron sources are relatively weak, with the flux of neutrons much lower than X-rays. X-ray photons with energy $\sim 10^4$ eV have wavelengths in the angstrom range; suitable for studying nanostructures. These X-rays can penetrate matter up to a millimeter and therefore provide "bulk" information. X-ray scattering also requires much lower sample amounts compared to neutron scattering. In addition, X-ray scattering experiments on biological macromolecules can be performed under near-physiological conditions, enabling us to examine their structural response to changes in a variety of parameters (e.g., pH, ionic strength, concentrations, temperature, etc.) Aside from studies of biomembrane-based systems, X-ray diffraction techniques can also be used to study other weakly scattering systems such as biopolymer-based systems under physiologically relevant conditions (Purdy *et al.*, 2007; Sanders *et al.*, 2005, 2007; Wong, 2006).

2.1. X-ray diffraction of weakly ordered systems

Most biological systems are not ordered into crystals. For example, membrane-based systems differ from conventional solid-state crystalline materials in that they are often weakly ordered, exhibiting only one-dimensional or two-dimensional periodicity rather than the three-dimensional periodicity of crystals. In this sense, they are analogous to liquid crystals. These systems are fluid and their periodic density distributions have much greater contributions from thermal fluctuation compared to crystalline samples. In the characterization of lipid-based systems by diffraction, two regions of the diffraction pattern are used to identify the structure. The small-angle region identifies the symmetry and long-range organization of the phase, while the wide-angle region gives information on the molecular packing or short-range organization of the phase (Seddon and Templer, 1995). The diffraction signals from these systems are in general weaker than those from crystals. Moreover, the molecular constituents of biological molecules are generally composed of combinations of mostly low Z elements, such as carbon, hydrogen, and oxygen, which make the electron density contrast between the constituent components low and thus lead to weak diffraction intensities. The collection of interpretable data from such systems, therefore, is facilitated by X-ray sources with high intensity and high resolution.

2.2. Synchrotron X-ray sources

Synchrotron radiation is produced by charged particles traveling at relativistic speeds forced to travel along curved paths by applied magnetic fields. High speed electrons circulating at constant energy in synchrotron storage rings produce X-rays. X-rays can also be produced in insertion devices, like wigglers or undulators, situated in the straight sections of storage rings. Alternating magnetic fields in these devices force the electrons along oscillating paths in the horizontal plane instead of straight lines, greatly enhancing the intensity of radiation. A wiggler consists of a series of magnets that force the electrons to turn in alternating in-plane directions for a fixed number of spatial periods. The intensity of the radiation from each wiggle is added up and the resultant intensity is proportional to the number of wiggles. Undulators also have a series of magnets; however, the radiation emitted from one undulation is in phase with the radiation from subsequent undulations. The resultant intensity is therefore proportional to the square of the number of undulations. The coherent addition of amplitudes is only valid at one particular wavelength; hence, radiation from undulators is quasi-monochromatic.

2.3. Theory of X-ray diffraction

In the classical description, X-rays are transverse electromagnetic waves, where the electric and magnetic fields are perpendicular to each other and to the direction of propagation. It is characterized by its wavelength λ, or its wavenumber $k = 2\pi/\lambda$. From a quantum mechanical perspective, the X-rays can be viewed as a beam of photons, with each photon having an energy $\hbar\omega$ and momentum $\hbar\mathbf{k}$. The intensity of the beam is given by the number of photons passing through a given area per unit time. When X-rays interact with a free scatterer with charge q and mass m, the scattered intensity I_{sc} at distance R from the scatterer is

$$I_{sc} = I_0 \frac{q^4}{m^2 c^4 R^2} \left(\frac{1 + \cos^2 2\theta}{2} \right) \tag{4.1}$$

where I_0 is the incident beam intensity, c is the velocity of light, and 2θ is the scattering angle. Protons and electrons have the same charge, but the mass of a proton is 1836 times larger than that of an electron. The scattered intensity by a proton is, therefore, $(1836)^2$ times smaller than that by an electron. Hence, the X-ray scattering pattern is predominantly contributed by the interactions between X-rays and electrons, and the scattering contrast is due to the electron density difference within the system. When electrons scatter X-rays, if the wavelength of the scattered wave is the same as that of the incident one, the scattering process is called *elastic*. However, if energy is transferred to the electron, the scattered photon has a longer wavelength relative to that of the incident photon, and the scattering process is *inelastic*. In this chapter, we are primarily concerned with elastic scattering.

In an X-ray diffraction experiment, the detectors usually count the number of scattered photons. The measured intensity, I_{sc}, is the number of photons per second recorded by the detector. The differential cross-section ($d\sigma/d\Omega$) can be defined as

$$\frac{d\sigma}{d\Omega} = \frac{(\text{number of X-ray photons scattered per second in } d\Omega)}{(\text{incident flux})d\Omega} \tag{4.2}$$

where σ is the scattering cross-section, $d\Omega$ is the solid angle subtend by the detector, and incident flux is the incident beam intensity (I_0) divided by its cross-section area (A_0). The measured intensity, I_{sc}, is related to the differential cross-section $d\sigma/d\Omega$ by

$$\left(\frac{d\sigma}{d\Omega}\right) = \frac{I_{SC}}{(I_0/A_0)d\Omega} \tag{4.3}$$

or

$$I_{sc} = \frac{I_0}{A_0}\left(\frac{d\sigma}{d\Omega}\right)(d\Omega) \tag{4.4}$$

From the theory of X-ray scattering (Als-Nielsen and McMorrow, 2001; Guinier, 1994; Warren, 1990), the differential cross-section for a system at thermal equilibrium:

$$\frac{d\sigma}{d\Omega} \propto P(q)|f(q)|^2 S(q) \tag{4.5}$$

where $P(q)$ is the polarization factor, $f(q)$ is the form factor of the scatterer, and $S(q)$ is the structure factor for the scatterer lattice. $q = \boldsymbol{k_s} - \boldsymbol{k_i}$, is the scattering vector which measures the photon momentum transfer ($\boldsymbol{k_i}$ and $\boldsymbol{k_s}$ are the wavevectors of the incident and scattered waves, respectively). For elastic X-ray scattering, the scattering vector q is given by

$$|q| = \frac{4\pi \sin\theta}{\lambda} = \frac{2\pi}{d} \tag{4.6}$$

where λ is the wavelength of the incident X-ray, θ is the half of the scattering angle, and d is the periodicity of the electron density fluctuation.

The polarization factor $P(q)$ depends on the X-ray source. In a synchrotron source, the electrons orbit in the horizontal plane and hence, the emitted X-rays are linearly polarized in the orbit plane but elliptically polarized when viewed out of that plane. So, for synchrotron source, polarization factor $P(q) = 1$ in the vertical scattering plane, while $P(q) = \cos^2 2\theta$ in the horizontal scattering plane where 2θ is the scattering angle. For an unpolarized X-ray source, the polarization factor $P(q) = (1 + \cos^2 2\theta)\ /2$. For the SAXS ($2\theta < 10°$), the polarization factor is not significantly different from $P(q) \sim 1$.

The static structural factor $S(q)$ accounts for the geometry of scatterer and contains the structural information of the biomolecular system.

$$S(q) = \frac{1}{N}\sum_{i,j}^{N}\left\langle e^{iq[ri(0)-rj(0)]}\right\rangle = \frac{V}{N}\int drG(r)e^{-iqr} \tag{4.7}$$

and,

$$G(r) = \frac{1}{V}\int dr'\langle\rho(r).\rho(r+r')\rangle \tag{4.8}$$

$G(r)$ is the density–density correlation function, with $\rho(r)$ being the electron density distribution of the system.

The form factor $f(q)$ of the scatterer is the Fourier transform of its electron density. For an atom, $f(q) = -\ r_0 f^0(q)$, where r_0 is the Thomson scattering length ($r_0 = 2.82 \times 10^{-5}$Å) and $f^0(q)$ is the atomic form factor given by

$$f^0(q) = \int \rho(r)e^{iq\cdot r}dr = \begin{cases} z, & \text{for}\, q \to 0 \\ 0, & \text{for}\, q \to \infty \end{cases} \tag{4.9}$$

where $\rho(r)$ is the number density of electron at position r around the nucleus in the atom and z is the total number of electrons in the atom.

For a molecule, the form factor $f^{mol}(q)$ is

$$f^{mol}(q) = -r_0\sum_{r_j} f_j(q)e^{iq\cdot r_j} \tag{4.10}$$

where $f_j(q)$ is the atomic form factor of the jth atom in the molecule. If $|f^{mol}(q)|^2$ can be determined experimentally with sufficient values of scattering vectors q, the position r_j of the jth atom in the molecule can be known.

For an ordered arrangement of atoms or molecules (e.g., a crystal), the form factor $f^{\mathrm{crystal}}(q)$ is

$$f^{\mathrm{crystal}}(q) = -r_0 \left(\sum_{r_j} f_j^{\mathrm{mol}}(q) \mathrm{e}^{iq \cdot r_j} \right) \left(\sum_{R_n} \mathrm{e}^{iq \cdot R_n} \right) \tag{4.11}$$

where the first term is the scattering amplitude from the basis of the molecules or atoms contained in the unit cell and is known as the "unit cell structure factor", in which r_j is the position of jth molecule or atom in the unit cell, and the second term is a sum over lattice sites and is known as the "lattice sum." All the terms in the lattice sum are phase factors located on the unit circle in a complex plane. This lattice sum and as a result, the crystal's form factor f^{crystal} is nonvanishing if and only if the scattering vector q coincides with a reciprocal lattice vector G which satisfies $G \cdot R_n = 2\pi \times m$, where m is an integer. This is the Laue condition for the observation of X-ray diffraction. Scattering from a crystal is confined to distinct points in the reciprocal space. The scattering signature can therefore be used to deduce structural information.

3. Preparation of Peptide–Lipid Complexes for X-Ray Measurements

3.1. Preparation of recombinant α-defensins

Recombinant α-defensins are expressed in *Escherichia coli* as N-terminally linked, 6×-histidine-tagged fusion peptides as described (Figueredo *et al.*, 2010; Satchell *et al.*, 2003b; Shirafuji *et al.*, 2003). α-Defensin cDNA coding sequences are amplified and directionally subcloned from cDNAs into a pET28a protein expression vector (Novagen, Inc., Madison, WI, USA). For example, Crp4 was amplified from mouse cDNA corresponding to nucleotides 182–274, and directionally subcloned into the *Eco*RI and *Sal*I restriction sites of pET28a. The Crp4 cDNA sequence was amplified using the forward primer, 5′-GCG CGA ATT CCA TCG AGG GAA GGA TGG GTT TGT TAG CTA TTG T, and paired with the reverse primer, 5′-ATA TAT GTC GAC TCA GCG ACA GCA GAG CGT GTA CAA TAA ATG. For expression of peptides that lack Met, forward primers incorporate a Met codon immediately 5′- of the peptide coding sequence, providing a unique cyanogen bromide (CNBr) cleavage site for subsequent separation of the defensin molecule from the pET-28-encoded, His-tag fusion partner. For example, Crp4 is Met-free, so the CNBr reaction does not cleave within the polypeptide chain. However, for Met-containing α-defensins, the CNBr cleavage site is replaced with enzymatic cleavage sites for enterokinase or thrombin.

α-Defensin constructs are transformed into *E. coli* BL21(DE3)-CodonPlus-RIL cells (Stratagene, La Jolla, CA, USA), whose additional Arg codons help to minimize *E. coli* codon bias for Arg-rich defensins. For induction purposes, 6 L Terrific Broth (TB) culture medium is prepared as follows: 12 g of BactoTryptone (BD biosciences, San Jose, CA, USA), 24 g of BactoYeast Extract (BD), and 4 mL of glycerol are combined with 900 mL of deionized water, and all solids are dissolved before autoclaving. The following sterile solution components are added to the medium: 100-mL phosphate buffer (0.17 M KH_2PO_4 and 0.72 M K_2HPO_4), 5-mL 30% dextrose, and kanamycin to a final concentration of 70 μg/mL. *E. coli* are grown to mid-log phase (OD_{600} nm = 0.6–0.9) at 37 °C TB and induced by addition of isopropyl-β-D-1-thiogalactopyranoside to 0.1 m*M*. The bacterial cells are harvested after 4–6 h or after growth, transferred to 1-L wide-mouth polycarbonate centrifuge bottles, and deposited by centrifugation at 5500×g for 10 min at 4 °C in a FIBERLite F8-4X1000Y rotor (FIBERLite Centrifuge, Santa Clara, CA, USA) in a Sorvall RC-26 Plus Superspeed centrifuge. Deposited cells are stored at −20 °C. Bacterial cell pellets resuspended in 6-*M* guanidine–HCl and 100-m*M* Tris–HCl (pH 8.1) are sonicated (70% power, 50% duty cycle for 2 min using a Branson Sonifier 450). Lyzates are clarified by centrifugation at 25,000×g for 30 min at 4 °C in a FIBERLite F21-8X50 rotor using rotor code SA-600 in a Sorvall RC-26 Plus superspeed centrifuge.

3.2. Purification of recombinant α-defensins

His-tagged fusion peptides are purified using nickel-nitrolotriacetic acid (Ni-NTA, Qiagen) resin affinity chromatography. Cell lyzates are incubated with resin at 4 °C overnight at a ratio of 25:1 (v/v) in 6-*M* guanidine–HCl in 100-m*M* Tris–HCl (pH 8.1). The His-tagged fusion peptides are eluted with approximately 10 bed vol of 1-*M* imidazole, 6-*M* guanidine–HCl, and 20-m*M* Tris–HCl (pH 6.5), dialyzed in SpectraPor 3 (Spectrum Laboratories, Inc., Rancho Dominquez, CA, USA) membranes, using three exchanges of 4 L of 5% acetic acid. The dialyzed peptides are lyophilized, dissolved in 80% formic acid, and solid CNBr is added to 10 mg/mL. The peptides are placed in polypropylene tubes, gently purged by a stream of N_2 gas, sealed, foil-wrapped, and incubated at ambient temperature overnight. The cleavage reaction is terminated by the addition of 10 vol of H_2O and lyophilized. Peptide lyophilate is dissolved in 5% acetic acid, centrifuged 15 min at 15,000×g in a microcentrifuge, and sterilized through a 0.22-μm filter.

Recombinant α-defensins are purified to homogeneity by reverse-phase high performance liquid chromatography (RP-HPLC). The peptide is initially purified from the 36 amino acid 6×-histidine-tag fusion partner on a semipreparative C18 column (Vydac 218TP510) at a 15–45% linear gradient of acetonitrile (0.1% trifluoroacetic acid (TFA) is used as the ion-pairing agent in the mobile phase). The peptides are further purified through an analytical C18 column (Vydac 218TP54), and homogeneity is assessed by using acid-urea polyacrylamide gel electrophoresis (AU-PAGE). AU-PAGE is a superior method for determining defensin homogeneity, because small cationic peptides are resolved on the basis of their electropositive charge-to-size ratios. Misfolded peptide variants display reduced mobilities relative to native peptides, providing an index of peptide quality. Peptide masses are confirmed by MALDI-TOF mass spectrometry.

3.3. Refolding of recombinant and synthetic peptides

Certain recombinant α-defensins such as Crp4 are produced within *E. coli* as properly folded peptides with correct disulfide pairings. Occasionally, however, certain recombinant and synthetic α-defensin peptides require reduction and refolding procedures to eliminate disulfide mispairings and insure proper disulfide linkages (Cys^{I-VI}, Cys^{II-IV}, Cys^{III-V}). Assessing proteolytic sensitivity provides a rapid test of whether an expressed α-defensin is correctly folded or not. α-Defensins are inherently resistant to proteolysis by trypsin, and disulfide mispairings render the peptide sensitive to tryptic cleavage. Lyophilized peptides dissolved in 6-*M* guanidine HCl, 0.2-*M* Tris base, and 2-m*M* sodium EDTA (pH 8.2) at peptide concentrations ranging from 0.5 to 2 mg/mL are purged under N_2, sealed, and denatured at 50 °C for 30 min. Following denaturation, 5-mol dithiothreitol per mol polypeptide Cys is added to the peptide solution. The reduction reaction mixture is purged briefly with N_2, incubated at 50 °C for 4 h, and then purified through RP-HPLC on a semipreparative C18 column. The extent of peptide reduction is confirmed by MALDI-TOF MS as an increase in peptide mass of six atomic mass units (a.m.u.). Reduced peptides are concentrated to ~1–6 mL by vacuum centrifugation in a SpeedVac® SC210A and diluted to concentrations of 0.1–0.3 mg/mL peptide with 0.1-*M* NH_4HCO_3, 2.0-m*M* EDTA, 0.1-mg/mL cysteine, 0.1-mg/mL cysteine (pH 7.8–8.0), and purged with N_2. The peptide refolding mixture is adjusted to pH 7.8–8.0 by dropwise addition of NH_4OH, purged again with N_2, sealed, and gently stirred at 4 °C. Samples (~0.1%) of the refolding mixture are assessed for correct folding at intervals by analytical RP-HPLC. As α-defensins fold, exposure of hydrophobic residues is reduced and the retention time on analytical RP-HPLC decreases. When refolding peptide mixtures are separated on semipreparative C18 RP-HPLC, the first peptide peak to elute is the correctly folded peptide in our experience.

3.4. Microbicidal peptide assays

α-Defensins are tested for microbicidal peptide activity against a panel of microorganisms. Exponential-phase bacteria are grown in trypticase soy broth (TSB). Microbes are deposited by centrifugation at 10,000×g for 3 min and washed three times with 10-m*M* PIPES (pH 7.4), supplemented with 1% (v/v) of respective growth medium (10-m*M* PIPES-TSB) and resuspended in PIPES-TSB). Approximately 1 5 × 10^6 CFU/mL of bacteria or fungi are incubated with peptides at various concentrations in a total volume of 50 μL. For assays of peptides other than Crp4, the most bactericidal known mouse α-defensin, is used as a positive control peptide, and cells suspended in 10-m*M* PIPES-TSB without added peptide provide negative controls. The test samples are incubated at 37 °C with shaking for 1 h, and 20-μL samples of incubation mixtures are diluted 1:2000 in 10-m*M* PIPES (pH 7.4) and plated on TSB agar plates using an Autoplate 4000 (Spiral Biotech Inc., Bethesda, MD, USA). After incubation overnight at 37 °C, bacterial cell survival is determined by counting CFU.

Results from a typical bactericidal peptide assay are shown in Fig. 4.2. Here, the activity of mouse Crp4 against *E. coli* ML35 cells was compared with that of its inactive precursor, proCrp4. The data show that Crp4 exposure reduces *E. coli* cell survival by 99.9% at concentrations of 1.5 μM or less, and it shows

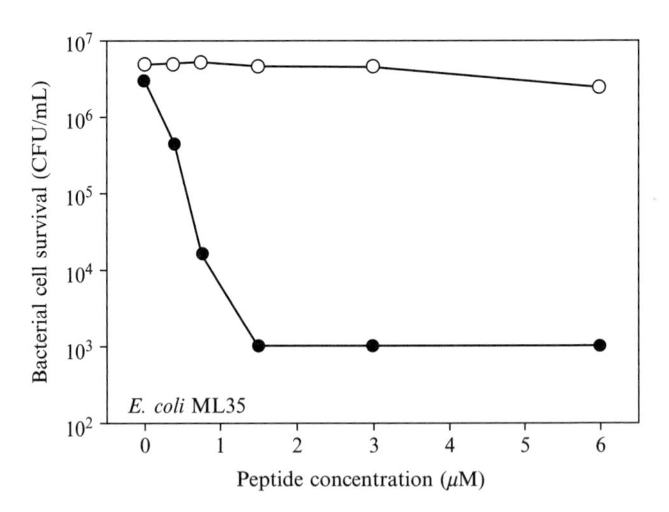

Figure 4.2 Bactericidal peptide activity of recombinant Crp4 (•) and the inactive pro-α-defensin, proCrp4 (○). *E. coli* ML35 cells were incubated with peptides in 50 μL of 10-m*M* PIPES, pH 7.4, 1% TSB (TSB, v/v) for 1 h at 37 °C at the concentrations shown. Following peptide exposure, incubation mixtures were plated on a semisolid medium and incubated for 16 h at 37 °C. Surviving bacterial cells were quantitated as colony-forming units (CFU)/mL. Cell survival values of 1 × 10^3 CFU/mL or less indicate that no colonies were detected after overnight growth.

minimal cytotoxic effects on mammalian cells in culture at levels of 100-μ g/mL peptide (data not shown). In contrast, the proCrp molecule lacks bactericidal peptide activity under these conditions due to the inhibitory effects of its anionic proregion. Previous studies have shown that Crp4 bactericidal effects occur by an undisclosed membrane disruptive mechanism, and consistent with its lack of activity, proCrp4 does not interact with or disrupt model membranes. Thus, the Crp4 molecule provides a useful peptide for studying peptide–lipid interactions using SAXS approaches.

3.5. Vesicle preparation

To investigate how AMPs interact with membranes, we examine the structure and interactions of corresponding peptide–lipid complexes. Small unilamellar vesicles (SUVs) of different compositions are used for X-ray diffraction experiments. The lipids 1,2-dioleoyl-*sn*-glycero-3-[phospho-L-serine] (sodium salt) (DOPS), 1,2-dioleoyl-*sn*-glycero-3-phosphocholine (DOPC), 1,2-dioleoyl-*sn*-glycero-3-phosphoethanolamine (DOPE), and 1,2-dioleoyl-*sn*-glycero-3-[phospho-*rac*-(1-glycerol)] (sodium salt) (DOPG) are purchased from Avanti Polar Lipids and used without further preparation. Mixtures of DOPG and DOPE can be used as an approximate model for bacterial membranes, and DOPS, DOPC, and DOPE mixtures can be used to model eukaryotic membranes. Stock solutions of lipids in chloroform are mixed at the desired ratios, dried under N_2, and desiccated under vacuum overnight. The dried lipids are rehydrated with Millipore H_2O (Tris/HEPES buffer may also be used) to a final concentration of 30 mg/mL and incubated at 37 °C overnight. This solution is sonicated to clarity and extruded through a 0.2-μm Nucleopore filter to make liposomes. Freshly prepared liposomes may be stored at 4 °C and should be used within a week.

The peptides are dissolved in Millipore H_2O at 5 mg/mL. The peptide solutions are mixed with liposomes at different peptide–lipid molar ratios and sealed in 1.5-mm quartz capillaries, which typically have 10-μm thick walls. These samples are typically incubated at least 24 h before data collection.

3.6. Data collection

The sample-containing X-ray capillaries are placed in the incident X-ray beam and the scattered X-rays are collected using a two-dimensional detector. Figure 4.3 shows a typical layout of a SAXS system. X-rays generated at the source are focused and collimated, passing through the presample flightpath. The length of the postsample flightpath depends on the q range of interest in the experiment. SAXS data at Stanford Synchrotron Radiation Laboratory (Palo Alto, CA) (BL4-2) Advanced Light Source (Berkeley, CA BL-7.3.3) and Advanced Photon Source (Argonne, IL, BESSRCAT BL-12ID) are collected using 9-, 10- and 12-keV X-rays,

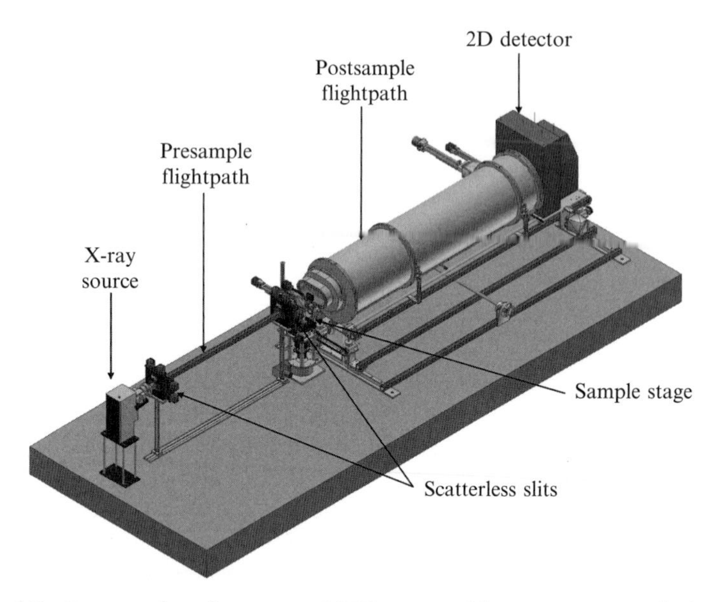

Figure 4.3 Layout of a microsource SAXS system. The spectrometer depicted is a Forvis custom-built instrument at UCLA CNSI. The microfocus sources, both sealed tube and rotating anode, give a brighter beam with lower total power than conventional rotating anode sources. The scatterless slits, made of a rectangular single-crystal substrate, for example, Si or Ge, bonded to a high-density metal base with a large taper angle (>10°), increase the usable flux by several folds (Li *et al.*, 2008).

respectively. The scattered intensity is collected using a MAR-Research (Hamburg) charge-coupled device detector (pixel size 79 μm). For in-house SAXS experiments, incident Cu Kα radiation ($\lambda = 1.54$ Å) from a Rigaku rotating-anode generator is monochromatized and focused using Osmic confocal multilayer optics and collimated to a final beam size of $\sim 0.8 \times 0.8$ mm^2. Scattered radiation is collected on a Bruker two-dimensional multiwire detector (pixel size 105 μm). All experiments are conducted at room temperature. No evidence of radiation damage to the samples is observed under the X-ray exposure levels used. For calibration, a capillary containing Millipore H_2O and an empty X-ray capillary are measured to assess the contribution of the solution and of the capillary itself. X-ray capillary containing dry silver behenate powder is measured to accurately and directly determine the sample to detector distance.

3.7. Translation of two-dimensional X-ray image to diffraction data

The detectors record the number of incident photons at each pixel position and generate two-dimensional diffraction images. The reciprocal lattice vector, q, is related to the pixel number by $q = (4\pi\ /\ \lambda)\sin\,\theta$, where 2θ is the scattering angle and λ is the wavelength of the photon. The two-dimensional diffraction images are calibrated for beam center and detector angular tilt using a standard silver behenate sample. Silver behenate is strongly scattering with several well-defined diffraction peaks. The peak position of the first diffraction peak is used to calculate the sample to detector distance. The two-dimensional images are then radially integrated using Nika 1.2 (usaxs.xor.aps.anl.gov/staff/ilavsky/nika.html) data reduction package or FIT2D (www.esrf.eu/computing/scientific/FIT2D/). The diffracted intensity is plotted against q. The appearance of one or more sharp (Bragg) peaks in the low-angle region of the diffraction pattern helps identify the phase of the complex. The positions of the diffraction peaks are related to periodicity in the phase and the width of the peaks is related to the extent of this periodicity.

For detailed descriptions of X-ray diffraction data analysis procedures, we refer the reader to specialized references on the topic (Cullity, 1956; Guinier, 1994; Ladd and Palmer, 1994; Warren, 1990; Woolfson, 1997). A number of simple structures often occur in lipid mesophases. For example, the $L\alpha$ lamellar phase consists of alternating layers of lipid bilayers and peptides, and its diffraction pattern has concentric equidistant rings centered at the origin. This quasi-one-dimensional periodic structure shows a series of peaks described by

$$q_n = \frac{2\pi}{d} n \tag{4.12}$$

where $n = 1, 2, 3$, etc. and d is the lamellar repeat distance of the one-dimensional lattice. The inverted hexagonal phase has cylindrical water channels, coated by inverse membrane monolayers, packed in a two-dimensional hexagonal lattice. The diffraction peaks are positioned at

$$q = \frac{4\pi}{\sqrt{3}a} \sqrt{h^2 + k^2 + hk} \tag{4.13}$$

where a is the distance between the centers of two neighboring water channels and h and k are the miller indices.

Cubic lipid phases have a more complex architecture. The lattice type can be identified by the characteristic ratios of the q positions of the Bragg peaks, given by

$$q = \frac{2\pi}{a} \sqrt{h^2 + k^2 + l^2} \tag{4.14}$$

where a is the cubic lattice constant and h, k, and l are the miller indices. The crystallographic space group to which the phase belongs is determined from the systematic absences of peaks in the diffraction pattern. However, this is often not trivial, as only a few low-angle Bragg peaks are usually detected. This is due to the large thermal disorder inherent in liquid-crystalline phases, which strongly damps the intensities at larger diffraction angles. From unaligned samples, it is sometimes only possible to identify the cubic aspect from the systematic absences, leaving an ambiguity about the precise space group (Seddon and Templer, 1995; Winter and Jeworrek, 2009).

Figure 4.4 shows the X-ray diffraction data for the α-defensin Crp4 complexed with a 20:80 DOPS:DOPE membrane bilayer. The diffraction peaks have ratios $\sqrt{2}:\sqrt{3}:\sqrt{4}:\sqrt{6}$, which indicate the formation of a cubic Pn3m "double-diamond" lattice. The fitted slope of the plot between the measured peak positions and the corresponding Pn3m cubic indexation, $\sqrt{h^2 + k^2 + l^2}$, gives the lattice parameter of the cubic phase. The presence of the cubic phase indicates that Crp4 is able to induce negative Gaussian curvature in membranes, which is topologically required for pore formation.

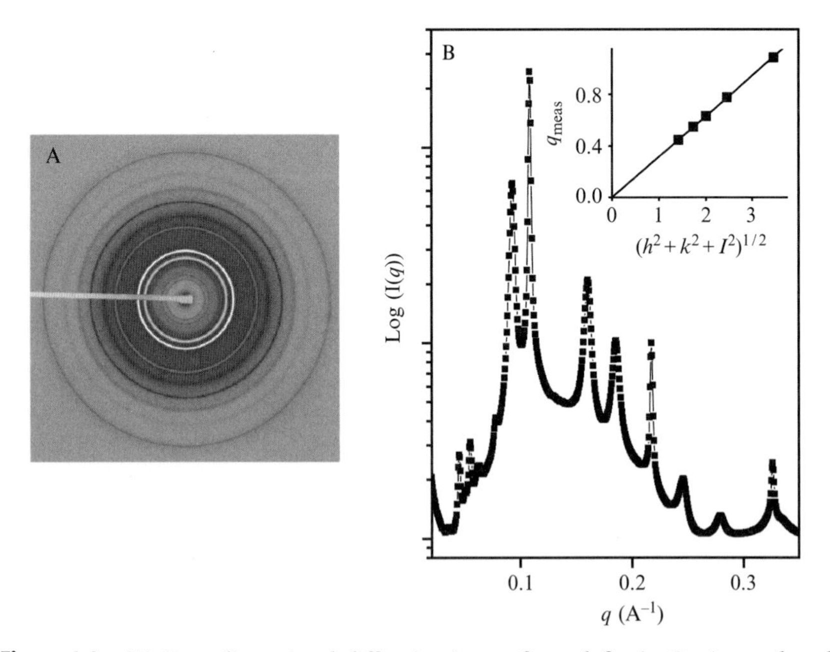

Figure 4.4 (A) Two-dimensional diffraction image for α-defensin Crp4 complexed with a 20:80 DOPS:DOPE membrane at a peptide–lipid molar ratio of 1/90. (B) Diffracted intensity $I(q)$ plotted against reciprocal lattice vector, q. The ratios of diffracted peak positions indicate formation of the Cubic *Pn*3*m* phase. The measured peak positions show good agreement with the corresponding *Pn*3*m* cubic indexation, $\sqrt{h^2 + k^2 + l^2}$ (inset).

4. Summary

In this chapter, we show how SAXS methods, used to study soft condensed matter systems, can be adapted to investigate peptide–membrane interactions. Examining the structures of corresponding peptide–membrane complexes can help elucidate the mechanism of actions of AMPs and CPPs. Here, Crp4 is used as an illustrative example, but the method can be generalized to a broad range of membrane-active peptides.

Acknowledgments

X-ray work was performed at the Stanford Synchrotron Radiation Lab (SSRL), the Advanced Light Source (ALS), the Advanced Photon Source (APS), and at the Fredrick Seitz Materials Research Laboratory (FS-MRL, Urbana, IL). This work is supported by NIH grants R01DK044632 and R01AI059346 (A. J. O.), NIH grant 1UO1 AI082192-01, and NSF grants DMR-0409769 and WaterCAMPWS (G. C. L. W.).

References

Aley, S. B., Zimmerman, M., Hetsko, M., Selsted, M. E., and Gillin, F. D. (1994). Killing of *Giardia lamblia* by cryptdins and cationic neutrophil peptides. *Infect. Immun.* **62,** 5397–5403.

Als-Nielsen, J., and McMorrow, D. (2001). Elements of Modern X-ray Physics. John Wiley & Sons, Inc., New York.

Borenstein, L. A., Ganz, T., Sell, S., Lehrer, R. I., and Miller, J. N. (1991). Contribution of rabbit leukocyte defensins to the host response in experimental syphilis. *Infect. Immun.* **59,** 1368–1377.

Brogden, K. (2005). Antimicrobial peptides: Pore formers or metabolic inhibitors in bacteria? *Nat. Rev. Microbiol.* **3,** 238–250.

Chaikin, P., and Lubensky, T. (1995). Principles of Condensed Matter Physics. Cambridge University Press, Cambridge.

Cullity, B. (1956). Elements of X-Ray Diffraction. Addison-Wesley, Reading.

Cummings, J. E., Satchell, D. P., Shirafuji, Y., Ouellette, A. J., and Vanderlick, T. K. (2003). Electrostatically controlled interactions of mouse Paneth cell alpha-defensins with phospholipid membranes. *Aust. J. Chem.* **56,** 1031–1034.

Cummings, J. E., and Vanderlick, T. K. (2007). Kinetics of cryptdin-4 translocation coupled with peptide-induced vesicle leakage. *Biochemistry* **46,** 11882–11891.

Diamond, G., Zasloff, M., Eck, H., Brasseur, M., Maloy, W. L., and Bevins, C. L. (1991). Tracheal antimicrobial peptide, a cysteine-rich peptide from mammalian tracheal mucosa: Peptide isolation and cloning of a cDNA. *Proc. Natl. Acad. Sci. USA* **88,** 3952–3956.

El-Sayed, A., Futaki, S., and Harashima, H. (2009). Delivery of macromolecules using arginine-rich cell-penetrating peptides: Ways to overcome endosomal entrapment. *AAPS J.* **11,** 13–22.

Elliott, G., and O'Hare, P. (1997). Intracellular trafficking and protein delivery by a herpes virus structural protein. *Cell* **88,** 223–233.

Figueredo, S., Mastroianni, J. R., Tai, K. P., and Ouellette, A. J. (2010). Expression and purification of recombinant alpha-defensins and alpha-defensin precursors in *Escherichia coli*. *Methods Mol. Biol.* **618,** 47–60.

Frankel, A. D., and Pabo, C. O. (1988). Cellular uptake of the tat protein from human Immunodeficiency virus. *Cell* **55,** 1189–1193.

Ganz, T. (2003). Defensins: Antimicrobial peptides of innate immunity. *Nat. Rev. Immunol.* **3,** 710–720.

Ganz, T., Selsted, M. E., Szklarek, D., Harwig, S. S., Daher, K., Bainton, D. F., and Lehrer, R. I. (1985). Defensins. Natural peptide antibiotics of human neutrophils. *J. Clin. Invest.* **76,** 1427–1435.

Green, M., and Loewenstein, P. M. (1988). Autonomous functional domains of chemically synthesized human immunodeficiency virus tat trans-activator protein. *Cell* **55,** 1179–1188.

Guinier, A. (1994). X-Ray Diffraction in Crystals, Imperfect Crystals, and Amorphous Bodies. Dover, New York.

Hancock, R. E. W., and Sahl, H.-G. (2006). Antimicrobial and host-defense peptides as new anti-infective therapeutic strategies. *Nat. Biotechnol.* **24,** 1551–1557.

Hickel, A., Danner-Pongratz, S., Amenitsch, H., Degovics, G., Rappolt, M., Lohner, K., and Pabst, G. (2008). Influence of antimicrobial peptides on the formation of nonlamellar lipid mesophases. *Biochim. Biophys. Acta* **1778,** 2325–2333.

Hristova, K., Selsted, M. E., and White, S. H. (1996). Interactions of monomeric rabbit neutrophil defensins with bilayers: Comparison with dimeric human defensin HNP-2. *Biochemistry* **35,** 11888–11894.

Huang, H. W. (2000). Action of antimicrobial peptides: Two-state model. *Biochemistry* **39,** 8347–8352.

Joliot, A., Pernelle, C., Deagostini-Bazin, H., and Prochiantz, A. (1991). Antennapedia homeobox peptide regulates neural morphogenesis. *Proc. Natl. Acad. Sci. USA* **88,** 1864–1868.

Keller, S., Gruner, S., and Gawrisch, K. (1996). Small concentrations of alamethicin induce a cubic phase in bulk phosphatidylethanolamine mixtures. *Biochim. Biophys. Acta* **1278,** 241–246.

Ladd, M., and Palmer, R. (1994). Structure Determination by X-Ray Crystallography. Plenum Press, New York.

Lehrer, R. I. (2004). Primate defensins. *Nat. Rev. Microbiol.* **2,** 727–738.

Lehrer, R. I. (2007). Multispecific myeloid defensins. *Curr. Opin. Hematol.* **14,** 16–21.

Lehrer, R. I., Barton, A., and Ganz, T. (1988). Concurrent assessment of inner and outer membrane permeabilization and bacteriolysis in *E. coli* by multiple-wavelength spectrophotometry. *J. Immunol. Methods* **108,** 153–158.

Lehrer, R. I., Selsted, M. E., Szklarek, D., and Fleischmann, J. (1983). Antibacterial activity of microbicidal cationic proteins 1 and 2, natural peptide antibiotics of rabbit lung macrophages. *Infect. Immun.* **42,** 10–14.

Li, Y., Beck, R., Huang, T., Choi, M. C., and Divinagracia, M. (2008). Scatterless hybrid metal-single-crystal slit for small-angle X-ray scattering and high-resolution X-ray diffraction. *J. Appl. Crystallogr.* **41,** 1134–1139.

Liang, H., Harries, D., and Wong, G. (2005). Polymorphism of DNA-anionic liposome complexes reveals hierarchy of ion-mediated interactions. *Proc. Natl. Acad. Sci. USA* **102,** 11173–11178.

Maemoto, A., Qu, X., Rosengren, K. J., Tanabe, H., Henschen-Edman, A., Craik, D. J., and Ouellette, A. J. (2004). Functional analysis of the {α}-defensin disulfide array in mouse cryptdin-4. *J. Biol. Chem.* **279,** 44188–44196.

Matsuzaki, K. (1999). Why and how are peptide-lipid interactions utilized for self-defense? Magainins and tachyplesins as archetypes. *Biochim. Biophys. Acta* **1462,** 1–10.

Matsuzaki, K., Sugishita, K.-I., Ishibe, N., Ueha, M., Nakata, S., Miyajima, K., and Epand, R. M. (1998). Relationship of membrane curvature to the formation of pores by magainin 2. *Biochemistry* **37,** 11856–11863.

Mishra, A., Gordon, V., Yang, L., Coridan, R., and Wong, G. (2008). HIV TAT forms pores in membranes by inducing saddle-splay curvature: Potential role of bidentate hydrogen bonding. *Angew. Chem. Intl. Ed.* **47,** 2986–2989.

Mitchell, D. J., Kim, D. T., Steinman, L., Fathman, C. G., and Rothbard, J. B. (2000). Polyarginine enters cells more efficiently than other polycationic homopolymers. *J. Pept. Res.* **56,** 318–325.

Ouellette, A. J. (2006). Paneth cell alpha-defensin synthesis and function. *Curr. Top. Microbiol. Immunol.* **306,** 1–25.

Ouellette, A. J., and Bevins, C. L. (2001). Paneth cell defensins and innate immunity of the small bowel. *Inflamm. Bowel Dis.* **7,** 43–50.

Prenner, E. J., Lewis, R. N. A. H., Neuman, K. C., Gruner, S. M., Kondejewski, L. H., Hodges, R. S., and McElhaney, R. N. (1997). Nonlamellar phases induced by the interaction of gramicidin S with lipid bilayers. A possible relationship to membrane-disrupting activity. *Biochemistry* **36,** 7906–7916.

Purdy Drew, K. R., Sanders, L. K., Culumber, Z. W., Zribi, O., and Wong, G. C. L. (2008). Cationic amphiphiles increase activity of aminoglycoside antibiotic tobramycin in the presence of airway polyelectrolytes. *J. Am. Chem. Soc.* **131,** 486–493.

Purdy, K. R., Bartles, J., and Wong, G. (2007). Structural polymorphism of the actin–espin system: A prototypical system of filaments and linkers in stereocilia. *Phys. Rev. Lett.* **98,** 058105.

Rädler, J., Koltover, I., Salditt, T., and Safinya, C. (1997). Structure of DNA-cationic liposome complexes: DNA intercalation in multilamellar membranes in distinct inter-helical packing regimes. *Science* **275,** 810–814.

Rajabi, M., de Leeuw, E., Pazgier, M., Li, J., Lubkowski, J., and Lu, W. (2008). The conserved salt bridge in human alpha-defensin 5 is required for its precursor processing and proteolytic stability. *J. Biol. Chem.* **283,** 21509–21518.

Rosengren, K. J., Daly, N. L., Fornander, L. M., Jonsson, L. M., Shirafuji, Y., Qu, X., Vogel, H. J., Ouellette, A. J., and Craik, D. J. (2006). Structural and functional characterization of the conserved salt bridge in mammalian paneth cell alpha-defensins: Solution structures of mouse Cryptdin-4 and (E15D)-Cryptdin-4. *J. Biol. Chem.* **281,** 28068–28078.

Rothbard, J. B., Jessop, T. C., and Wender, P. A. (2005). Adaptive translocation: The role of hydrogen bonding and membrane potential in the uptake of guanidinium-rich transporters into cells. *Adv. Drug Deliv. Rev.* **57,** 495–504.

Sanders, L., Guáqueta, C., Angelini, T., Lee, J., Slimmer, S., Luijten, E., and Wong, G. (2005). Structure and stability of self-assembled actin–lysozyme complexes in salty water. *Phys. Rev. Lett.* **95,** 108302.

Sanders, L., Xian, W., Guáqueta, C., Strohman, M., Vrasich, C., Luijten, E., and Wong, G. (2007). Control of electrostatic interactions between F-actin and genetically modified lysozyme in aqueous media. *Proc. Natl. Acad. Sci. USA* **104,** 15994–15999.

Satchell, D. P., Sheynis, T., Kolusheva, S., Cummings, J. E., Vanderlick, T. K., Jelinek, R., Selsted, M. E., and Ouellette, A. J. (2003a). Quantitative interactions between cryptdin-4 amino terminal variants and membranes. *Peptides* **24,** 1793–1803.

Satchell, D. P., Sheynis, T., Shirafuji, Y., Kolusheva, S., Ouellette, A. J., and Jelinek, R. (2003b). Interactions of mouse Paneth cell alpha-defensins and alpha-defensin precursors with membranes: Prosegment inhibition of peptide association with biomimetic membranes. *J. Biol. Chem.* **278,** 13838–13846.

Schmidt, N., Mishra, A., Lai, G., Davis, M., Sanders, L., Tran, D., Garcia, A., Tai, K., McCray, P., Ouellette, A., Selsted, M., and Wong, G. (2010a). Submitted.

Schmidt, N. W., Mishra, A., Lai, G. H., and Wong, G. C. L. (2010a). Arginine-rich cell-penetrating peptides. *FEBS Lett.* **584,** 1806–1813.
Schonwetter, B. S., Stolzenberg, E. D., and Zasloff, M. A. (1995). Epithelial antibiotics induced at sites of inflammation. *Science* **267,** 1645–1648.
Schutte, B. C., Mitros, J. P., Bartlettt, J. A., Walters, J. D., Jia, H. P., Welsh, M. J., Casavant, T. L., and McCray, P. B. (2002). Discovery of five conserved beta-defensin gene clusters using a computational search strategy. *Proc. Natl. Acad. Sci. USA* **99,** 2129–2133.
Seddon, J., and Templer, R. (1995). Polymorphism of lipid-water systems. *In* "Structure and Dynamics of Membranes: From Cell to Veiscles," (R. Lipowsky and E. Sackmann, eds.), Vol. 1. Elsevier, Amsterdam.
Selsted, M. E., Brown, D. M., Delange, R. J., and Lehrer, R. I. (1983). Primary structures of MCP-1 and MCP-2, natural peptide antibiotics of rabbit lung macrophages. *J. Biol. Chem.* **258,** 14485–14489.
Selsted, M. E., and Harwig, S. S. (1989). Determination of the disulfide array in the human defensin HNP-2. A covalently cyclized peptide. *J. Biol. Chem.* **264,** 4003–4007.
Selsted, M. E., and Ouellette, A. J. (2005). Mammalian defensins in the antimicrobial immune response. *Nat. Immunol.* **6,** 551–557.
Selsted, M. E., Tang, Y. Q., Morris, W. L., McGuire, P. A., Novotny, M. J., Smith, W., Henschen, A. H., and Cullor, J. S. (1993). Purification, primary structures, and antibacterial activities of beta-defensins, a new family of antimicrobial peptides from bovine neutrophils. *J. Biol. Chem.* **268,** 6641–6648.
Shai, Y. (1999). Mechanism of the binding, insertion and destabilization of phospholipid bilayer membranes by α-helical antimicrobial and cell non-selective membrane-lytic peptides. *Biochim. Biophys. Acta* **1462,** 55–70.
Shirafuji, Y., Tanabe, H., Satchell, D. P., Henschen-Edman, A., Wilson, C. L., and Ouellette, A. J. (2003). Structural determinants of procryptdin recognition and cleavage by matrix metalloproteinase-7. *J. Biol. Chem.* **278,** 7910–7919.
Som, A., Yang, L., Wong, G. C. L., and Tew, G. N. (2009). Divalent metal ion triggered activity of a synthetic antimicrobial in cardiolipin membranes. *J. Am. Chem. Soc.* **131,** 15102–15103.
Staudegger, E., Prenner, E. J., Kriechbaum, M., Degovics, G., Lewis, R. N. A. H., McElhaney, R. N., and Lohner, K. (2000). X-ray studies on the interaction of the antimicrobial peptide gramicidin S with microbial lipid extracts: Evidence for cubic phase formation. *Biochim. Biophys. Acta* **1468,** 213–230.
Warren, B. E. (1990). X-Ray Diffraction. Dover Publications, New York.
Wender, P. A., Galliher, W. C., Goun, E. A., Jones, L. R., and Pillow, T. H. (2008). The design of guanidinium-rich transporters and their internalization mechanisms. *Adv. Drug Deliv. Rev.* **60,** 452–472.
White, S. H., Wimley, W. C., and Selsted, M. E. (1995). Structure, function, and membrane integration of defensins. *Curr. Opin. Struct. Biol.* **5,** 521–527.
Winter, R., and Jeworrek, C. (2009). Effect of pressure on membranes. *Soft Matter* **5,** 3157.
Wong, G. (2006). Electrostatics of rigid polyelectrolytes. *Curr. Opin. Colloid Interface Sci.* **11,** 310–315.
Wong, G., Tang, J., Lin, A., Li, Y., Janmey, P., and Safinya, C. (2000). Hierarchical self-assembly of F-Actin and cationic lipid complexes: Stacked three-layer tubule networks. *Science* **288,** 2035–2039.
Woolfson, M. (1997). An Introduction to X-Ray Crystallography. Cambridge University Press, Cambridge.
Wu, Z., Li, X., de Leeuw, E., Ericksen, B., and Lu, W. (2005). Why is the Arg5-Glu13 salt bridge conserved in mammalian alpha-defensins? *J. Biol. Chem.* **280,** 43039–43047.

Yang, L., Gordon, V. D., Mishra, A., Som, A., Purdy, K. R., Davis, M. A., Tew, G. N., and Wong, G. C. L. (2007). Synthetic antimicrobial oligomers induce a composition-dependent topological transition in membranes. *J. Am. Chem. Soc.* **129,** 12141–12147.

Yang, L., Gordon, V. D., Trinkle, D. R., Schmidt, N. W., Davis, M. A., DeVries, C., Som, A., Cronan, J. E., Tew, G. N., and Wong, G. C. L. (2008). Mechanism of a prototypical synthetic membrane-active antimicrobial: Efficient hole-punching via interaction with negative intrinsic curvature lipids. *Proc. Natl. Acad. Sci. USA* **105,** 20595–20600.

Yang, L., Liang, H., Angelini, T. E., Butler, J., Coridan, R., Tang, J. X., and Wong, G. C. L. (2004). Self-assembled virus-membrane complexes. *Nat. Mater.* **3,** 615–619.

Zanetti, M. (2004). Cathelicidins, multifunctional peptides of the innate immunity. *J. Leukoc. Biol.* **75,** 39–48.

Zasloff, M. (2002). Antimicrobial peptides of multicellular organisms. *Nature* **415,** 389–395.

Zhu, S. (2008). Discovery of six families of fungal defensin-like peptides provides insights into origin and evolution of the CSalphabeta defensins. *Mol. Immunol.* **45,** 828–838.

Zweytick, D., Tumer, S., Blondelle, S. E., and Lohner, K. (2008). Membrane curvature stress and antibacterial activity of lactoferricin derivatives. *Biochem. Biophys. Res. Commun.* **369,** 395–400.

CHAPTER FIVE

Synergy of Molecular Dynamics and Isothermal Titration Calorimetry in Studies of Allostery

Rebecca Strawn,* Thomas Stockner,† Milan Melichercik,‡ Lihua Jin,§ Wei-Feng Xue,¶ Jannette Carey,* *and* Rüdiger Ettrich‡

Contents

* Chemistry Department, Princeton University, Princeton, New Jersey, USA
† Center for Physiology and Pharmacology, Institute of Pharmacology, Medical University of Vienna, Waehringerstrasse, Vienna, Austria
‡ Department of Structure and Function of Proteins, Institute of Systems Biology and Ecology, Academy of Sciences of the Czech Republic, and University of South Bohemia, Nove Hrady, Czech Republic
§ Chemistry Department, DePaul University, Chicago, Illinois, USA
¶ Astbury Centre for Structural Molecular Biology, University of Leeds, Leeds, United Kingdom

Methods in Enzymology, Volume 492
ISSN 0076-6879, DOI: 10.1016/B978-0-12-381268-1.00017-3
© 2011 Elsevier Inc.
All rights reserved.

Abstract

Despite decades of intensive study, allosteric effects have eluded an intellectually satisfying integrated understanding that includes a description of the reaction coordinate in terms of species distributions of structures and free energy levels in the conformational ensemble. This chapter illustrates a way to fill this gap by interpreting thermodynamic and structural results through the lens of molecular dynamics simulation analysis to link atomic-level detail with global response. In this synergistic approach molecular dynamics forms an integral part of a feedback loop of hypothesis, experimental design, and interpretation that conforms to the scientific method.

1. Allostery

Allostery is a manifestation of the global response of a macromolecule to ligand binding. The molecular origins of this fascinating emergent phenomenon have been a holy grail of biochemistry ever since Monod and Jacob named the concept (Jacob and Monod, 1961) and generalized it to include homotropic cooperativity (Monod *et al.*, 1963); Monod is said (Perutz, 1989) to have considered allostery to be the second secret of life. Studies of ligand-binding kinetics, thermodynamics, dynamics, and structures have led to many levels of description, but despite decades of intensive study, allosteric effects have eluded an integrated understanding that includes a description of the reaction coordinate in terms of species distributions of structures and free energy levels in the conformational ensemble. In this chapter we illustrate the power of molecular dynamics (MD) simulations to fill this gap by linking atomic-level detail with global response, with the aim of promoting the use of MD in concert with thermodynamic studies.

The possible role of protein dynamics as a likely source of allosteric effects was formalized soon after Monod (Weber, 1972); in recent years results for many systems have re-popularized the view that allosteric ligands redistribute protein conformational ensembles (Gunasekaran *et al.*, 2004; Goodey and Benkovic, 2008). As described in this chapter, it was recently possible in one system to trace allosteric activation from specific atomic details to global conformational response by interpreting puzzling thermodynamic and structural results through the lens of MD simulation analysis (Strawn *et al.*, 2010). The proposed allosteric mechanism resulting from this analysis is simple and direct, and it expresses in remarkable detail many features anticipated by the concerted model of Monod, Wyman, and Changeux (MWC) (Monod *et al.*, 1965). The mechanism makes structurally and thermodynamically explicit predictions for homotropic and heterotropic ligand binding and subunit assembly in this system that can be evaluated by clearly defined experiments. As such, these predictions serve as independent tests of the MD results.

In this example extremely complex thermodynamic observations were mirrored in unexpected detail by results from standard MD approaches, supporting a higher level of confidence in the MD methods than if the experimental observations had been simple, and suggesting an unanticipated degree of complementarity and synergy between thermodynamic and molecular dynamics approaches for understanding allostery. Thus, one subtext of this chapter is that analysis of similar examples using the MD strategies illustrated here may lead to a broader view of allostery. Many other allosteric proteins present highly complex, even similar, thermodynamic data (Jin *et al.*, 2005), yet thermodynamics is characteristically — fundamentally — silent about molecular details of mechanism. As well, many allosteric proteins with structures in the Protein Data Bank, like the one described here, present scarcely any conformational difference in the presence and absence of their effectors, offering little or no hint about allosteric mechanism, whereas other examples display differences that do not accurately reflect the allosteric process (Zhang *et al.*, 1987; Zhao *et al.*, 1993). Furthermore, as illustrated by the present example, X-ray crystal structures can specifically mask certain features that can play a critical role in allostery. Thus, structural and thermodynamic analyses may provide before-and-after pictures, but not critical pathway information needed for an intellectually satisfying understanding of the molecular origins of allostery. The example presented here strongly suggests this gap can be filled by MD. Thus, nearly fifty years after MWC, a comprehensive understanding of Monod's second secret from atomic detail to global response might be within reach through the combination of quantitative studies integrating structural, dynamic, and thermodynamic information.

Yet despite the potential of molecular dynamics for adding molecular detail to featureless thermodynamic state functions and motion to static structures, simulation has not been widely accepted among experimental biophysical chemists who have generated most of the available data on allosteric systems. Experimentalists may understandably view MD results with suspicion as something of a *post hoc* rationalization. However, MD is a thermodynamically rigorous method that can represent thermodynamically-defined states at equilibrium in a closed system of dilute solution with conservation of mass and energy and constant temperature and pressure; in principle MD is simply the mechanically detailed counterpart of thermodynamic data, provided that the solution states of the system are accurately mimicked. As such, thermodynamic principles, such as definitions of thermodynamic states, can be applied rigorously in interpreting the observed dynamic behaviors, as will be exemplified here. Thus another subtext of this chapter is to illustrate that MD results can lead to hypotheses that are structurally detailed and thermodynamically rigorous, as well as specifically falsifiable by experiment, bringing the approach squarely into the realm of scientific method by using it as a tool for experimental design;

this is the synergy of the title. We aim to demonstrate several ways in which MD can be used to establish a nested iterative feedback cycle in which analysis of MD simulations suggests experiments, new analysis of experimental data, and further MD analysis.

One novel feature of the work summarized here is that it was conducted as a collaboration among biophysical chemists and simulations experts, neither of whom appreciated fully at the outset the details of the other's approaches, viewpoints, and historical background. This unintentional double-blind approach led to extensive cross-checking and mutual education. One fortunate and unexpected outcome was that during analysis of the simulations in light of the thermodynamic results, which was done truly and intensively together, features emerged that provided unforeseeable corroboration of details that previously had been understood fully by only part of the team. Thus both sides acquired a much higher level of confidence in the interpretations than might have been otherwise possible. A lesson from this experience, and another subtext, is that an intentional double-blind approach has the potential to increase biochemists' confidence in MD results.

The explicit purpose of this chapter is to illustrate how MD simulations can be applied to elucidate allosteric mechanisms with the desired range of detail from atomic to global, aiming to outline for typical practicing biochemists how MD might provide a clearer view of allostery in other systems as well. Unlike typical chapters in these volumes, this one does not aim to present a recipe-like approach to MD or its application, nor basic training in MD itself or in its physical or computational underpinnings; for background of this kind aimed at a biochemical audience see the excellent recent chapter in these volumes by Schleif (Schleif, 2004) and references therein. Rather, the present aim is to illustrate how the results of MD simulations can yield the kind of information about allostery that is of interest to biochemists by using the strategies employed in the recent example from our own work to analyse, interpret, and relate the results to thermodynamic data, and by using the experimental and MD results iteratively to guide further design and analysis of both computational and wet experiments.

2. Arginine Repressor

Arginine repressor (ArgR) is the master regulator of the arginine regulon in a wide variety of bacteria, acting as direct sensor and transcriptional transducer of intracellular L-arginine (L-arg) concentrations to provide feedback control over biosynthesis and catabolism of L-arg (Maas, 1994). The co-effector L-arg binds deep within a central hexamerization domain (Fig. 5.1), altering affinity and specificity of DNA binding (Szwajkajzer *et al.*, 2001) by peripheral domains (Grandori *et al.*, 1995;

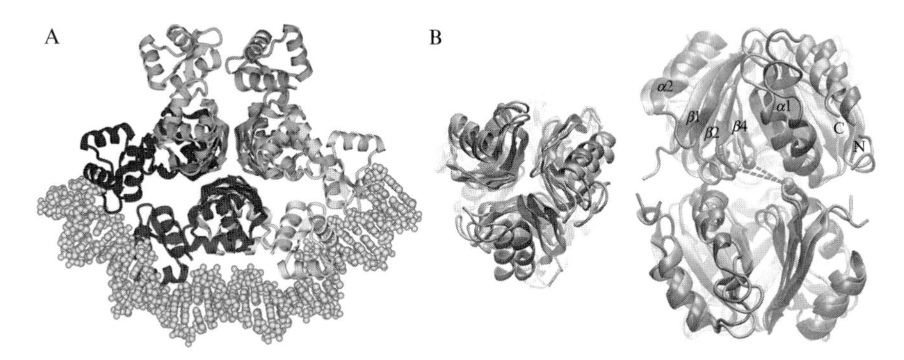

Figure 5.1 ArgR structure. A. Model of DNA complex. Intact ArgR viewed down the three-fold axis with central ArgRC domains and peripheral ArgRN domains docked with bent B-form DNA. Subunits A, yellow; B, green; C, magenta; D, cyan; E, blue; and F, red correspond to those in Fig. 5.2. Protein structure prepared from PDB ID 1B4A (apoBstArgR (Ni *et al.*, 1999)), DNA from PDB ID 1J59 (cAMP receptor) as described in (Sunnerhagen *et al.*, 1997). B. ArgRC rotation. Overlay of average hexamer structures from the equilibrated part of the simulations showing the conformational shift from the holoArgRC (orange) structure that occurs uniquely in apoArgRC (blue). The bottom trimer (ABC of panel A) was used for C_α RMSD minimization. Left, top view. The size and viewpoint are chosen to match panel A. Right, side view. Selected secondary structure elements and N- and C-termini are labeled for orientation. CPK spheres mark C_α atoms of two Gly103-Asp128 residue pairs whose interatomic distances (dashed) are measured to quantify rotation. (See Color Insert.)

Sunnerhagen *et al.*, 1997). Thus ArgR is an apparent example of a true action-at-a-distance allosteric protein (Monod *et al.*, 1963). The structural organization of ArgR into N- (ArgRN) and C-terminal (ArgRC) domains, and the functional division of labor between them, are conserved even among distant homologs that display a bewildering diversity of reported biochemical properties, notably the L-arg dependence of their hexamerization and DNA-binding equilibria (Czaplewski *et al.*, 1992; Lu *et al.*, 1992; Dion *et al.*, 1997; Morin *et al.*, 2003), hinting that allosteric mechanisms could vary.

An allosteric mechanism for ArgR was previously inferred by comparison between the crystallized intact unliganded ArgR apoprotein from the thermophile *Bacillus stearothermophilus* and its liganded C-terminal domain fragment ArgRC (Ni *et al.*, 1999). The two structures differ by $\sim 15°$ rotation about the trimer-trimer interface that was ascribed to L-arg binding and was presumed to be transmitted to the DNA-binding domains. However, rotation could as well reflect crystal packing or crystallization conditions, or the differential presence of the N-terminal domains in the two crystals, and no structure of apoBstArgRC has been reported. A similar degree of rotation was later observed between apo- and holoArgRC of

Mycobacterium tuberculosis (Cherney *et al.*, 2008), suggesting that rotation might be an inherent response of ArgRC to L-arg, but offering no hint if or how this response might be involved in allosteric activation. Pursuit of the allosteric mechanism of *E. coli* K-12 ArgR (EcArgR), the most thoroughly studied ArgR, has been motivated by the wealth of physiological, genetic, biochemical, and biophysical knowledge (Jin *et al.*, 2005; Maas, 1994; Szwajkajzer *et al.*, 2001; Grandori *et al.*, 1995; Sunnerhagen *et al.*, 1997; Lim *et al.*, 1987; Tian *et al.*, 1992, 1994; Tian and Maas, 1994, Van Duyne *et al.*, 1996; Niersbach *et al.*, 1998) that is unavailable for any homolog and is expected to constrain activation models.

Unlike the *Bacillus* and *Mycobacterium* proteins, crystal structures of the *E. coli* ArgR C-terminal domain with (holoEcArgRC) and without (apoEcArgRC) bound L-arg are essentially identical, with maximum local shifts of $<\sim$0.5 Å except at poorly-ordered termini (Van Duyne *et al.*, 1996) and overall C_α RMSD $\sim$ 0.76 Å for hexamers. The maximum difference in Stokes' radius between intact ($\sim$100,000 Da) apo- and holoEcArgR that would be consistent with extensive analytical ultracentrifugation data is less than 1 Å (Jin *et al.*, 2005), severely constraining possible allosteric mechanisms. Crystalline apo- and holoEcArgRC hexamers are also entirely symmetric, and intact apoEcArgR and ArgRC hexamers are symmetric as judged by the number of NMR HSQC crosspeaks (J. Carey, unpublished), yet symmetric hexamers are seemingly incongruent with the complex thermodynamics of L-arg binding. Isothermal titration calorimetry (ITC) confirms that both EcArgR and EcArgRC hexamers bind six equivalents of L-arg, but with a multiphasic binding isotherm, slow endothermic plus fast exothermic heat flows, and $\sim$100-fold stronger binding for the first L-arg than for the subsequent five (Jin *et al.*, 2005). This negative cooperativity might be physiologically useful to ArgR because such distinct affinities create two action levels in the protein's response to L-arg. It is not yet known if or how two action levels are exploited, but they could make sense for multifunctional ArgR, which not only governs the arginine regulon both positively and negatively, but also autoregulates its own expression, and acts as a required cofactor for resolution of ColE1 plasmid multimers during replication; all these activities require L-arg (Maas, 1994). Asymmetric L-arg binding is an apparent paradox for the symmetric ArgR and ArgRC hexamers, and it reflects a second, separate allosteric transition that occurs even in the absence of DNA binding or DNA-binding domains.

Further interpretation of this complex system required an understanding of the protein's conformational landscape. As described here, MD simulations with the ArgRC domain were used to provide a structural and energetic picture of populated conformations and their response to the presence of the ligand. The analysis offered unexpected insight into the ITC results, resolving the apparent symmetry paradox, explaining the slow endotherm, and suggesting a detailed mechanism for ArgR—L-arg allostery

(Strawn *et al.*, 2010). In brief, conserved Arg and Asp sidechains of each L-arg binding pocket promote oscillations of apoArgRC trimers by engagement and release of salt bridges. Exogenous L-arg shifts the dynamic quaternary ensemble to favor a higher-energy state with broken Arg-Asp bonds. A single equivalent of L-arg per hexamer is necessary and sufficient to accomplish this shift, with the bond-breaking assigned to the ITC endotherm. A single bound ligand arrests trimer oscillation and promotes intense monomer motions that generate an entropic driving force while maintaining hexamer symmetry. Despite its lower effective concentration, an L-arg ligand competes successfully with a resident Arg residue because its C_α substituents engage multiple subunits *via* an extensive network of interactions that exploit the hexamer's symmetry. The results provide a first view of the symmetric relaxed state predicted by the MWC model, revealing that this state is achieved by exploiting the dynamics of the protein-ligand assembly and the distributed nature of its cohesive free energy.

3. Preparation for Simulations

Preparation of structure files deposited in the PDB that originate from X-ray diffraction or NMR involves deleting irrelevant atoms detected in the experiment and rebuilding essential undetected atoms. In the case of ArgR, multiple alignment of 500 homologs was used (Strawn *et al.*, 2010) to ensure that the crystal structures of *E. coli* apo- and holoArgR domain ArgRC (PDB 1XXC and 1XXA, respectively (Van Duyne *et al.*, 1996)), which contain residues 80-156 of intact ArgR, comprise the entire C-terminal domain. These PDB files were then prepared as simulation templates using standard methods (Schleif, 2004; Lindahl *et al.*, 2001; Krieger *et al.*, 2002). Each prepared template was then placed in a box of water molecules to mimic the dilute solution state, and with added cations or anions sufficient to achieve electric neutrality. The size of the box is governed by the need to prevent the protein from interacting with its virtual copies in periodic boundary conditions, typically requiring 10 Å beyond the protein. Virtual-site hydrogens (Feenstra *et al.*, 1999) were employed to facilitate longer simulation times for this large system by avoiding the need to calculate the vibrational and rotational motions of hydrogen atoms, which are the fastest and most numerous motions in protein simulations, but are not expected to be significant for a global process like allostery. The virtual-site approach enables a larger step size during the simulations (here 5 fs), thereby enabling longer simulations by reducing the number of repetitions of the basic MD step and greatly reducing the computational burden of data collection for full-atom simulations.

Although the basic MD step is calculated every 5 fs, a snapshot of the structure is captured along the trajectory and saved at a frequency that must be chosen based on the nature of the analyses to be conducted and the available computational capacity. For energy calculations it is essential to capture a sufficient number of observations to ensure that rare species are adequately represented in the population distribution. For proteins a typical interval adequate for this purpose, determined empirically, is 1 ps. For standard analysis calculations including root-mean-square deviation, root-mean-square fluctuations, and radius of gyration, if the trajectory contains frames captured every 1 ps, then every fiftieth frame contains adequate information. For calculations of entropy, however, the number of frames captured must equal or exceed the number of degrees of freedom of the atoms for the entropic process being considered. For example, for a protein of N atoms a typical calculation of conformational entropy using the quasi-harmonic approximation would require 3N-6 frames to account for motion of each atom in three dimensions (minus translational and rotational degrees of freedom of the whole protein). Entropy calculations are thus computationally expensive, and are typically carried out with the minimum number of frames adequate to yield an entropy value that is independent of frame number. Thus, the output of a typical protein simulation is essentially a movie of the structure comprising frames collected every 0.1 to 50 ps, with a typical 20-ns trajectory containing 20,000 to 200,000 frames, making data analysis a formidable task.

At the time these simulations were initiated (2004) the ~ 50,000-Da *E. coli* ArgRC system was near the upper limit of accessible computational size, with 20-ns full-atom simulations requiring ~one month of computation time. The technical configuration included a typical Beowulf parallel CPU cluster and nodes communicating via high-speed fiber-optic cable connections. At that time the optimal balance of calculation speed and internode communication per job for molecular dynamics calculations peaked with eight processors in the cluster, limiting overall simulation lengths. Advances since then have eliminated the reduced efficiency at higher processor numbers so that computation time now scales linearly with the overall number of processors and nodes up to at least several hundred (Hess *et al.*, 2008); thus, availability of nodes and processors has now become rate-limiting. With the computational resources now available simulations are underway with intact ArgR, and preliminary results indicate that the dynamic behaviors observed for ArgRC are also observed for ArgR, consistent with their essentially identical ITC behaviors (Jin *et al.*, 2005). Total simulation length is limited in practice by the precision of force fields, with cumulative errors in numerical integration leading to distortions that can be minimized with proper selection of algorithms and empirical force field parameters, but not eliminated entirely. One-microsecond all-atom simulations are the approximate current maximum, and even longer simulations will become possible as empirical force fields are continually improved. Simulations for ArgR used the GROMACS program package (Lindahl *et al.*, 2001) with the GROMOS 87 force field (van Gunsteren and Berendsen, 1987), approaches that are standard for systems of proteins and amino acids in aqueous solution. Different empirical force fields are designed for specific kinds of molecules, properties, or processes, and are specifically parameterized using sets of experimental values that are known to be appropriate for each case. Force fields accommodate systems containing more than one type of molecule, e.g., protein and DNA. A comparison of force fields can be found in Ponder & Case (2003).

For apoArgRC, independent runs reproducibly produced a very large early conformational shift corresponding to rotation of one trimer about the other in the clockwise direction only as judged by visual inspection of the trajectories, and the starting state based on the apoArgRC crystal structure was never visited again during any simulation. The clockwise direction of rotation is opposite of that detected between the crystals of apo- and holoArgRC from *M. tuberculosis* (Cherney *et al.*, 2008), and intact hexameric apoBstArgR (Ni *et al.*, 1999) is also rotated counterclockwise relative to EcArgRC. Thus, in an initial effort to exclude artifactual causes of this early conformational shift, numerous control runs were carried out (Strawn *et al.*, 2010) using a range of force fields and corresponding water models, as well as different initial velocities; all showed the same large, early, clockwise rotational shift between apoEcArgRC trimers. Furthermore, multiple independent runs with holoEcArgRC showed no comparable change, greatly limiting the potential kinds of artifactual causes and thus suggesting the conformational change might be related to the properties of apoArgRC. As discussed below, these results and others suggest that crystals trap a high-energy conformation of apoArgRC, and the apparently conflicting directions of rotation in homologous ArgR systems can be explained by the allosteric mechanism derived from the analysis.

4. Sampling of States

A broad series of simulations was then initiated representing the various liganded states of ArgRC. In addition to multiple independent replicates of apo- and holoArgRC, additional starting structures were derived by removing all six L-arg ligands from 1XXA (holoArgRC-6) and by adding six L-arg to 1XXC (apoArgRC+6); six more by adding one L-arg in turn to each monomer of 1XXC (apoArgRC+1); six by removing five L-arg from 1XXA in all permutations (holoArgRC-5); and fifteen by adding two L-arg to apoArgRC in all permutations (apoArgRC+2). Based on the results from these simulations, further incremental additions or deletions of L-arg ligands appeared unnecessary. The apparent redundancies of these simulations are in fact used directly to probe a system's free energy landscape along the reaction coordinate in both directions and to compensate in some measure for the fact that each simulation operates on only one protein molecule for times that may be short relative to its full range of possible motions. The multiple minima of typical protein conformational landscapes present barriers that may not be crossed sufficiently often to obtain an equilibrium distribution of the conformations accessible within one thermodynamic state in available computation times. Thus, with present limits on total simulation time, any one simulation might under-represent conformations that are accessed only rarely within an ensemble. By conducting more than one independent replicate it is possible to verify the ergodic behavior of the system in accessible computation times, as some independent simulations might initiate near rare conformations, permitting observation of their behaviors or transitions.

Independent simulations of the same system might also evolve differently after equilibration due to the different initial velocities or other differences in starting conditions. However, the initial velocities that are assigned to each atom randomly in each simulation satisfy a Maxwell-Boltzmann distribution at the chosen temperature, and should not affect the region of conformational space that is sampled upon equilibration in the ergodic limit. Results that depend on initial velocities indicate non-ergodic behavior and should not be used. On the other hand, slight differences in starting states, as in the apoArgRC+1 and holoArgRC−5 simulations considered here, which have distinct atom coordinates, may give rise to different system behavior because they may encounter barriers of different heights when approaching the same state from opposite directions on the reaction coordinate. Thus, apparently redundant simulations that behave distinctly do not necessarily indicate artifactual results as long as ergodicity is met, but rather are useful to broaden the sampling of states and to probe energy levels and barriers along the reaction coordinate.

5. Equilibration

ArgRC simulations were allowed to proceed to equilibration and beyond for at least 20 ns. C_α root-mean-square deviations (RMSDs) comparing each frame of a trajectory with its starting structure are used to assess whether the system has equilibrated. RMSDs tend toward values of $\sim$ 2 Å (Fig. 5.2), typical for systems of this size (Speranskiy *et al.*, 2007); stable plateau values are reached by $\sim$10 ns except for holoArgRC-6 and apoArgRC+6 that drift slightly even at 70 ns, indicating that equilibration has not yet been achieved. The significance of equilibration is that it implies the observed ensemble represents a set of conformations that is accessible to the system within one thermodynamic state, with the above caveat that any one simulation is likely too short to sample all accessible states. Simulations that do not equilibrate indicate that the system did not yet find a minimum on the energy

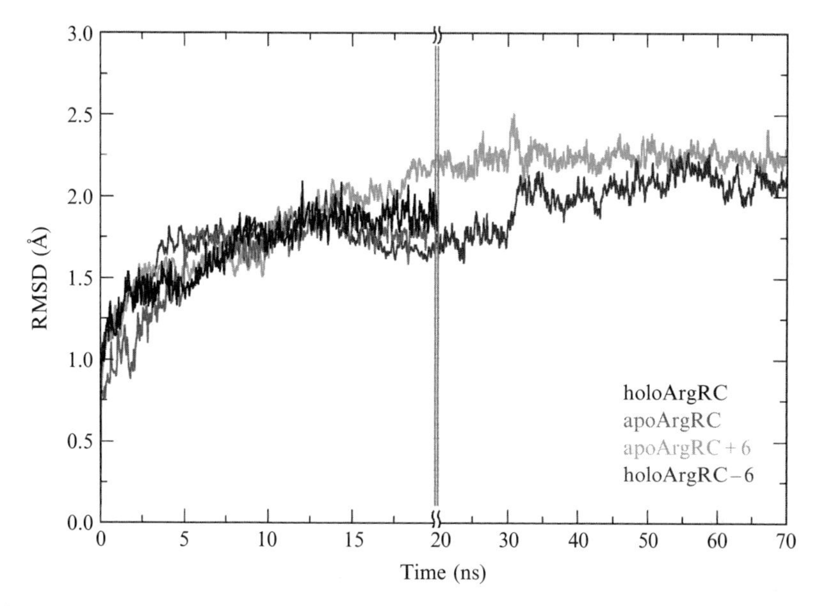

Figure 5.2 Equilibration. Every 50 ps during the trajectory, each simulation structure is compared to the corresponding initial simulation structure after overlaying by C_α superposition to derive root-mean-square deviation (RMSD, Å) of C_α positions. Each color represents the individual simulation indicated. Note the compressed time scale after 20 ns (vertical line). (See Color Insert.)

landscape. The relevance of sampled conformations cannot be presumed in simulations that do not equilibrate because the results do not describe a "state" according to the thermodynamic definition. Thus, the importance of analysing only equilibrated simulations cannot be overemphasized.

Three additional measures were used to assess the equilibrated status of the system at a coarse level of structural detail. C_α fluctuations (RMSFs) relative to the structure averaged over the equilibrated phase report on chain excursions and can be compared with crystal structures. For ArgRC, C_α fluctuations during the last 10 ns follow closely the pattern of crystallographic B factors, with maximal values ~ 1.5 Å for peripheral residues and minimal values ~ 0.3 Å for internal chain segments (Fig. 5.3), indicating that secondary and tertiary structures are maintained. Monomer mass distributions and radii of gyration, recovered from the trajectories using standard approaches incorporated in the program packages (Lindahl *et al.*, 2001), equilibrate by ~5 ns for ArgRC (not shown). All three measures reflect the nature and range of motions at the secondary, tertiary, and quaternary structure levels, and thus constrain the interpretation of observed motions. For ArgRC the results from these three measures taken together exclude the occurrence during the simulation of significant chain excursions within monomers, and also significant excursions of monomers within the hexamer.

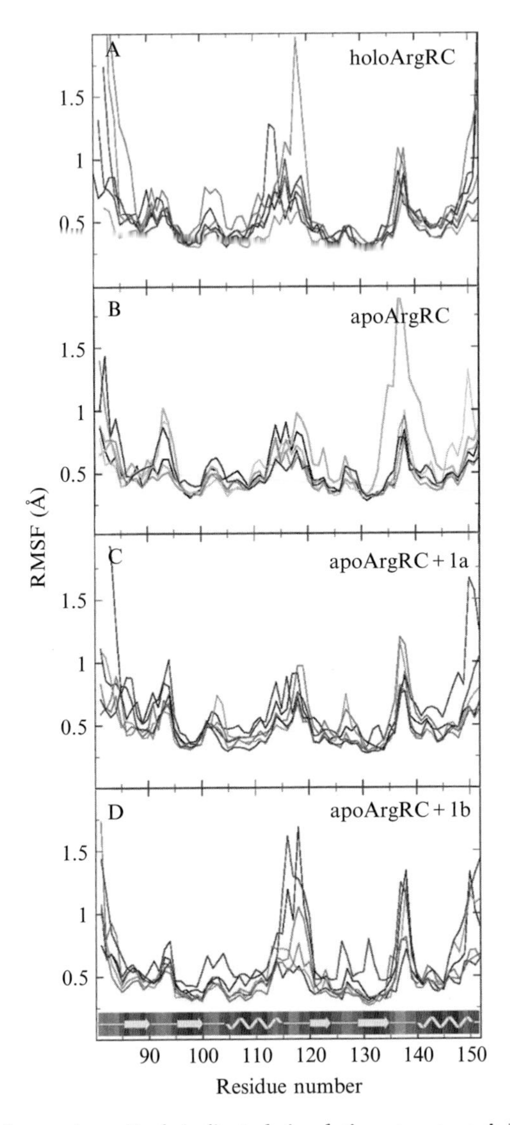

Figure 5.3 C_α fluctuations. Each indicated simulation structure A-D was compared every 50 ps with the corresponding average structure calculated during the last 10 ns of the trajectory. The maximum displacement of each C_α during the last 10 ns is shown (RMSF, Å). Colors indicate individual monomers as in Fig. 5.1 except monomer A is black for clarity; the outlier referred to in the text for apoArgRC is monomer E, shown in grey. The secondary structure of ArgRC is indicated below: arrows, strands; folded tape, helices. (See Color Insert.)

6. Observing System Motions

The first level of observation is examination of the trajectories by eye, i.e., watching the movie of molecular motions that is the output of a simulation. This may seem to be an inexplicably low-tech approach given the high-level nature of the computations, but it is an essential step because some observations simply cannot be codified in any useful way. In fact the reporting of parameter values averaged over a trajectory can mask important discrete effects. For example in the present case it was only by direct human observation that the essential sidechain interactions were detected that govern the global motions of the system, as discussed further below. An effective way to screen the results is to overlay selected snapshots from the trajectories. Nevertheless, the majority of analysis must be handled by automated routines that return only averaged parameter values, simply because the enormous amount of molecular information present in the trajectories cannot be assimilated by human observation. Some automated routines yield data that can be used to narrow the range of human observation. For example, RMSD as a function of atom number can identify potentially interesting regions of the structure, and H-bond number or specific inter-atomic distances as a function of time can identify subtle conformational processes.

7. Correlated Motions

All motions of the system must be analysed to determine which ones are significant for the process under investigation. This is a daunting task because the motions of each atom must be resolved from and correlated with all others in each of the 20,000 or more frames. The immense quantity of data produced by large-scale molecular dynamics simulation is handled through principal-components analysis (PCA) (Lindahl *et al.*, 2001), which is used to detect correlations that enable discrimination between relevant conformational changes and the background of atomic fluctuations; to identify systematic atomic displacements within individual trajectories; and to compare motions in separate MD trajectories. PCA is a standard statistical method for pattern recognition in data of high dimensionality, and involves calculating the eigenvalue decomposition of a data covariance matrix to determine which components account for which fraction of the variance. As applied to MD, the data covariance matrix represents the products obtained when the vector describing the magnitude and direction of the deviation for each atom relative to a reference position is multiplied by that of each other atom. Positive vector products indicate motions in a common direction (correlated), negative vector products indicate motions in opposite directions (anti-correlated), and vector products near zero indicate random motions, and not necessarily absence of motion (uncorrelated). The reference position used to define the atomic deviations is typically the averaged structure over the equilibrated part of the trajectory, which is assumed to lie in the middle of the free energy basin of the state sampled in an equilibrated MD simulation. Deviations are interpreted as describing fluctuations up the sidewalls of the basin, and are typically calculated only for C_α atoms to reduce the computational burden, as the procedure involves matrix inversion, a slow calculation that scales as the square of the number of atoms. The covariance matrix thus contains a highly condensed summary of all motions of C_α atoms occurring during the trajectory, presented as the extent of their pairwise correlations.

Covariance analysis of correlated motions during the last 2 ns of each ArgRC simulation is shown in Fig. 5.4, where each panel presents two covariance matrices fused along the diagonal to facilitate comparison and avoid redundancy. Correlations between pairs of ArgRC monomers are represented by the full matrix elements, and correlations within individual monomers by elements along the diagonal. Each matrix element is itself a matrix of monomer residues 80–156 arrayed along each axis of the element. When the hexamer structure averaged over the last 2 ns is used as the reference structure for analysis of each corresponding trajectory (Fig. 5.4A), the pattern of correlations within each monomer is mostly, though weakly, positive, with a few interspersed regions of negatively, but again weakly, correlated motions; negative correlations are in approximately the same locations in each monomer, in regions that correspond to loops in the tertiary structure. This pattern indicates that monomers move largely as units. Although near-zero vector products indicate absence of correlation rather than the absence of motion *per se*, the interspersion of narrow white regions between areas of positive and negative correlation likely identifies pivot points for the observed local motions, i.e., regions that are not moving. The absence of substantial negatively correlated motions within monomers excludes secondary and tertiary rearrangements as a source of the correlated motions, consistent with the RMSFs. Among monomer pairs no clear pattern of correlated motions can be discerned; the random distribution of mostly small or negative vector products indicates that their motions are not correlated.

8. Structural Features of Correlated Motions

The covariance matrix itself reveals nothing about the geometric features of motions, only the extent to which they are correlated. To convert the quantitative MD results into useful structural information

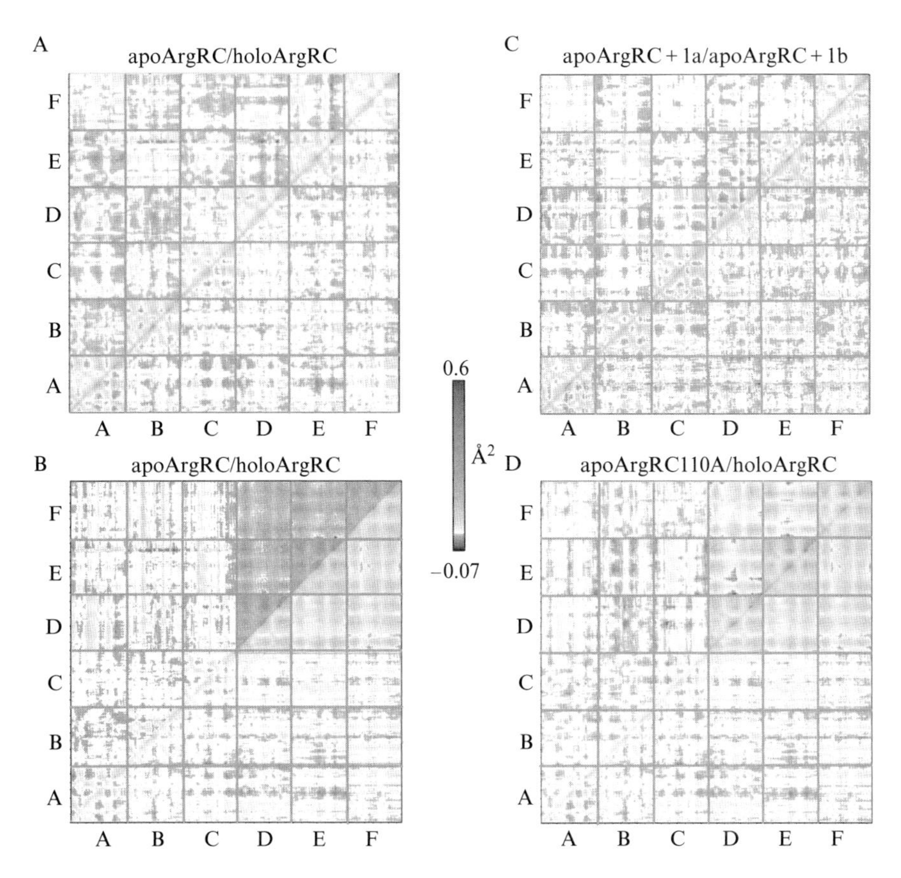

Figure 5.4 Correlated motions. Vector products representing the maximum extent of correlated motion (Å^2) for each C_α pair are plotted. The color scale indicates the degree of correlation: red, positively correlated; blue, negatively correlated; white, uncorrelated. Each panel (A)-(D) displays the six subunits of each hexamer, A-F, arrayed along the axes. In each panel two covariance matrices are fused along the diagonal to facilitate comparison of the simulations indicated above the panel; the simulation presented on the upper left half is indicated first, followed by a slash (/) representing the diagonal, followed by the simulation on the lower right half. The reference state for calculating the extent of C_α motion is the average hexamer for panels A and C; for B and D the reference state is the trimer composed of ABC monomers. (See Color Insert.)

requires geometric evaluation of atomic motions. Although much of the desired geometric data can be acquired by automated routines applied to the MD trajectory, such as recording of dihedral angles, secondary structures, or interatomic distances as a function of time, a surprisingly large part of the analysis must be conducted by human beings with thorough knowledge of protein properties, through detailed visual examination of the trajectories themselves and of the outputs of automated geometric routines. In this and the following two sections, the ArgR example is used to illustrate how insights about allosteric processes can be derived from MD results in systems that have been properly prepared and analysed.

To assign the correlated motions to structural attributes of ArgRC, the program DynDom (Hayward and Berendsen, 1998; Hayward *et al.*, 1997) was used that clusters correlated residue displacements consistent with screw motions and assigns them to individual motional units, thus enabling visualization of large-scale relative motions within the protein. Analysis of apoArgRC structures along the trajectory reveals that the dominant motion consists of rotational oscillation of one apoArgRC trimer with respect to the other trimer about the inter-trimer interface, with little or no other significant global motion. Thus in addition to the early rotational shift, which DynDom quantifies as ~13°, the main motion of apoArgRC during the simulation is rotational oscillation across the trimer interface that returns the system occasionally to a less-rotated conformer but never quite reaches the non-rotated starting conformer. When one trimer (subunits ABC) was then used as reference structure for covariance analysis of the ArgRC trajectory, the striking pattern in Fig. 5.4B upper emerged, indicating dramatically correlated motions between trimers reflecting this rotational oscillation. This result indicates that an appropriate reference structure for covariance analysis may not be the starting structure or the average structure as is usually the case, and that prior analysis of motions can suggest other relevant reference structures.

When the holoArgRC trajectory is analyzed with trimer referencing, a very similar, but less intensely positively correlated, pattern is found (Fig. 5.4B lower). The similarity in covariance pattern between apo- and holoArgRC in Fig. 5.4B need not reflect similar motions *per se*; rather it reflects only that in both cases correlated motions occur between trimers. In fact the relative motion between holoArgRC trimers has no rotational component as judged by DynDom; the trimer basis for the weaker correlation detected in this case reflects instead mainly small variations in inter-trimer distance along the three-fold axis. Trimer referencing also reinforces the underlying pattern of traces reflecting the tertiary structure (nine faint blocks within each monomer square, particularly notable where monomer E straddles the diagonal in Fig. 5.4B), consistent with the RMSF analysis, indicating apoArgRC monomers remain folded despite the extreme relative motions of trimers; tertiary traces are observed also for holoArgRC.

In the equilibrated part of the apoArgRC trajectory the structure oscillates freely between more rotated and less rotated conformers; one entire cycle occurs over ~200–300 ps. Parameter values recovered by automated routines for atom position, atom velocity, and energy change continuously during oscillation of apoArgRC, indicating there is no barrier between the two conformers and thus suggesting they represent extremes of natural hexamer motion driven by thermal flux. This result demonstrates that large, synchronized global motions of a rather large system can be observed in a full-atom simulation on a reasonable computational timescale. The interpretation of this dynamic behavior from the viewpoint of a free energy landscape is of a system sampling a broad potential well or basin, with one conformer occupying the bottom of the basin and the other populating the ascending sidewalls. The non-rotated conformer of apoArgRC is the one more similar to the starting state based on the apoArgRC crystal structure, which is not rotated and is never visited again during any simulation, suggesting it lies at higher energy. Thus, the bottom of the apoArgRC basin is most likely occupied by the rotated conformer, and the sidewalls by the non-rotated conformer, as confirmed below. An energy barrier presumably separates this basin from higher-energy conformations like the starting state, which must represent an accessible state of apoArgRC even if it is rare outside crystals, where lattice energy and/or crystallization conditions may stabilize it. Because the crystal structure of apoArgRC is nearly identical to that of holoArgRC, the state of apoArgRC in the crystal may be most closely represented by the holoArgRC-6 simulations, which did not equilibrate, perhaps because the simulation conditions do not mimic the crystal. Simulations are in progress with holoArgRC-6 in high salt to test this hypothesis.

9. Arg Residues Promote Rotation and Oscillation

The unexpected finding that a large rotational shift occurs early in the simulation, in one direction only, and not in holoArgRC prompted detailed examination of the trajectories to identify the structural correlates of this shift and understand its underlying causes. Because this shift occurs across the trimer interface, attention was focused on the interface as a likely locus of atomic interactions involved in the shift. Visual inspection of the trajectories identified a major alteration at the Arg110 sidechain in each L-arg binding pocket and only very minor changes elsewhere. Rotation is associated with a flip of these sidechains that extends them almost completely, permitting each Arg110 guanidino group to make a bidentate, doubly hydrogen-bonded salt bridge across the trimer interface with an Asp128 carboxylate lying diagonally opposite within each L-arg binding pocket. The interaction is equivalent to that made by the ligand guanidino group with Asp128 in holoArgRC crystals and simulations (Fig. 5.5A and B). This finding illustrates the importance of detailed human visual inspection of the trajectories for discovering possible atomic origins of global effects.

To clarify whether Arg-Asp interactions are a cause or a consequence of apoArgRC rotation, an energy-minimized starting structure in which Ala replaced Arg110 in every subunit was created from the apoArgRC crystal structure; 20-ns simulations were conducted and the equilibrated

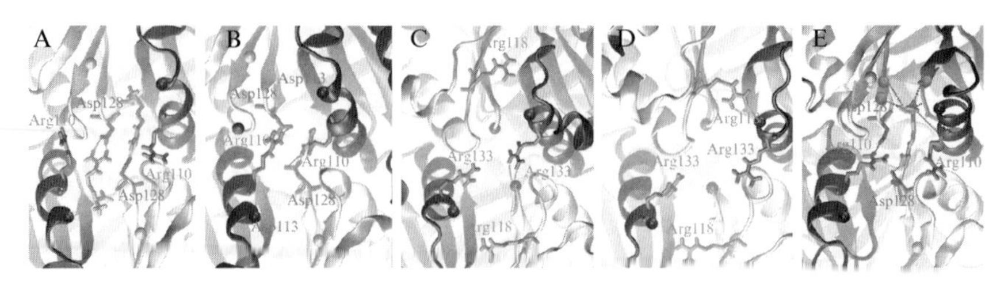

Figure 5.5 L-Arg binding sites. The view is approximately that of panel B but zoomed in on the trimer interface, showing four subunits in front with the two in back faded for clarity. The viewpoint is fixed in all panels by minimizing RMSD of the dark grey helix shown at lower left. Each panel is one snapshot from the indicated simulation resembling the mean state. A. HoloEcArgRC. The guanidino group of each L-arg (cyan) forms a doubly-H-bonded salt-bridge with Asp128 (dashed lines). Approximate locations of His99 (yellow) and Asp113 (purple) are marked by unlabeled dots for clarity. B. ApoEcArgRC. Each Arg110 forms a doubly-H-bonded salt-bridge with Asp128. Gly103 used for distance measurements to Asp128 is marked by unlabeled blue dot. C. ApoMtArgRC clockwise. Rotation is promoted by Arg132-Asp146 salt bridges, the equivalent of EcArgRC Arg110-Asp128. Asp132 and Asp146 positions marked by purple CPK spheres. D. ApoMtArgRC counterclockwise. Rotation is promoted by Arg118-Asp132 salt bridges; Asp132 is equivalent to EcArgRC Asp113 but Arg118 corresponds to EcArgRC His99. E. EcArgRC, singly-bound L-arg. Some residues and their interactions with L-arg C_α substituents are marked by unlabeled cyan dots and dashes; others are omitted for clarity. (See Color Insert.)

phase analysed. No initial trimer rotation occurred, and only crystal-like, non-rotated states were populated after equilibration; the covariance matrix of Fig. 5.4D indicates that apoArgRCArg110Ala presents a pattern very similar to that of holoArgRC. Thus Arg110 promotes apoArgRC trimer rotation *via* interaction with Asp128. Inspection of the apoArgRC trajectory indicated that repetitive engagement and disruption of one or more of the Arg-Asp salt bridges of the hexamer is responsible for rotational oscillation as well. These findings indicate that global conformational changes can indeed be traced to specific atomic interactions. Therefore, effort should be directed at attempting to do so.

These results suggested a general functional role of Arg residues in ArgR rotational dynamics that might be related to the static rotations observed in crystals of homologous ArgRs. To test this hypothesis, simulations were carried out for apoMtArgRC (PDB ID 2ZFZ (Cherney *et al.*, 2008)) using parameters and preparation as for apoEcArgRC. Two different rotated conformations, rotated in opposite directions (Fig. 5.5C and D), are sampled during the equilibrated phase, promoted by two pairs of Arg-Asp salt bridges originating from opposite sides of the L-arg binding pocket, the second of which has no equivalent in EcArgRC. Extrapolation of this result to apoBstArgRC predicts that rotation in the counterclockwise direction engages its Arg-Asp pair equivalent to the second Arg-Asp pair of MtArgRC, and that apoBstArgRC is incapable of clockwise rotation because no Arg residue is present near the position of *E.coli* Arg110.

10. Structural Correlates of Rotational Oscillation

Hydrogen bonds can be directly counted during each trajectory using automated routines, and even the resulting averaged values can provide structural and thermodynamic insight. The rotated and non-rotated conformers that populate the apoEcArgRC basin differ in the average number of H-bonds between trimers, with ~10-11 in the rotated conformation and ~7-8 in the non-rotated conformation. This difference suggests that the rotated conformer is indeed the more stable of the two and populates the bottom of the basin as earlier assigned. Visual inspection of the apoArgRC trajectory indicates that in the non-rotated conformer on average approximately one hydrogen-bonded Arg110-Asp128 salt bridge is broken with the Arg110 residue flipped outward from the binding pocket, and one other salt bridge lacks one or both of its H-bonds due to slightly increased distance between the functional groups, though electrostatic interaction is preserved. In the rotated conformation all six salt bridges of the hexamer remain intact, and are almost fully H-bonded because rotation permits a slightly closer approach distance of the salt-bridging functional groups. These findings further imply that access of the L-arg ligand to its binding pocket differs in the two conformations of apoArgRC, indicating that thermal flux drives rotational oscillation of apoArgRC to transiently sample a conformation with one open L-arg binding site per hexamer. Although the ligand might occasionally sample such an open site, the rotated conformation does not represent a binding-competent state because it is not a state by the thermodynamic definition, but only one conformer among the ensemble constituting one state; however, another, binding-competent, state must exist.

11. Single-Arginine Simulations

Artificial placement or removal of a ligand in a structure can lead to drifting under the applied force field to accommodate such relatively large changes. In eight of the 27 such simulations that were initiated, L-arg maintained a crystal-like binding geometry after equilibration, and by this criterion only these eight were judged suitable for further analysis: two apoArgRC+1 (a and b), three holoArgRC-5 (a, b, c), and three apoArgRC+2 (a, b, c) simulations. At least 10 ns of each apoArgRC+1 and holoArgRC-5 simulation were equilibrated and at least 5 ns of the apoArgRC+2 simulations (due to longer time to reach equilibration). None experienced rotation, but covariance analysis (referenced therefore only to the hexamer to reveal motions of individual monomers) reveals distinct patterns (Fig. 5.4C). ApoArgRC+1a (upper) displays large regions within most monomers with no correlated motion, interspersed with regions of negatively correlated motions, suggesting that subunit motions become more random when L-arg binds, particularly for subunits BCF that in this simulation do not contact L-arg. The prominent tertiary traces despite very different overall extents of correlation within monomers indicate that internal motions are correlated with monomer motion and that no unfolding occurs. ApoArgRC+1b (4C lower) displays an approximately uniform and equal distribution of correlated and uncorrelated motions over all monomers, with less prominent tertiary traces, indicating intense local motions uncorrelated with monomer motion.

In all five single-ligand simulations the L-arg guanidino group salt-bridges to Asp128, leaving Arg110 to make random motions (Fig. 5.5E). The ligand conformation was also analysed, and was found to be as in holoArgRC cocrystals except in the apoArgRC+1a simulation, where it is more extended and has slightly altered frequency of H-bond contacts. In every singly-liganded simulation all five unliganded subunits retain their Arg110-Asp128 interactions, though with variable extents of H-bonding, and all other pocket residues experience only very minor changes. The ability of the ligand to displace a residue that presumably has far higher effective concentration in the binding site is undoubtedly due to the extensive network of contacts the ligand uniquely can make with its α-amino and α-carboxylate substituents (Fig. 5.5E).

12. Rotational Ensembles

Quantifying the populations of rotated and non-rotated apoArgRC conformers provides information about their relative energies and enables visualization of the ensemble species distribution. Despite its high information content, PCA analysis does not report the frequency with which each conformer is sampled during the trajectory. To quantify rotational populations, snapshots of the apoArgRC structure along the trajectories were inspected visually to identify atom pairs that could serve as accurate metrics. Pairs of atoms lying along a line that is approximately perpendicular to the axis of rotation and as far as possible from the center should report on rotation with maximum sensitivity. To monitor global motions separately from local motions requires atom pairs experiencing relatively small and infrequent intra-monomer and intra-trimer motions. Suitably chosen atom pairs can also report on excursions of monomers and trimers that may shed further light on the nature of global motions even in the absence of rotational oscillation as, e.g., in the holoArgRC simulations. The details of this analysis are presented below to illustrate the nature of the information that can be deduced from it.

The atom pair found to best fulfill these somewhat conflicting requirements comprises C_α of Gly103 in the β2-α1 turn of one monomer and C_α of binding-site residue Asp128 in the β3-β4 turn of the monomer diagonally across the binding pocket (Fig. 5.1B, right panel). Each hexamer presents six such C_α-C_α distances; in most cases these distances report uniquely on the extent of inter-trimer motion, but one apoArgRC subunit presents a minor local conformational change that moves the β3-β4 turn toward the center of the binding pocket, obscuring measurement of rotation and accounting for the one outlier among calculated RMSFs in Fig. 5.3B; this one distance was not used for analysis of species distributions. In both apo- and holoArgRC crystal structures the mean of the six Gly103-Asp128 pair distances is ~ 9.8–9.9 Å, reflecting that the two structures are scarcely different. The six Gly103-Asp128 pair distances of each hexamer were measured during the equilibrated last 10 ns of each trajectory (Fig. 5.6).

In holoArgRC the mean of six Gly103-Asp128 pair distances is ~ 9.7 Å with a narrow range, indicating that holoArgRC samples mainly conformations resembling those in the crystal structure (lower red line) and with relatively little variation among subunits, i.e., a relatively high degree of symmetry of the hexamer. The small distance fluctuations within each monomer reflect motion along rather than perpendicular to the threefold axis, consistent with the covariance matrix. The species distribution is approximately Gaussian, with deviations due to small (~ 0.3 Å) local motions of the two residues. No sampling of rotated conformations occurs, which would move the two atoms apart by maximally ~1.6 Å (upper red line); the change in distance appears small only because the residues used for measurement lie relatively close to the center of the hexamer. In apoArgRC (omitting the one outlier with local conformational change shown in grey) the mean Gly103-Asp128 distance is ~ 11.4 Å, indicating that the weighted average is dominated by the rotated conformer, verifying for the first time during the analysis that it is the more stable of the two. The slightly larger range of distances suggests greater freedom of movement for apo- than for holoArgRC. The range of apoArgRC distances is not so large as to include significant sampling of holo-like rotational conformations, further indicating that the non-rotated state is distinct from holoArgRC. A small part of the increased range (estimated as ~10% from the ratio of smallest to largest C_α fluctuations, ~0.3Å/~3Å) is due to local motions, which fall into two classes and thus account for the slightly bimodal distribution. Thus, most of the increased distance range reflects greater freedom in the rotational

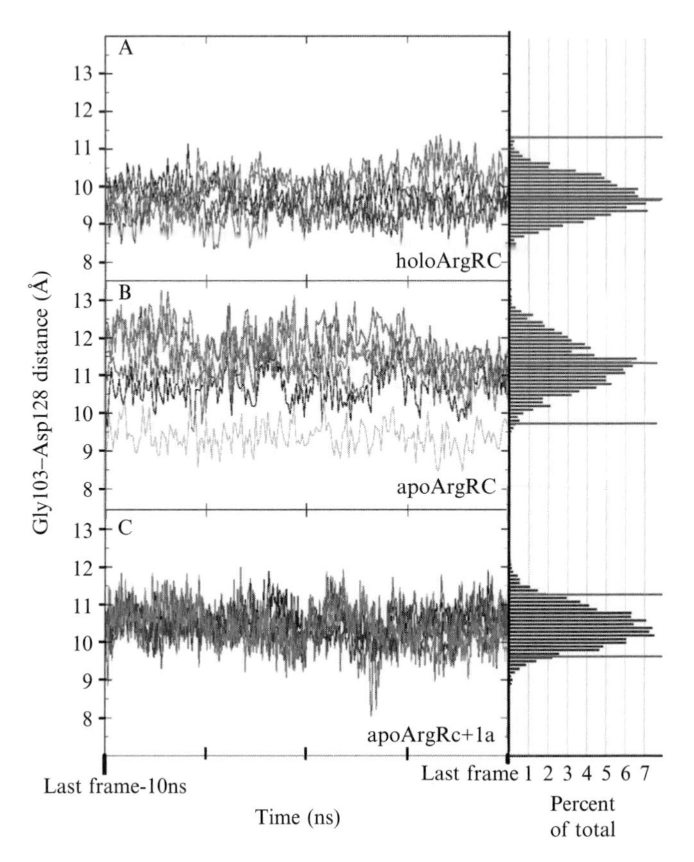

Figure 5.6 Species distribution. Left, distances between Gly103 and Asp128 residues. Distances are measured for each of the six residue pairs in the hexamer every 50 ps during the final 10 ns of each indicated simulation. Monomer colors, defined by the subunit housing Gly103, correspond to those of Fig. 5.3. Right, frequency histograms of Gly–Asp distances shown on left panels. Distances are binned by size in groups of 0.1 Å. On each histogram, lower red line marks the mean crystal distance; upper line marks mean distance in the rotated ensemble of apoArgRC. (See Color Insert.)

motion of trimers with respect to each other and of monomers with respect to each other.

In both apoArgRC+1 simulations, mean Gly103-Asp128 distances are smaller than in apoArgRC, indicating that both fluctuate about a common mean structure approaching that of holoArgRC. The narrow Gaussian distribution of the apoArgRC+1a simulation indicates that the highly random, intense monomer motions detected by covariance analysis do not occur at the trimer interface, and that, as in apoArgRC, no subunit shows unique behavior, indicating a very high degree of symmetry. This finding suggests that all inter-subunit and inter-ligand contacts are optimized simultaneously. The distribution for the apoArgRC+1b simulation is broader but otherwise similar (not shown). The symmetry of the +1 state is corroborated by center-of-mass distances between the hexamer and each monomer, which vary over the simulation by less than the length of one covalent bond (~18 ± 0.5 Å, not shown). Addition of a second L-arg ligand causes little further shift in the mean or range of Gly-Asp distances (Strawn *et al.*, 2010), nor in the symmetry according to measurements of inter-residue distances, hexamer radius of gyration, and center-of-mass distances (not shown); further L-arg additions were not examined.

13. Energetic Contributions

Per-ligand binding enthalpies and free energies, and configurational entropy contributions to the total system free energy (Andricioaei and Karplus, 2001), can be calculated directly from the simulations by standard methods implemented in MD packages [e.g., Lindahl *et al.*, 2001], and H-bond numbers can be counted over the trajectories. These values contain information that can be used to develop an energetic picture of the species distribution and thus map the conformational landscape. We recently introduced (Strawn *et al.*, 2010) a way to use these data that involves combining the entropy and H-bond number to roughly estimate relative free energy levels for each state, and ordering the states to produce a free energy reaction coordinate that accounts for all available information (Fig. 5.7). The reaction coordinate separates the conformational equilibria of apoArgRC from the binding of ligands. Although the free energy estimates are very approximate, and necessarily omit many potentially significant energetic contributions such as differential solvation, the relative magnitudes of such other contributions can to some extent be judged from the simulations. When other contributions are small as for solvation in the present case, it is possible to infer relative trends along the reaction coordinate based on the contributions that present the largest differences between states. Although the absolute values of the calculated energies might be doubted, the relative values can be compared with more confidence.

Binding enthalpies per L-arg are essentially the same in each single-ligand simulation as for each of the six ligands of holoArgRC, suggesting equal enthalpy increments for binding of all six L-arg, consistent with the interpretation of ITC data (Jin *et al.*, 2005); these values (tabulated in (Strawn *et al.*, 2010)) are not shown as they make no differential contribution. Per-ligand binding affinities (tabulated in (Strawn *et al.*, 2010)) range widely among singly-bound simulations, but all are significantly more favorable than the average value for holoArgRC, implying that free energy

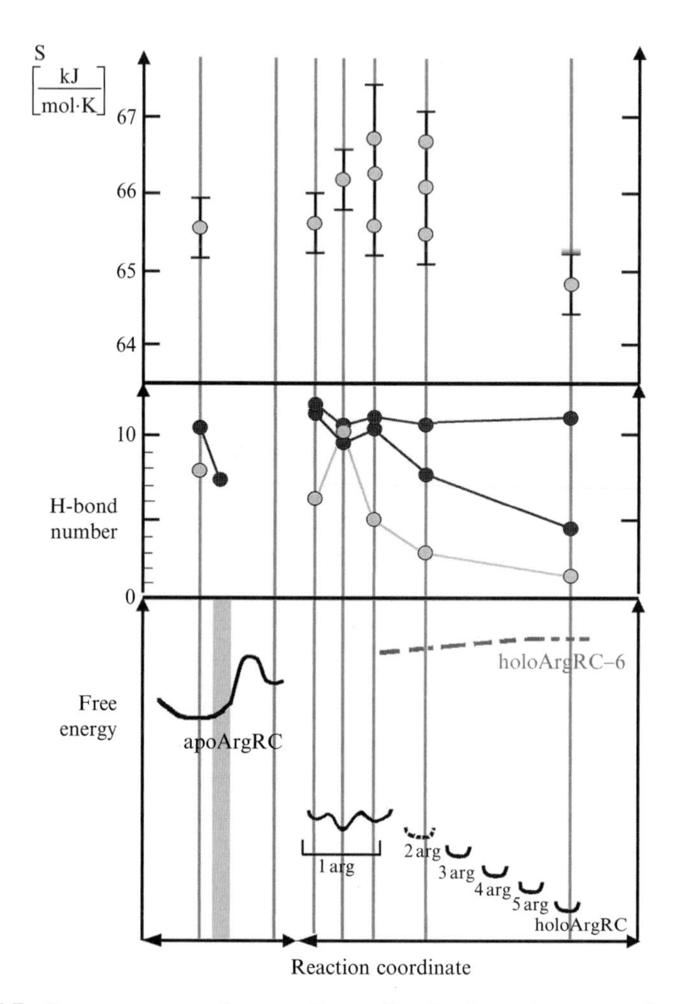

Figure 5.7 Free energy reaction coordinate. *X*-axis, discontinuous conformational and ligand-binding reaction coordinates; vertical lines mark coordinates for which values are plotted based on calculations from a corresponding simulation. *Y*-axis, energy contributions: top, system configurational entropy *S*, kJ/mol K, calculated from each covariance matrix; middle, inter-trimer H-bond number counted from each simulation structure (filled circles; lines are only to guide the eye: light blue, persistent bonds present >50% of time; dark blue, total not including L-arg; green, total including L-arg); bottom, relative free energy levels estimated by combining values for entropy and hydrogen bond number with average enthalpy and per-ligand binding free energy values calculated for each simulation in (Strawn *et al.*, 2010) (or where unavailable, the values calculated for holoArgRC). On the apoArgRC conformational coordinate, the grey zone identifies the higher-energy, non-rotated conformer of apoArgRC with one open L-arg binding site and most other H-bonded salt bridges intact; the second vertical line from the left identifies the high-energy, binding-competent conformation. From left to right on the binding coordinate, the three lines for the 1arg state represent simulations apoArgRC+1b, apoArgRC+1a, and holoArgRC-5,

lowering is substantially greater in the step from zero to one bound L-arg than in subsequent steps, consistent with the greater ligand affinity found by ITC for the first L-arg (Jin *et al.*, 2005). Entropies are similar for apo- and holoArgRC, and values for apoArgRC+1 and holoArgRC-5 are surprisingly undiminished despite the absence of rotation, which is a major contributor to the calculated configurational entropy of apoArgRC. These relatively high entropy values reflect the greatly increased monomer fluctuations in the +1arg state detected by covariance analysis, which is the direct source of data for the configurational entropy calculation.

Both total and persistent (present >50% of the time) H-bonds can be calculated for each state, and H-bonds involving ligand can also be counted separately. Persistent H-bonds cannot be resolved separately for the two apoArgRC conformers; only the total number of persistent H-bonds can be calculated for the single state of apoArgRC, and the mean number over the trajectory is plotted in Fig. 5.7 as if for the more populated rotated conformer, which dominates the calculation. The average number of persistent inter-trimer H-bonds is highly variable in simulations with one L-arg: apoArgRC+1a has ten, 1b six, and three holoArgRC-5 simulations have five each; the three +2 states have three each. The relative contribution of the ligand to total H-bond number also varies over these partially-liganded simulations. The landscape resulting from the combination of these energetic contributions is quite rough on the left side of the coordinate, with double minima for apoArgRC and multiple minima for singly-liganded states; ligation states beyond +1 lie at progressively lower energy levels, reflecting the cumulative free energy lowering of successive ligand additions.

The +1a simulation yields the seemingly self-contradicting result that entropy is high even though all H-bonds are persistent; the latter fact implies the existence of a cooperative H-bond network unique to the +1a conformer. Furthermore, all the H-bonds involve the ligand, implying that increasingly random subunit motions in the +1a conformer enable access to improved H-bond partners within the mesh of interactions surrounding the C_{α} substituents of L-arg. These results can be understood together with the other unexpected result for the +1a state, that its symmetry is as high as or higher than that of apo- or holoArgRC, as judged from

with average values plotted for the holoArgRC-5 states except for entropy. The energy level for the +2 state is dashed to reflect present uncertainty; for ligands three, four, and five estimated energy levels are obtained by interpolation between the values calculated for the flanking coordinates with two and six bound L-arg ligands. The red dashed line represents the non-equilibrated holoArgRC-6 simulation, which is inferred to approach the energy barrier from the right. (For interpretation of the references to color in this figure legend, the reader is referred to the Web version of this chapter.)

the distributions of Gly103-Asp128 distances, hexamer radius of gyration (not shown), and center-of-mass distances (not shown). Such a symmetric state can be visualized as resulting from bonding constraints with the ligand that limit monomer motions close to the binding site while transferring momentum to the peripheral parts of each subunit. This picture is corroborated by atomic B factors that can be calculated from the simulations, showing low values at the center of the hexamer and higher values at the periphery (Strawn *et al.*, 2010); this result does not reflect inconsistency between the RMSFs, which are compared on a monomer-to-monomer basis, and the covariance analysis, which compares hexamers to hexamers, as does the B-factor analysis. Thus the +1 state is conceptually asymmetric, but it is structurally symmetric, and it has higher affinity than subsequent ligation states because interactions among all subunits and between the ligand and subunits can be optimized simultaneously in some conformers within the +1 manifold.

The free energy reaction coordinate of Fig. 5.7 can be interpreted in terms of the structural, kinetic, and energetic events and their manifestation in ITC as L-arg binds to a population of oscillating apoArgRC hexamers. L-arg occasionally encounters a hexamer in the high-energy, non-rotated conformation (grey zone) in which the ligand-binding site of one subunit presents a broken Arg-Asp salt bridge and the remaining five salt bridges remain largely H-bonded. L-arg can enter this site and promote conversion to a holo-like, binding-competent conformation of apoArgRC lying at even higher energy (upper plateau, second vertical line from left), in which nearly all H-bonds of the salt bridges are broken because the functional groups are slightly too distant on average, though electrostatic interactions are maintained. The mechanism by which L-arg promotes conversion to this state is presently under investigation by simulations in which L-arg is allowed to approach the binding site. Breaking of the intertrimer H-bonds constitutes an energy barrier between the apoArgRC basin and the binding-competent conformation, and is assigned to the slow ITC endotherm. Binding of one L-arg to the binding-competent conformation acts as a brake on rotation, transducing momentum from the oscillatory motions of trimers into intense random motions of monomers, enabling optimization of H-bonds with the ligand while maintaining hexamer symmetry. In the language of the MWC model (Monod *et al.*, 1965), the constraints arising from subunit assembly are relaxed via interactions with the ligand, with maintenance of symmetry as the model predicts.

The rate of L-arg dissociation from the singly-bound state is presumably slow relative to the time required for redistribution of the conformational ensemble; estimates from surface plasmon resonance (Jin *et al.*, 2005) suggest an aggregate L-arg off-rate constant of ~0.1 to 1.0 sec^{-1}. Thus, in the presence of less than one full equivalent of L-arg, as during the early stages of the ITC titration, part of the rotating apoArgR population slowly crosses the barrier to the binding-competent state with broken Arg-Asp H-bonds, giving rise to a slow endotherm, and becomes trapped by L-arg binding; additions of further aliquots of L-arg repeat the cycle of barrier-crossing, endotherm evolution, and trapping until binding of the first equivalent of L-arg per hexamer is complete. A single equivalent of L-arg is thus necessary and sufficient to accomplish the shift of the dynamic quaternary ensemble. A second L-arg forces a compromise in the optimized interactions of the +1 state, as inferred from the absence within the +2 manifold of any conformers with a +1a-like cooperative H-bond network. This compromise reduces the free energy of the system by an unknown amount (shown therefore as dashed on the reaction coordinate), but is accompanied by no further endothermic heat flow, and symmetry is maintained according to structural measures. Analysis of additional +2 simulations is required to set the relative energy level of that state more accurately and perhaps guide improved modeling of the ITC data, as discussed below in the section on complementarity of MD and ITC. Binding energies are presumably the same or similar for each of the remaining binding events after the second, shown as equivalent free energy increments.

14. Reconciliation with Crystallographic Data

The holo-like, non-rotated state of apoArgRC observed in crystals with all Arg-Asp H-bonds and salt bridges broken must lie at even higher energy than the binding-competent conformation represented by the upper plateau in Fig. 5.7 (second vertical line from left). In the absence of crystal factors like high salt and/or packing forces that may trap this state, it would lie at even higher energy and thus is not shown. The MD results predict that L-arg binding to this state would display no endothermic heat flow because the H-bonded salt bridges are already broken. ITC experiments are under way to test the possibility that crystal-like salt concentrations eliminate the slow endotherm. Although the effect of crystal packing cannot be mimicked in ITC experiments, its relative contribution would have to be limited if salt alone eliminates the endotherm. Even though the binding sites in apoArgRC crystals are already in a binding-competent conformation, the crystals crack when L-arg is soaked in, due to their slightly smaller distance between C_α atoms of the salt-bridge pair than in holoArgRC (Van Duyne *et al.*, 1996). This observation furthermore implies that it is crystal packing alone that enforces the slightly smaller distances in apoArgRC. Thus an even higher-energy state of apoArgRC must also exist with Arg-Asp distances equal to those of holoArgRC (also not shown). High-salt simulations are under way with holoArgRC-6, which did not equilibrate in simulations at ordinary salt concentrations, to test the hypothesis that holoArgRC-6 may represent or lie near one of the high-energy, binding-competent states that are presumably unstable at low salt. This is an example of the kind of iterative feedback cycle that can be established between MD analysis, experimental data, and further MD analysis.

Significantly, *none* of the critical inter-trimer Arg-Asp salt bridges is detected in any of the relevant ArgR or ArgRC crystal structures. In apoEcArgRC crystals the Arg110 and Asp128 functional groups point away from each other, yet their alpha carbons are close enough so the sidechains could span the inter-trimer distance to form a salt bridge; furthermore, neither makes any other interactions in the crystals but each is instead surrounded by solvent density. These findings suggest that the high ion concentrations used in crystallization (50 m*M* NaHepes, 100 m*M* NaCl, 20 m*M* $CaCl_2$; Van Duyne *et al.*, 1996) interfere with salt-bridging and stabilize a high-energy conformation of apoEcArgRC that is unrotated and lacks salt bridges. In apoMtArgRC crystals as well, the functional groups of the two Arg-Asp pairs are within salt-bridging distance, yet each of the four sidechains instead adopts random orientations as in apoEcArgRC, again suggesting that high salt used in crystallization (0.1 *M* Hepes, 0.1 *M* NaCl (Cherney *et al.*, 2008)) may interfere. To examine this possibility for apoEcArgRC, a simulation was prepared in an effort to mimic the crystallization conditions by adding additional ions beyond those required to achieve electric neutrality. The preliminary results suggest that at high concentrations cations can compete with Arg110 for interaction with Asp128, and they coordinate also with Asp129, altering the orientation of the tandem Asp residues and remaining bound during the simulation, with Arg110 displaced in random orientations and rotation damped. ITC experiments are also under way with crystal-like salt concentrations to test the hypothesis that breaking the Arg-Asp hydrogen bonds is the process associated with the ITC endotherm.

The MD results suggest an interpretation of a very puzzling crystallographic observation: apoArgR hexamers from both *B. stearothermophilus* (Ni *et al.*, 1999) and *B. subtilis* (Dennis *et al.*, 2002) present good electron density for only five of the six equivalent subunits, a seemingly unusual coincidence that might reflect trapping by crystallization of a state that is not quite holo-like but resembles more the binding-competent upper-plateau state in which one salt bridge per hexamer is broken and the other five salt-bridge distances are variable because they lack H-bonds. Such a structure might be unresolvable in fitting the experimental electron density, not only because of the motions of the sidechains and the presence of solvent density, but also due to the rotational degeneracy of the hexamer. Preliminary analysis of ongoing apoArgRC simulations at high salt indicates that one subunit is indeed slightly more mobile than the other five, apparently due to a differential influence of salt on isologous (intra-trimer) and heterologous (inter-trimer) subunit interactions. One of our original motivations for comparative study of ArgR with respect to the dimeric tryptophan repressor (Lavoie & Carey, 1994) was the symmetry principle articulated in the MWC model (Monod *et al.*, 1965) that a hexamer is the minimal multimeric assembly in which both kinds of subunit interactions are used.

15. Complementarity and Synergy of MD and ITC

Thermodynamic results for the ArgR system offered several sources of complexity (Jin *et al.*, 2005) that both motivated and enabled detailed molecular interpretation through the lens of MD simulations to uncover its allosteric mechanism. Such complexities are likely to be more common among other allosteric systems than is presently recognized, thus offering the promise of a more general understanding of allostery by applying the approaches outlined here. An unforeseeably lucky feature of ArgR was the slowly-evolving endotherm that directly implies the existence of a high energy barrier associated with bond-breaking, an unexpected feature for a binding process that is observed to be both fast and overall highly exothermic. Of course one's first instinct is that a small offsetting heat flow is probably an artifact, and on this presumption observations of such a seemingly minor effect may have been dismissed in other cases, or even deleted from the raw data. But exhaustive experimental analyses diminished the possibility of artifact for ArgR, eventually demonstrating quantitative association of the endotherm with the first ligand-binding event only, under a wide range of conditions with both ArgR and ArgRC and with a Cys68Ser ArgR mutant that can be studied in simplified buffer conditions, as well as for binding of not only L-arg, but also L-canavanine with almost 1000-fold weaker affinity (Jin *et al.*, 2005).

Many nominally symmetric macromolecules present multiphasic ITC isotherms for ligand binding even in the absence of detectable endothermic heat flows. Multiphasic isotherms in well-controlled experiments constitute *prima facie* evidence for unequal binding enthalpies, and thus imply cooperative ligand binding. The molecular origins of cooperativity should be identifiable in such systems by analysis of the atomic-to-global response as described here. Any multiphasic ITC isotherm should be carefully evaluated through well-controlled experiments that systematically examine all variables to establish or rule out the presence of a cryptic offsetting heat flow, despite the frustrations inevitably associated with such painstaking work. Furthermore, systematic exploration of initial conditions for ArgR in ITC experiments revealed that the ability to detect an explicit endotherm in multiphasic ITC data depends heavily, but predictably, on experimental variables (Jin *et al.*, 2005). Thus, offsetting heat flows may be a more common underlying contributor to multiphasic isotherms than is presently recognized; we have even suggested they may constitute a thermodynamic signature of global conformational processes (Jin *et al.*, 2005). Far more examples must be analysed to evaluate this suggestion, and MD offers a powerful tool to do so in combination with ITC.

The structures and energies determined in the MD analysis suggested the possibility that the second L-arg binding event is substantially distinct from the remaining four, implying that L-arg binding might follow a more complex model than the original 1+5 model, which yielded qualitatively adequate fit to the ITC data (Jin *et al.*, 2005). The MD results thus prompted further analysis of the published ITC data using a more complex binding model in which the first, second, and remaining four binding events acquire three distinct affinities due to cooperativity, a so-called "1+1+4" model. Available dedicated fitting routines for analysis of calorimetric data cannot flexibly link affinities in this way. Equations based on an Adair scheme (Adair, 1925) relating the three binding constants to the population distribution of protein-ligand occupancy states as a function of ligand concentration for the 1+1+4 model were solved numerically using Matlab, with the affinity for the second binding event fixed at one of three values relative to the first: ten-fold stronger, equal, or ten-fold weaker. The 1+1+4 model was then fit globally to two titration datasets (Jin *et al.*, 2005) for 22 μ*M* ArgR hexamer with 2 m*M* and with 15 m*M* L-Arg (Fig. 5.8 and Table 5.1). All three fits improve upon the fit of the 1+5 model, indicating that a more complex model like the one suggested by the MD results is also consistent with the ITC data. However, despite the additional constraints arising from simultaneous fitting of titrations at low and high ligand concentrations, all three fits are very similar (thus only one of the three is shown on the figure) due to covariation between affinity and enthalpy, which is an indication that the information content of the data is insufficient to provide accurate parameter values. Nevertheless, the statistical robustness of the two models determined using the so-called Akaike information criterion (Hurvich & Tsai, 1989), which assesses the goodness of fit of a model with respect to its complexity (number of independent adjustable parameters), indicates that the improved fit of the 1+1+4 binding model to the ITC data is not due simply to the increased number of parameters (five with stoichiometry and second binding constant fixed vs. four) but is a statistically significant improvement relative to the 1+5 model.

The improved fit of the 1+1+4 binding model indicates that analysis of additional +2arg MD simulations is required to set the relative energy level of the +2 state more accurately, and to guide improved modeling of the ITC data. It is important to note that interaction energies calculated from MD data represent average values on a per-site basis, which is not always true for the binding affinities derived from fitting to ITC binding data. Depending on the binding model used, binding constants derived from analysis of ITC data can be represented on a per-site or a per-molecule basis.

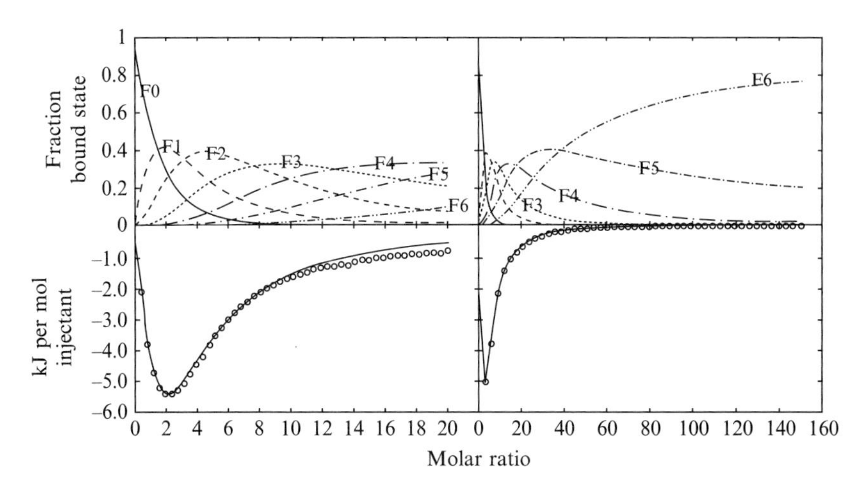

Figure 5.8 Global analysis of ITC binding data. Top, population distribution of ligand-bound protein species during the course of titration. *F0* to *F6* indicate the fractions of species with 0 to 6 L-arg equivalents bound. Bottom, integrated heat flow from raw ITC data (circles; taken from (Jin *et al.*, 2005)) and global fit (lines) to the 1+1+4 binding model with the affinity of the second binding event fixed at the same value as the first event; fits with the affinity of second event ten times higher or lower than the first event are essentially coincident with the solid lines shown. Left, ArgR at 22 μ*M* hexamer in the reaction cell titrated with L-Arg at 2 m*M* in the syringe at 25 °C. Right, 22 μ*M* ArgR hexamer titrated with 15m*M* L-Arg. The binding constants and reaction enthalpies obtained from the global analysis are listed in Table I.

The distinction between these two cases is that per-molecule binding constants take into account the multivalency of the target molecule and thus the reduced availability of empty sites on each hexamer as titration progresses. Thus for a multivalent target like the ArgR hexamer, the filling of empty sites is governed by the availability of unfilled sites, i.e., by occupancy levels in the hexamer population, and not only by site affinities *per se*. Both cases are considered in Table 5.1, and the species distribution is plotted in Fig. 5.8.

The improved fit deriving from global analysis of only two ITC datasets suggests that inclusion of additional datasets may further refine the binding model and shed light on the second event. Furthermore, one of these two titrations (bottom left in Fig. 5.8) did not reach completion as indicated by the substantial residual heat at the end of the injection series, which is much greater than the heat of L-arg dilution (Jin *et al.*, 2005). Standard recommendations for ITC protocols emphasise optimization of experimental conditions aimed at acquiring all thermodynamic parameters (affinity, enthalpy change, and molar ratio) from a single complete dataset (Wiseman *et al.*, 1989). The ArgR case and other evidence addressed in (Freyer & Lewis, 2007) suggest, however, that even incomplete or

Table 5.1 Parameter values from global fitting

	K_d, M (binding site)			K_d, M (hexameric protein)			$\Delta H°$, kJ/mol		
Binding event	*A*	*B*	*C*	*A*	*B*	*C*	*A*	*B*	*C*
1	3.7×10^{-4}	8.8×10^{-5}	1.5×10^{-5}	6.1×10^{-5}	1.5×10^{-5}	2.5×10^{-6}	−0.9	0.9	−1.9
2	3.7×10^{-5}	8.8×10^{-5}	1.5×10^{-4}	1.5×10^{-5}	3.5×10^{-5}	6.0×10^{-5}	−24.2	−22.4	−27.6
3	1.2×10^{-4}	1.4×10^{-4}	1.6×10^{-4}	8.7×10^{-5}	1.0×10^{-4}	1.2×10^{-4}	−8.5	−9.4	−7.4
4	1.2×10^{-4}	1.4×10^{-4}	1.6×10^{-4}	1.5×10^{-4}	1.8×10^{-4}	2.2×10^{-4}	−8.5	−9.4	−7.4
5	1.2×10^{-4}	1.4×10^{-4}	1.6×10^{-4}	2.9×10^{-4}	3.4×10^{-4}	4.0×10^{-4}	−8.5	−9.4	−7.4
6	1.2×10^{-4}	1.4×10^{-4}	1.6×10^{-4}	7.0×10^{-4}	8.1×10^{-4}	9.7×10^{-4}	−8.5	−9.4	−7.4

The 1+1+4 binding model was fitted globally to the two ITC datasets shown in Fig. 5.8. Column one lists the six binding events required to fill the hexamer; columns 2 to 5 list the per-site affinities, K_d, in units of monomer M^{-1}; columns 6 to 8 list the per-hexamer affinities, K_d, in units of hexamer M^{-1}; columns 9 to 11 list the enthalpy change, ΔH, per binding event in units of kJ/mol. The letters *A*, *B*, and *C* in row two indicate parameter values recovered from fits in which the affinity of the second binding event was fixed to a value ten times stronger than the first event (A), equal to the first event (B), or ten times weaker (C).

otherwise suboptimal ITC datasets can provide useful constraints for model-fitting by global analysis, and for data simulation to guide experimental design. For example, extensive simulation of ITC data (Turnbull & Daranas, 2003) has very nicely demonstrated that in many cases concentrations can be much lower than standard recommendations. As well, very high or very low concentration regimes can provide limiting values of the thermodynamic parameters, and in complex cases the absence of heat flow can also be meaningful. Simulation of binding data can and should be used to design each experiment so as to recover maximally useful datasets, even partial datasets, for global analysis. It should be noted that fitting and simulation of binding data use a shared set of mathematical and computational tools, and simulation of experimental results using alternative binding models is a very powerful approach that should be incorporated into ITC experimental design. In the present case, simulation of ITC results using alternative binding models may aid the design of ITC experiments aimed at defining the thermodynamic parameters of the second event, although the values in Table 5.1 suggest that appropriate conditions may be difficult to achieve.

The working model derived from the MD results has also opened an entirely new avenue of experimental approach for EcArgR, and guides its design. Despite the widespread popularity of mutational analysis in contemporary studies of protein properties including allostery, with ArgR there had been until now no clear rationale for making mutants to gain insight into its allosteric mechanism because no experiments were found able to report uniquely on the unknown allosteric transition. The results of the MD analysis suggest that the ITC endotherm corresponds structurally to disruption of H-bonded Arg-Asp salt bridges that are not detected in crystal structures (and thus likely would not have been thought to influence allostery); elimination of these interactions is now predicted to eliminate the endotherm. As seen in MD, an Arg110Ala mutant lacks Arg-Asp hydrogen bonds, is incapable of rotation, and is apparently trapped in a holo-like and/or binding-competent state. Its L-arg binding isotherm detected by ITC is thus predicted to display no endothermic heat flow; it is unclear whether the mutant would still display L-arg binding cooperativity of any kind, and ITC analysis is expected to offer further insight. As well, the His99Arg substitution in MD enables rotation in both directions (not shown), although the prediction for its binding isotherm is unclear. However, an Arg110Ala, His99Arg double mutant rotates in MD, and is predicted to regain the endothermic heat flow lost in the Arg100Ala mutant and conform to the wildtype binding model. Investigation of all three mutants is now in progress. Importantly, study of these mutations by ITC has the power to falsify the motivating hypothesis. Falsification is an essential criterion in the design of any study according to the scientific method, but mutational studies often fail to conform to this requirement. Although several of the experimental tests of the allosteric model are predicted to have negative outcomes in the sense that the endotherm is expected to disappear, other tests have stronger predictions because observation of the endothermic heat flow characteristic of the first L-arg equivalent negates the hypothesis. Nevertheless, well-controlled ITC experiments should enable adequate testing of even negative predictions.

The consistency between the ArgR.C ITC data and the more complex binding model suggested by the MD results implies that the structural picture available from MD, which assigns H-bond breakage to the slow endotherm, indeed reflects this distinctive thermodynamic signature. Future work will be needed, however, to determine if the 1+1+4 binding model is an accurate physical model and not simply a model consistent with the data. It should be borne in mind that goodness of fit is neither a necessary nor a sufficient condition to judge a model as the best physical model; independent physical information is always required to distinguish accurate models from incorrect ones with equivalent fits (Johnson, 1992). MD simulation can be a useful tool in this regard because it can predict structural features that can be subjected to independent experimental tests to establish their relationship to thermodynamic observables.

16. Prospects

Rapid developments in MD will continue to extend the useful time scales of simulations for systems of increasing sizes. A major challenge remains the calculation of absolute binding constants, although as the ArgR example shows, even relative affinities may provide useful insight. The ArgR example reveals no residue-to-residue network or pathway for transmission of the allosteric signal, although analysis of additional allosteric systems is needed to establish how general is this aspect of the ArgR case. But if pathways prove to be common, as is widely expected (and deliberately sought at present by numerous experimental and computational methods (Rousseau & Schymkowitz, 2005; Shi *et al.*, 2006), detailed analysis of MD data as outlined here may also play an important role in elucidating the mechanism of information transfer along those pathways. In that case another experimental avenue, vibrational spectroscopy, which is synergistic with both MD and ITC (Kopecký *et al.*, 2004), could also be important. Information transfer along intramolecular pathways is presumably mediated by propagated changes in vibrational modes, which in addition have a thermodynamic manifestation in the heat capacity change of the ligand-binding process (Sturtevant, 1977). Global changes in vibrational modes accompany many ligand-binding processes, and study of the pathways of energy distribution by MD should further illuminate allosteric mechanisms. Another complementary application of vibrational spectroscopy is for determining the minimum composition of buffers, salts, etc. compatible with native protein structure, to optimize, simplify, and match the solution conditions to those used in MD. Recent advances in sample preparation, e.g., durable glassy-state native samples requiring very small protein quantities (Kopecký & Baumruk, 2006), should particularly facilitate such applications.

Advances in ITC technology have been aimed largely at reducing sample requirements without compromising sensitivity; current instruments can operate with only ~0.2 mL of protein. Sample concentration requirements depend ultimately on the affinity and enthalpy change of ligand binding. Many biomolecular interactions are accompanied by substantial change in heat capacity (Sturtevant, 1977), which can be exploited to alter the accessible sample concentration regime by shifting measurements to a temperature range with larger or smaller enthalpy change. However, the greatest advantage of ITC - its generality based on the ability to detect heat flow, which accompanies virtually every reaction as well as many intramolecular processes that accompany binding - is also its greatest disadvantage for the unwary, because all heat flows are detected in the experiment, regardless of their source or relevance to the process under study. Thus there is no substitute for incisive design of experiments and controls, using simulation as an additional tool, and for extensive manual intervention in data analysis as illustrated by the ArgR case, including very careful inspection of the raw heat spikes before integration, which may bear the traces of offsetting or slow heat flows that can report on unusual processes. Systematic variation of experimental conditions including titrant and titrate concentrations, salt, pH, and buffer identity can help to distinguish artifacts from *bona fide* reaction heats, and MD can be used to derive clues about how solution conditions may affect the system.

An iterative approach to the design and analysis of ITC experiments and MD simulations, in combination with parallel advances in computerized global analysis and modeling tools capable of rigorous evaluation of complex binding scenarios, may eventually enable the full deconvolution of multiphasic ITC data and shed light on complex allosteric processes like that of ArgR. The findings with ArgR illustrate the power of this approach that exploits the synergy between ITC experiments and MD simulations. We have suggested (Jin *et al.*, 2005) that the binding of any species, even a simple ion that might not otherwise be regarded as a true ligand, has the potential to shift the conformational ensemble that is a characteristic feature of all dynamic proteins (Gunasekaran *et al.*, 2004), producing enthalpic and/or entropic consequences that can be manifested without structural identification in measurements of ligand-binding thermodynamics. It is thus likely that careful analysis of virtually any protein-ligand binding process by the combination of MD and ITC can broaden and deepen our understanding of Monod's second secret.

ACKNOWLEDGEMENT

Access to METACentrum supercomputing facilities was provided under the research intent MSM6383917201. We thank Curt Hillegas, Director of Research Computing at the Princeton University Office of Information Technology, for access to the high-performance computing center. We gratefully acknowledge support from the Ministry of Education, Youth and Sports of the Czech Republic (MSM6007665808, LC06010), Academy of Sciences of the Czech Republic (AVOZ60870520), Grant Agency of the Czech Republic (P207/10/1934 and 203/08/0114 to R.E.), and joint Czech - US National Science Foundation International Research Cooperation (INT03-09049 and OISE08-53423 to J.C., and ME09016 to R.E.).

REFERENCES

Adair, G. S. (1925). The hemoglobin system. VI. The oxygen dissociation curve of hemoglobin. *J. Biol. Chem.* **63,** 529–545.

Andricioaei, I., and Karplus, M. (2001). On the calculation of entropy from covariance matrices of the atomic fluctuations. *J. Chem. Phys.* **115,** 6289–6292.

Cherney, L. T., Cherney, M. M., Garen, C. R., Lu, G. J., and James, M. N. (2008). Structure of the C-domain of the arginine repressor protein from *Mycobacterium tuberculosis. Acta Crystallogr.* **D64,** 950–956.

Czaplewski, L. G., North, A. K., Smith, M. C., Baumberg, S., and Stockley, P. G. (1992). Purification and initial characterization of AhrC: The regulator of arginine metabolism genes in *Bacillus subtilis. Mol. Microbiol.* **6,** 267–275.

Dennis, C. A., Glykos, N. M., Parsons, M. R., and Phillips, S. E. V. (2002). The structure of AhrC, the arginine repressor/activator protein from *Bacillus subtilis. Acta Crystallogr.* **D58,** 421–430.

Dion, M., Charlier, D., Wang, H., Gigot, D., Savchenko, A., Hallet, J.-N., Glansdorff, N., and Sakanyan, V. (1997). The highly thermostable arginine repressor of *Bacillus stearothermophilus*: Gene cloning and repressor–operator interactions. *Mol. Microbiol.* **25,** 385–398.

Feenstra, K. A., Hess, B., and Berendsen, H. J. C. (1999). Improving efficiency of large time-scale molecular dynamics of hydrogen-rich systems. *J. Comput. Chem.* **20,** 786–798.

Freyer, M. W., and Lewis, E. A. (2007). Isothermal titration calorimetry: Experimental design, data analysis and probing macromolecule ligand binding interactions. *In* "Biophysical Tools for Biologists," (J. J. Correia and H. W. Detrich, eds.), pp. 79–113. Academic Press, San Diego, CA.

Goodey, N. M., and Benkovic, S. J. (2008). Allosteric regulation and catalysis emerge via a common route. *Nat. Chem. Biol.* **4,** 474–482.

Grandori, R., Lavoie, T. A., Pflumm, M., Tian, G., Niersbach, H., Maas, W. K., Fairman, R., and Carey, J. (1995). The DNA-binding domain of the hexameric arginine repressor. *J. Mol. Biol.* **254,** 150–162.

Gunasekaran, K., Ma, B., and Nussinov, R. (2004). Is allostery an intrinsic property of all dynamic proteins? *Proteins* **57,** 433–443.

Hayward, S., and Berendsen, H. J. C. (1998). Systematic analysis of domain motions in proteins from conformational change; new results on citrate synthase and T4 lysozyme. *Protein* **30,** 144–154.

Hayward, S., Kitao, A., and Berendsen, H. J. C. (1997). Model-free methods of analyzing domain motions in proteins from simulation: A comparison of normal mode analysis and molecular dynamics simulation of lysozyme. *Protein* **27,** 425–437.

Hess, B., Kutzner, C., van der Spoel, D., and Lindahl, E. (2008). GROMACS 4: Algorithms for highly efficient, load-balanced, and scalable molecular simulation. *J. Chem. Theory Comput.* **4,** 435–447.

Hurvich, C. M., and Tsai, C. L. (1989). Regression and time-series model selection in small samples. *Biometrika* **76,** 297–307.

Jacob, F., and Monod, J. (1961). On the regulation of gene activity. *Cold Spring Harb. Symp. Quant. Biol.* **26,** 193–211.

Jin, L., Xue, W. F., Fukayama, J. W., Yetter, J., Pickering, M., and Carey, J. (2005). Asymmetric allosteric activation of the symmetric ArgR hexamer. *J. Mol. Biol.* **346,** 43–56.

Johnson, M. L. (1992). Why, when, and how biochemists should use least squares. *Anal. Biochem.* **206,** 215–225.

Kopecký, V., Jr., and Baumruk, V. (2006). Structure of the ring in drop coating deposited proteins and its implication for Raman spectroscopy of biomolecules. *Vib. Spectrosc.* **42,** 184–187.

Kopecký, V., Ettrich, R., Hofbauerova, K., and Baumruk, V. (2004). Vibrational spectroscopy and computer modeling of proteins: Solving structure of alpha(1)-acid glycoprotein. *Spectrosc. Int. J.* **18,** 323–330.

Krieger, E., Koraimann, G., and Vriend, G. (2002). Increasing the precision of comparative models with YASARA NOVA: A self-parameterizing force field. *Proteins* **47,** 393–402.

Lavoie, T. A., and Carey, J. (1994). Adaptability and specificity in DNA binding by *trp* repressor. *Nucleic Acids Mol. Biol.* **8,** 184–196.

Lim, D., Oppenheim, J. D., Eckhardt, T., and Maas, W. K. (1987). Nucleotide sequence of the argR gene of *Escherichia coli* K-12 and isolation of its product, the arginine repressor. *Proc. Natl. Acad. Sci. USA* **84,** 6697–6701.

Lindahl, E., Hess, B., and van der Spoel, D. (2001). GROMACS 3.0: A package for molecular simulation and trajectory analysis. *J. Mol. Model.* **7,** 306–317.

Lu, C.-D., Houghton, J. E., and Abdelal, A. T. (1992). Characterization of the arginine repressor from *Salmonella typhimurium* and its interactions with the carAB operator. *J. Mol. Biol.* **225,** 11–24.

Maas, W. K. (1994). The arginine repressor of *Escherichia coli. Microbiol. Mol. Biol. Rev.* **58,** 631–640.

Monod, J., Changeux, J.-P., and Jacob, F. (1963). Allosteric proteins and cellular control systems. *J. Mol. Biol.* **6,** 306–329.

Monod, J., Wyman, J., and Changeux, J.-P. (1965). On the nature of allosteric transitions: A plausible model. *J. Mol. Biol.* **12,** 88–118.

Morin, A., Huysveld, N., Braun, F., Dimova, D., Sakanyan, V., and Charlier, D. (2003). Hyperthermophilic *Thermotoga* arginine repressor binding to full-length cognate and heterologous arginine operators and to half-site targets. *J. Mol. Biol.* **332,** 537–553.

Ni, J., Sakanyan, V., Charlier, D., Glansdorff, N., and Van Duyne, G. D. (1999). Structure of the arginine repressor from *Bacillus stearothermophilus. Nat. Struct. Biol.* **6,** 427–432.

Niersbach, H., Lin, R., Van Duyne, G. D., and Maas, W. K. (1998). A superrepressor mutant of the arginine repressor with a correctly predicted alteration of ligand binding specificity. *J. Mol. Biol.* **279**(4), 753–760.

Perutz, M. F. (1989). Mechanisms of cooperativity and allosteric regulation in proteins. *Q. Rev. Biophys.* **22,** 139–236.

Ponder, J. W., and Case, D. A. (2003). Force fields for protein simulations. *Adv. Protein Chem.* **66,** 27–85.

Rousseau, F., and Schymkowitz, J. (2005). A systems biology perspective on protein structural dynamics and signal transduction. *Curr. Opin. Struct. Biol.* **15,** 23–30.

Schleif, R. (2004). Modeling and studying proteins with molecular dynamics. *Methods Enzymol.* **383,** 28–47.

Shi, Z., Resing, K. A., and Ahn, N. G. (2006). Networks for the allosteric control of protein kinases. *Curr. Opin. Struct. Biol.* **16,** 686–692.

Speranskiy, K., Cascio, M., and Kurnikova, M. (2007). Homology modeling and molecular dynamics simulations of the glycine receptor ligand binding domain. *Proteins* **67,** 950–960.

Strawn, R., Melichercik, M., Green, M., Stockner, T., Carey, J., and Ettrich, R. (2010). Symmetric allosteric mechanism for hexameric *E. coli* arginine repressor exploits competition between L-arginine ligands and resident arginine residues. *PLoS Comput. Biol.* **6,** e1000801.

Sturtevant, J. (1977). Heat capacity and entropy changes in processes involving proteins. *Proc. Natl. Acad. Sci. USA* **74,** 2236–2240.

Sunnerhagen, M., Nilges, M., Otting, G., and Carey, J. (1997). Solution structure of the DNA-binding domain and model for the complex of multifunctional hexameric arginine repressor with DNA. *Nat. Struct. Biol.* **4,** 819–826.

Szwajkajzer, D., Dai, L., Fukayama, J. W., Abramczyk, B., Fairman, R., and Carey, J. (2001). Quantitative analysis of DNA binding by the *Escherichia coli* arginine repressor. *J. Mol. Biol.* **312,** 949–962.

Tian, G., and Maas, W. K. (1994). Mutational analysis of the arginine repressor of *Escherichia coli. Mol. Microbiol.* **13,** 599–608.

Tian, G., Lim, D., Carey, J., and Maas, W. K. (1992). Binding of the arginine repressor of *Escherichia coli* K12 to its operator sites. *J. Mol. Biol.* **226**(2), 387–397.

Tian, G., Lim, D., Oppenheim, J. D., and Maas, W. K. (1994). Explanation for different types of regulation of arginine biosynthesis in *Escherichia coli* B and *Escherichia coli* K12 caused by a difference between their arginine repressors. *J. Mol. Biol.* **235**(1), 221–230.

Turnbull, W. B., and Daranas, A. H. (2003). On the value of *c*: Can low affinity systems be studied by isothermal titration calorimetry? *J. Am. Chem. Soc.* **125,** 14859–14866.

Van Duyne, G. D., Ghosh, G., Maas, W. K., and Sigler, P. B. (1996). Structure of the oligomerization and L-arginine binding domain of the arginine repressor of *Escherichia coli. J. Mol. Biol.* **256,** 377–391.

van Gunsteren, W. F., and Berendsen, H. J. C. (1987). Gromos-87 Manual. Biomos BV, Nijenborgh 4, 9747 AG Groningen, The Netherlands, pp. 331–342.

Weber, G. (1972). Ligand binding and internal equilibriums in proteins. *Biochemistry* **11,** 864–878.

Wiseman, T., Williston, S., Brandts, J. F., and Lin, L. N. (1989). Rapid measurement of binding constants and heats of binding using a new titration calorimeter. *Anal. Biochem.* **179,** 131–137.

Zhang, R. G., Joachimiak, A., Lawson, C. L., Schevitz, R. W., Otwinowski, Z., and Sigler, P. B. (1987). The crystal structure of trp aporepressor at 1.8 Å shows how binding tryptophan enhances DNA affinity. *Nature* **327,** 591–597.

Zhao, D., Arrowsmith, C. H., Jia, X., and Jardetzky, O. (1993). Refined solution structures of the *Escherichia coli* trp holo- and aporepressor. *J. Mol. Biol.* **229,** 735–746.

CHAPTER SIX

Using Tryptophan Fluorescence to Measure the Stability of Membrane Proteins Folded in Liposomes

C. Preston Moon *and* Karen G. Fleming

Contents

T.C. Jenkins Department of Biophysics, Johns Hopkins University, Baltimore, Maryland, USA

Methods in Enzymology, Volume 492
ISSN 0076-6879, DOI: 10.1016/B978-0-12-381268-1.00018-5

© 2011 Elsevier Inc.
All rights reserved.

Abstract

Accurate measurements of the thermodynamic stability of folded membrane proteins require methods for monitoring their conformation that are free of experimental artifacts. For tryptophan fluorescence emission experiments with membrane proteins folded into liposomes, there are two significant sources of artifacts: the first is light scattering by the liposomes; the second is the nonlinear relationship of some tryptophan spectral parameters with changes in protein conformation. Both of these sources of error can interfere with the method of determining the reversible equilibrium thermodynamic stability of proteins using titrations of chemical denaturants. Here, we present methods to manage light scattering by liposomes for tryptophan emission experiments and to properly monitor tryptophan spectra as a function of protein conformation. Our methods are tailored to the titrations of membrane proteins using common chemical denaturants. One of our recommendations is to collect and analyze the right-angle light scattering peak that occurs around the excitation wavelength in a fluorescence experiment. Another recommendation is to use only one tryptophan spectral parameter, the emission intensity at a specific wavelength, for the determination of membrane protein stability. Emission intensity is the only parameter we find to be linearly proportional to the protein conformational population, and we show that other spectral parameters lead to errors in protein stability measurements.

1. Introduction

The physical forces governing the thermodynamic stability of a membrane protein are important cues to the relationship of the protein's structure and its function. The most relevant measurements of a membrane protein's stability will come from studies that use thermodynamically stable mimetics of biological lipid bilayers. Stable mimetics include large unilamellar vesicles (LUVs) composed of glycerophospholipids. However, LUVs can pose challenges to membrane protein experiments with luminescent spectroscopy because of their tendency to scatter light.

One example of luminescent spectroscopy that is popular for membrane protein studies is tryptophan fluorescence emission spectroscopy. All three research groups that have so far reported reversible equilibrium thermodynamic information for the stability of membrane proteins in liposomes chose tryptophan fluorescence emission as their primary source of data on protein conformational states (Hong and Tamm, 2004; Huysmans *et al.*, 2010; Sanchez *et al.*, 2008). These three groups also employed titrations of chemical denaturants as a technique to perturb equilibrium between the folded and unfolded conformations of their respective membrane proteins. Denaturant titrations pose two problems that need to be solved for thermodynamic studies of membrane proteins: (1) the light scattering by LUVs depends on the denaturant concentration and (2) a mixture of protein conformations in a sample has a nonlinear relationship with most tryptophan spectral parameters. In this chapter, we propose solutions to overcome both of these problems that enable accurate measurements of tryptophan emission spectra and membrane protein stabilities.

2. Issues with Managing Light Scattering from Liposomes

Light scattering by liposomes can be ruinous to spectroscopic experiments if it is not appropriately managed. Ladokhin and coworkers have provided several extremely useful procedures to minimize or correct for the effects of light scattering on tryptophan spectrofluorometry (Ladokhin *et al.*, 2000). Here, we expand upon some of those techniques to tailor them for the study of membrane protein folding, especially folding perturbed by the titration of chemical denaturants, such as urea or guanidine HCl.

2.1. The contribution of light scattering to a tryptophan fluorescence emission scan can be divested from true tryptophan emission

We show in Fig. 6.1 how light scattering usually manifests in a typical fluorescence experiment. In this experiment, we folded the fatty acid transporter FadL from the outer membrane of *Escherichia coli* into LUVs of 1,2-dilauroyl-*sn*-glycero-3-phosphocholine (DLPC). We excited the folded protein sample with light at 295 nm and then collected its fluorescence emission data from 280–400 nm. By collecting emission data all the way down to 280 nm, we could observe the full peak caused by light scattering that was distributed around the excitation wavelength. The light scattering peak is easily discernable (Fig. 6.1) from the emission peak that came from the tryptophans in FadL.

The most common procedure to account for light scattering in an emission scan is to eliminate it by subtracting a scan of a blank sample of liposomes in buffer from the protein scan. However, we have chosen to keep the light scattering peaks in our emission data to understand how they change from one experiment to another. We can fully divest the light scattering and protein emission peaks and consider them separately if we fit both peaks with an expression expanded from the single log-normal distribution used by Ladokhin *et al.* (2000) and Permyakov (1993).

The single log-normal distribution describes the asymmetric shape of a tryptophan emission peak (Fig. 6.1). It can give the tryptophan emission intensity at a wavelength λ by

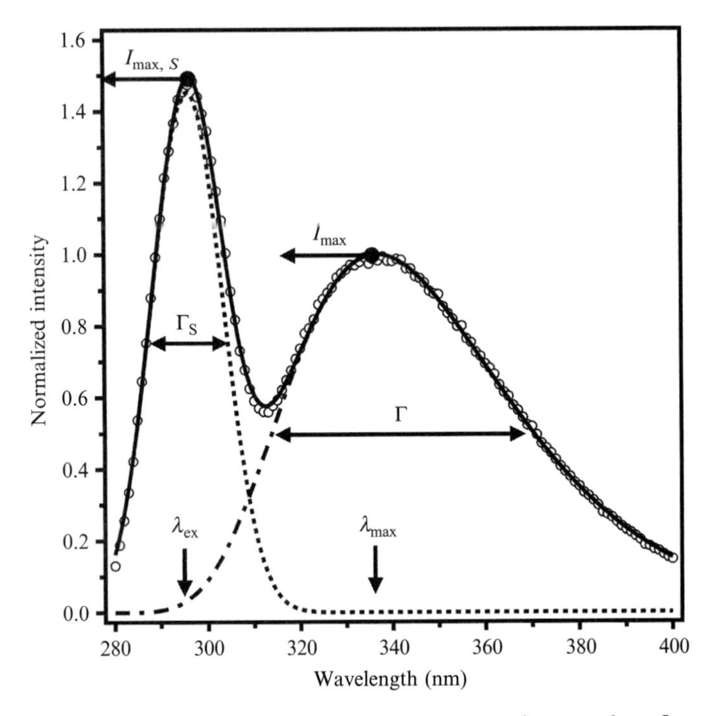

Figure 6.1 Relative contributions of light scattering and tryptophan fluorescence emission to the observed data from a typical fluorescence experiment. The observed data (open circles) came from a sample of 400 n*M* FadL folded into LUVs of DLPC at a lipid-to-protein ratio of 2000:1 and in 1.5 *M* guanidine HCl, 2 m*M* EDTA, and 100 m*M* citrate, pH 3.8. The sample was excited at 295 nm, the excitation polarizer was set at 90°, and the emission polarizer was set at 0°. The sample pathlength was 10 mm, the excitation slits were 2.4 mm wide, and the emission slits were 2.0 mm wide. The solid line represents a fit to the observed data by a sum of a normal distribution and a log-normal distribution as described by Eq. (6.3). Fit parameters from Eq. (6.3) are labeled next to the parts of the data that they describe. The dotted line is the normal distribution component of Eq. (6.2), which represents light scattering at the excitation wavelength. The dot/dashed line represents the log-normal distribution component of Eq. (6.1), which describes the fluorescence emission of the tryptophans in FadL.

$$I_{Trp}(\lambda) = I_{\max} \times \exp\left\{ -\frac{\ln(2)}{\ln(\rho^2)} \times \left[\ln\left(\frac{\left(\lambda_{\max} + \frac{\Gamma \times \rho}{\rho^2 - 1}\right) - \lambda}{\left(\lambda_{\max} + \frac{\Gamma \times \rho}{\rho^2 - 1}\right) - \lambda_{\max}} \right) \right]^2 \right\} \tag{6.1}$$

as long as

$$\lambda > \lambda_{\max} - \frac{\rho \times \Gamma}{\rho^2 - 1}$$

while for smaller λ,

$$I_{Trp}(\lambda) = 0$$

where I_{max} is the maximum emission intensity of the tryptophan peak, λ_{max} is the emission wavelength at I_{max}, Γ is the width of the peak at half of I_{max}, and ρ is a factor that describes the degree of asymmetry in the peak.

The shape of the light scattering peak can be adequately described by a normal distribution. The right-angle (90°) light scattering intensity at wavelength λ is given by

$$I_S(\lambda) = I_{\max,S} \times \exp\left\{-\left[\frac{\lambda - \lambda_{ex}}{\Gamma_S}\right]^2\right\} \tag{6.2}$$

where $I_{max,S}$ is the maximum emission intensity of the light scattering peak, λ_{ex} is the excitation wavelength, and Γ_S is the width of the peak at half of $I_{max,S}$.

We use the sum of Eqs. (6.1) and (6.2) to describe our observed data from emission scans of tryptophans and liposomes:

$$I(\lambda) = I_{\mathrm{Trp}}(\lambda) + I_S(\lambda). \tag{6.3}$$

It does not take much extra time to include the light scattering peak in a tryptophan fluorescence scan. Once collected, it can be removed directly from an existing set of data by using Eq. (6.3) to determine the log-normal parameters coming only from tryptophan emission.

2.2. Effects of light scattering on the tryptophan fluorescence from membrane proteins

Ladokhin and coworkers noted four problems that light scattering has for tryptophan fluorescence experiments: (1) it directly contributes to the observed emission signal, (2) it causes less of the excitation light to reach the tryptophans, (3) it causes less of the light emitted by the tryptophans to reach the detector, (4) it causes the observed emission spectrum to be red-shifted because it asymmetrically affects tryptophan's excitation band (Ladokhin *et al.*, 2000).

These problems are demonstrated in Fig. 6.2. We dissolved one sample of L-tryptophan zwitterion into a blank buffer solution and a second sample into the same buffer solution that also contained LUVs of DLPC. The tryptophan zwitterion should not have partitioned onto the LUVs (Ladokhin *et al.*, 2000). In principle then, the fluorescence emission should

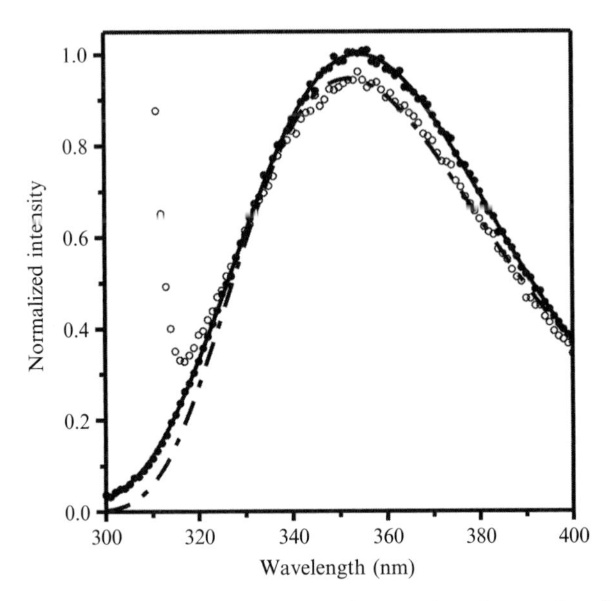

Figure 6.2 Light scattering changes the spectral properties of tryptophan fluorescence. Two samples of L-tryptophan zwitterion were dissolved at a concentration of 6.0 μ*M* in a background buffer of 2 m*M* EDTA and 100 m*M* citrate, pH 3.8. One sample included LUVs of DLPC (open circles) at a lipid concentration of 800 μ*M*. The dot/dashed line represents a fit of Eq. (6.1) to the open circles. The other sample had no LUVs (filled circles). The solid line represents the log-normal distribution component of Eq. (6.3) fit to the filled circles.

have been the same from both samples. However, Fig. 6.2 shows that there were spectral changes due to light scattering from the LUVs.

The first problem with light scattering (Ladokhin *et al.*, 2000) is apparent in both Figs. 6.1 and 6.2 where the experimentally observed emission at low wavelengths (295–325 nm) from a sample containing LUVs is higher than the true tryptophan emission component that resulted from a fit of Eq. (6.3) to the observed data. The light scattering directly added to the tryptophan signal to increase the observed emission at those low wavelengths. Eliminating the contribution from the light scattering, either by subtracting a blank scan or by using Eq. (6.3), can remove the first problem. Problems 2 and/or 3 are to blame for the peak emission intensity (I_{max}) of the sample with LUVs being lower than the peak intensity of the sample not containing LUVs. Problem 4 can be observed in Fig. 6.2 as a very subtle increase in the wavelength position of the peak emission (λ_{max}) going from the sample without LUVs to the sample with LUVs. Altogether, these observations suggest that inaccuracies in protein conformation measurements will arise if tryptophan emission data that were troubled by these problems were used to measure the thermodynamics of membrane protein folding. To prevent problems 2–4, Ladokhin and coworkers advocated for the use of a simple method to correct protein spectra using their L-tryptophan spectra as a reference (Ladokhin *et al.*, 2000).

2.3. Reducing light scattering with a spectrofluorometer

Besides correcting protein spectra or removing the direct contribution from light scattering, Ladokhin and coworkers also proposed several means of reducing light scattering to begin with by changing the experimental set-up with the spectrofluorometer itself (Ladokhin *et al.*, 2000). These means included using an excitation polarizer at 90° and an emission polarizer at 0°, using emission slits at or narrower than 5 mm and excitation slits at or narrower than 10 mm, and using an excitation wavelength of 295 nm. The emission polarizer also eliminates the Wood's anomaly that is an artifact of many monochromators and that would disrupt proper fitting of the spectra with Eq. (6.3) (Ladokhin *et al.*, 2000). The relatively long excitation wavelength at 295 nm (compared to 280 nm) avoids energy transfer from tyrosines to tryptophans and additionally prevents intertryptophyl transfer (Burstein *et al.*, 1973). Not having either of these energy transfer events allows the position-width analysis of tryptophan spectra that we discuss below. Ladokhin and coworkers finally recommend using a cuvette with a pathlength ≤ 4 mm to reduce light scattering (Ladokhin *et al.*, 2000). However, to prevent protein aggregation, we used protein concentrations so low that a 10 mm square cuvette was necessary to increase our signal-to-noise ratio. Otherwise, we followed all of the advice from Ladokhin and coworkers in all of our experiments presented here (see Section 6).

2.4. Reducing light scattering by refractive index matching

Light scattering by liposomes can also be reduced by matching their refractive index with solutes. As we describe more fully below, the ratio between the refractive index of a lipid bilayer and the refractive index of the background solution is one of the factors that influence how much light is scattered by liposomes (Matsuzaki *et al.*, 2000). If the refractive index of the background solution is raised by the addition of solutes, then the liposomes will scatter less light. Virtually any solute, including buffers, can raise the refractive index of the background solution. However, some solutes in high concentrations could affect the structure of lipid bilayers or the structure of membrane proteins. Therefore, solutes with high refractive indices coupled with high solubility would be the best candidates for refractive index matching. One group has used sucrose to make liposomes invisible to linear dichroism spectroscopy for the study of membrane pore forming peptides (Ardhammar *et al.*, 2002).

Importantly for membrane protein folding experiments, the most frequently used chemical denaturants (urea and guanidine HCl) are in fact good solutes for refractive index matching. In Fig. 6.3A, we show the effects of guanidine HCl on light scattering at three different concentrations of DLPC where it can be observed that the peak intensity of scattered light measured as right-angle emission in our spectrofluorimeter decays with increasing concentrations of guanidine HCl. At a lipid concentration of 400 μ*M*, the LUVs are essentially invisible in solutions having greater than 3.5 *M* guanidine. At higher lipid concentrations, it takes more guanidine to make the LUVs invisible because there are more LUVs scattering light. The data in Fig. 6.3A are fully reversible, and we recover the same amount of light scattering whether the LUVs are first put in concentrated guanidine and then diluted or whether they are first put in buffer and then titrated into guanidine. Also, if LUVs of DLPC are prepared by extrusion in 8.0 *M* guanidine, they do not scatter light until the guanidine is diluted.

If we had included a membrane protein in the titrations shown in Fig. 6.3A, we would also have needed a full set of blank samples of liposomes without protein for each concentration of guanidine HCl in order to adequately remove the direct contribution of scattering from the tryptophan emission signal from each protein sample. There are three key reasons why we do not favor this approach. First, preparing a full set of

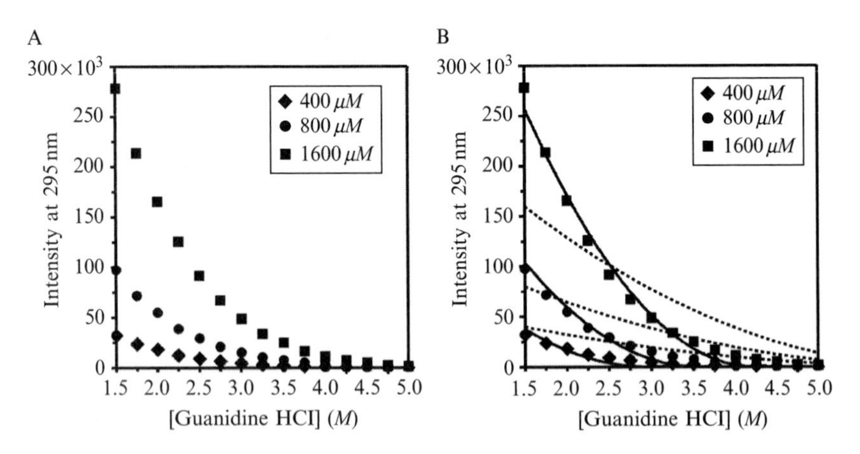

Figure 6.3 Denaturants can reduce light scattering from LUVs by matching the refractive index of the lipid bilayers. Peak intensities of RGD light scattering by LUVs of DLPC at an excitation wavelength of 295 nm are plotted. The background buffer for all samples was 2 m*M* EDTA and 100 m*M* citrate, pH 3.8. (A) Effect of the number of LUVs on their light scattering at different concentrations of the solute guanidine HCl. (B) Same as in (A) where the dotted lines represent fits to Eq. (6.9), and the solid lines represent fits to Eq. (6.11).

blanks for each protein titration is inefficient for a titration with a large number of points and it doubles the data collection time. Second, since light scattering is sensitive to the output of a spectrofluorimeter's excitation light source, one day's data may not align with another day's data. A typical xenon arc lamp, for example, can lose brightness over time as it ages, quickly making a set of data from blank samples obsolete. Third, since the peaks are well distinguished from each other, collecting the light scattering and explicitly divesting its contribution from that of the true fluorescence in the exact same sample work well. Therefore, we suggest using Eq. (6.3) as the primary method to remove the contribution of light scattering from observed fluorescence emission data when doing experiments with membrane proteins in different concentrations of denaturants. At the very least, we suggest collecting the full light scattering peak as a second stream of information about each sample. This information could reveal issues, such as the formation of membrane protein aggregates, which would not be readily apparent from just the tryptophan emission peak.

2.5. Rayleigh–Gans–Debye theory describes light scattering by liposomes

The right-angle light scattering by liposomes seen in fluorescence emission experiments can be described by the Rayleigh–Gans–Debye (RGD) theory of light scattering by particles around the same size as the wavelength of the incident light (Matsuzaki *et al.*, 2000). The RGD scattering is not the same as Rayleigh light scattering, which comes from particles much smaller than the wavelength of light.

Matsuzaki and coworkers found that RGD scattering by LUVs is proportional to the number of LUVs in the pathlength (N_p), the radius of the LUVs (R), the excitation wavelength (λ_{ex}), and the refractive indices of both the LUVs (n_{luv}) and the background solution (n_{back}) (Matsuzaki *et al.*, 2000). For our experiments, we assumed N_p is directly proportional to the lipid concentration. We also assumed that all of our LUVs that were extruded through membranes with 100 nm pores had a radius of 50 nm. These parameters can be combined in the RGD equation for the intensity of right-angle light scattering ($I(90°)$). If the LUVs are treated as optically homogenous spheres, then the RGD equation is (Matsuzaki *et al.*, 2000)

$$I(90^\circ) \propto \frac{1}{2} N_p \alpha^6 \left[\frac{(\lambda_{ex}/n_{back})^2}{4\pi^2} \right] \left[\frac{(n_{luv}/n_{back})^2 - 1}{(n_{luv}/n_{back})^2 + 2} \right]^2 \left[\frac{3(\sin u - u \cos u)}{u^3} \right]^2 \tag{6.4}$$

where the size parameter (α) is

$$\alpha = \frac{2\pi R n_{\mathrm{back}}}{\lambda_{\mathrm{ex}}} \tag{6.5}$$

and

$$u = 2\alpha \sin(45^\circ) \tag{6.6}$$

The exact value of $I(90^\circ)$ in a given experiment would also be governed by the intensity of the excitation light, and the orientation of any polarizers or filters that are in line with the incident beam or the detector.

To account for changing refractive index matching across a titration of guanidine HCl, we used the following polynomial, which relates the concentration of guanidine in a solution to the difference in the measured refractive indices of the guanidine solution (n_{GdnHCl}) and the blank buffer (n_{buffer}):

$$\begin{aligned}[\mathrm{GdnHCl}] = {} & 57.147(n_{\mathrm{GdnHCl}} - n_{\mathrm{buffer}}) + 38.68(n_{\mathrm{GdnHCl}} - n_{\mathrm{buffer}})^2 \\ & - 91.6(n_{\mathrm{GdnHCl}} - n_{\mathrm{buffer}})^3\end{aligned} \tag{6.7}$$

The guanidine becomes part of the background solution for liposomes. Our particular buffer (2 m*M* EDTA, 100 m*M* citrate, pH 3.8) has a refractive index (n_{buffer}) of 1.337 measured with an Abbe-3L refractometer at a wavelength of 589.3 nm. Refractive indices vary with wavelength according to the Sellmeir equation, but for a first approximation, we used the difference in refractive indices measured with our refractometer for Eq. (6.7). We used n_{buffer} and Eq. (6.7) to express the refractive index of the background solution (n_{back}) containing both buffer and guanidine as

$$n_{\mathrm{back}} = \frac{[\mathrm{GdnHCl}] + 80.803}{60.401} \tag{6.8}$$

Combining Eqs. (6.4) and (6.7) gives

$$\begin{aligned}I(90^\circ, [\mathrm{GdnHCl}]) \propto {} & \frac{1}{2} N_{\mathrm{p}} \alpha^6 \left[\frac{(60.401\lambda_{\mathrm{ex}}/[\mathrm{GdnHCl}] + 80.803)^2}{4\pi^2}\right] \\ & \left[\frac{(60.401 n_{\mathrm{luv}}/[\mathrm{GdnHCl}] + 80.803)^2 - 1}{(60.401 n_{\mathrm{luv}}/[\mathrm{GdnHCl}] + 80.803)^2 + 2}\right]^2 \times [3(\sin u - u \cos u)/u^3]^2\end{aligned} \tag{6.9}$$

In Fig. 6.3B, we show global fits of Eq. (6.9) to the same three sets of data that are shown in Fig. 6.3A, where we held n_{luv} to be the same value for each of the three lipid concentrations. The fits are shown as dotted lines. The fits gave the value of n_{luv} as 1.445, but they do not describe the data well. Nevertheless, the fits do accurately represent the general decay in light scattering intensity with increasing concentrations of guanidine. We then fit the data in Fig. 6.3B with a refined model that assumed that the guanidine had an effect on the lipid bilayer structure of LUVs of DLPC. This new model included the following simple linear decrease in the refractive index of the lipids with increasing concentrations of guanidine:

$$n_{\text{luv}}([\text{GdnHCl}]) = m_{\text{luv}}[\text{GdnHCl}] + n^{\circ}_{\text{luv}} \tag{6.10}$$

where n°_{luv} is the refractive index of the LUVs in the absence of guanidine and m_{luv} describes the linear decrease in the refractive index. The refractive index of a lipid bilayer depends on its thickness and on the spacing between the lipid headgroups (Ohki, 1968). As a chaotrope and a Hofmeister ion, guanidine HCl may affect both bilayer thickness and headgroup spacing (Aroti *et al.*, 2007; Feng *et al.*, 2002). Combining Eqs. (6.9) and (6.10) gives:

$$I(90^{\circ}, [\text{GdnHCl}]) \propto \frac{1}{2} N_{\text{p}} \alpha^{6} \left[\frac{(60.401\lambda_{\text{ex}}/([\text{GdnHCl}] + 80.803))^{2}}{4\pi^{2}} \right] \left[\frac{\left(60.401\left(m_{\text{luv}}[\text{GdnHCl}] + n^{\circ}_{\text{luv}}\right)/([\text{GdnHCl}] + 80.803)\right)^{2} - 1}{\left(60.401\left(m_{\text{luv}}[\text{GdnHCl}] + n^{\circ}_{\text{luv}}\right)/[\text{GdnHCl}] + 80.803\right)^{2} + 2} \right]^{2} \times \left[3(\sin u - u\cos u)/u^{3}\right]^{2}. \tag{6.11}$$

The solid lines in Fig. 6.3B are global fits of Eq. (6.11) to the three sets of titration data where we held n°_{luv} to be the same value for each of the three lipid concentrations. The refined model does reasonably describe the data considering we made such a simple assumption about the behavior of lipid bilayers in guanidine. The global fit value for n°_{luv} was 1.5008, which is very consistent with the expected refractive index of bilayers of phosphatidylcholine lipids at a wavelength of 295 nm (Chong and Colbow, 1976; Huang *et al.*, 1991).

3. Using Tryptophan Spectral Properties to Monitor Membrane Protein Folding into Liposomes

It is well understood that the spectral parameters of tryptophan fluorescence emission are sensitive to the environment of tryptophans and that a protein's tryptophans can report on environmental changes during events such as folding or unfolding. For example, the λ_{max} of many proteins' emission scans will red-shift upon unfolding as the tryptophans become more exposed to polar solvent. What is less understood is which spectral parameter is the best one to observe for a given experiment. We focus here on the choice for the best parameter to use for measurements of the equilibrium thermodynamic stability of a membrane protein when it is folded in lipid bilayers.

3.1. Position-width analysis

In 1973, Burstein and coworkers showed that a plot of peak spectral position (λ_{max}) versus width (Γ) can reveal the degree of heterogeneity of tryptophan microenvironments in a protein (Burstein *et al.*, 1973). The group compared the position-width data of proteins against position-width data for *N*-acetyl-L-tryptophanamide (NATA) and other indole derivatives that were dissolved in various solvents. We also prepared a similar set of NATA samples (see Section 6) and plotted their position-width values in Fig. 6.4A. The position-width data for NATA varied linearly with the polarity of the solvent solutions. When NATA was in apolar solutions, it had a blue-shifted λ_{max} and a smaller Γ. When it was in water or other polar solutions, it had a red-shifted λ_{max} and a larger Γ. A line fit to our NATA data (Fig. 6.4A) can be expressed by $\Gamma(\lambda_{max}) = 0.457\lambda_{max} - 96.8$. This line is characteristic of our ISS PC1 spectrofluorometer set-up. Ladokhin and coworkers also produced a similar position-width plot of NATA and found a line expressed by $\Gamma(\lambda_{max}) = 0.624\lambda_{max} - 156.7$ (Ladokhin *et al.*, 2000). Burstein and coworkers found an even steeper line fit to their indole data (Burstein *et al.*, 1973). The variation in these three lines suggests that some leeway is needed when comparing spectral parameters reported from different groups or from different spectrofluorometer set-ups.

These lines for indole behavior have significance for tryptophan fluorescence spectra of proteins. Burstein and coworkers concluded that protein emission spectra having a position-width pair falling on the indole line indicated that all of the protein's tryptophans in the sample were in equivalent environments (Burstein *et al.*, 1973). If more than one tryptophan environment were sampled, the position-width data would fall above the

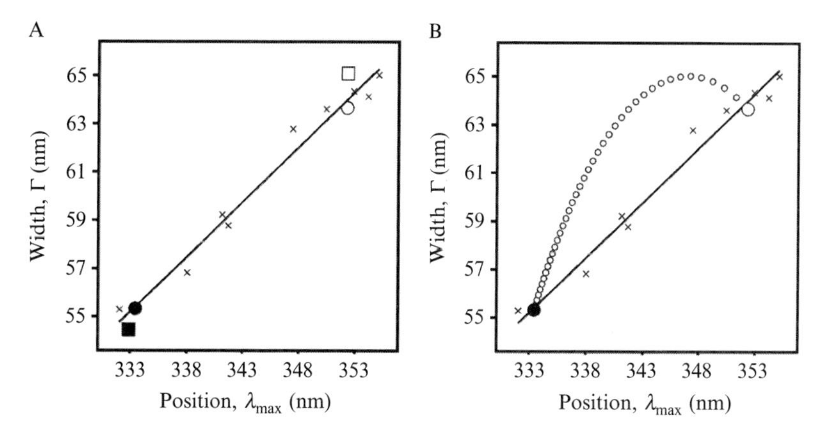

Figure 6.4 Position-width analysis of tryptophan fluorescence from folded and unfolded outer membrane proteins. Positions (λ_{max}) and widths (Γ) come from fits to Eq. (6.3). The solid line is a linear fit to data from spectra of *N*-acetyl-tryptophan-amide dissolved in different solvents or mixtures of solvents (crosses). (A) Two proteins, FadL (filled square) and OmpW (filled circle), were folded into LUVs of DLPC in 1.5 *M* guanidine HCl. The two proteins were also unfolded in the presence of the same LUVs but in 5.5 *M* guanidine HCl (open symbols of the same shapes). (B) Simulated data (small open circles) showing the expected heterogeneity of tryptophan environments when different fractions of an ensemble of OmpW molecules are folded versus unfolded. These different fractions could be obtained, for example, when a sample of OmpW is titrated into different concentrations of guanidine HCl. Each data point along the arc comes from Eq. (6.12) and represents a 2.5% step in a mixture of the folded and unfolded states. The endpoints of the arc (large circles) were taken from the actual data for OmpW shown in (A).

indole line. Multiple environments could occur in a single protein conformation for proteins that have more than one tryptophan. Multiple environments could also occur from a mixture of more than one conformation of the same protein being present at the same time. For example, Ladokhin and coworkers showed that a mixture of peptide oligomeric states could have a mixture of tryptophan environments and yield position-width data above their NATA line (Ladokhin *et al.*, 2000). Position-width analysis also allows the identification of classes of tryptophan environments (Burstein *et al.*, 1973). On our NATA line, class I tryptophans in apolar environments would have a λ_{max} around 333 nm and a Γ around 55 nm. Class III tryptophans in polar environments would have a λ_{max} around 352 nm and a Γ around 64 nm.

Along with the NATA data in Fig. 6.4A, we show position-width values for two outer membrane proteins from *E. coli*, FadL, and Outer Membrane Protein W (OmpW), in both their folded and unfolded conformations. The folded conformations were with LUVs of DLPC in 1.5 *M* guanidine HCl. The unfolded conformations were also with the LUVs but were in 5.5 *M* guanidine HCl. FadL has eleven tryptophans and OmpW has five. According to their crystal structures (Hong *et al.*, 2006; van den Berg *et al.*, 2004), if FadL and OmpW were folded, all of their tryptophans would either be buried inside their barrel structures or inside the membrane. The position-width data points for both proteins were consistent with that expectation since they fell on our NATA line in the class I region when the proteins were folded. When the proteins were unfolded, they produced position-width data points that fell on the NATA line in the class III region. We concluded that the unfolded proteins were not embedded in membranes and had all of their tryptophans fully accessible to water.

3.2. Variation of tryptophan spectral properties with fractional populations of folded protein

We considered what the position-width data for OmpW would look like if there were a heterogeneous mixture of its folded and unfolded conformations. In Fig. 6.4B, we show a simulated transition between the actual data for the folded and unfolded states of OmpW shown in Fig. 6.4A. To create this transition, we generated a set of spectra for discrete steps in the fraction of folded protein (f_{fold}) in the mixture. In Fig. 6.5A, we show an example of

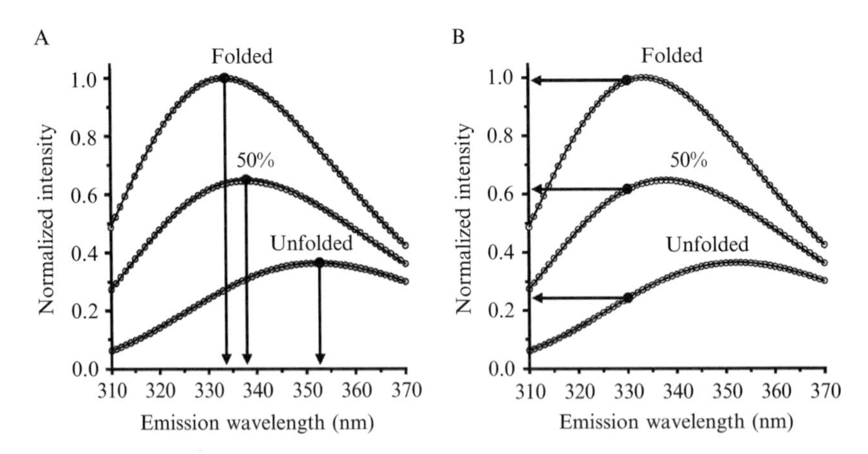

Figure 6.5 Comparison of the dependence of two spectral parameters to the fraction of folded OmpW in a sample. Tryptophan fluorescence emission intensity is normalized to the peak intensity (I_{max}) of a folded sample of OmpW in LUVs of DLPC in 1.5 *M* guanidine HCl. The unfolded sample of OmpW is in the same LUVs but instead is in 5.5 *M* guanidine HCl. The curve for 50% folded protein was generated using Eq. (6.12) and the spectra from folded and unfolded protein samples. (A) Emission intensity at a fixed wavelength (e.g., 330 nm) linearly depends on the fraction of folded protein. (B) The position of maximum emission (λ_{max}) does not linearly depend on the fraction of folded protein.

the generated spectra where f_{fold} is 0.5. The generated spectra were calculated from the equation

$$I_{step}(\lambda) = f_{fold}I_{fold}(\lambda) + (1 - f_{fold})I_{unfold}(\lambda); 0 \leq f_{fold} \leq 1 \qquad (6.12)$$

where $I_{step}(\lambda)$ is the tryptophan emission intensity at a given wavelength at each step in f_{fold}, $I_{fold}(\lambda)$ is the emission intensity taken from a fit of Eq. (6.3) to the data from the folded sample of OmpW in 1.5 *M* guanidine HCl, and $I_{unfold}(\lambda)$ is the emission intensity taken from a fit of Eq. (6.3) to the data from the unfolded sample of OmpW in 5.5 *M* guanidine HCl. We fit all of our generated spectra to Eq. (6.3) to extract the simulated values for λ_{max} and Γ for each step. In Fig. 6.4B, we show position-width values for 0.025 steps in f_{fold} as small circles. The small circles form an arc above the NATA line. This arc implies that mixtures of folded and unfolded OmpW contain mixtures of tryptophan environments and would produce spectral widths that are broader than the widths from either conformation alone.

It is also notable to observe that the small circles in Fig. 6.4B are also more closely spaced along the left-side of the arc for lower values of f_{fold} than they are on the right side of the arc for higher values of f_{fold}. This result is because of a characteristic of tryptophan spectra resulting from mixtures of tryptophan environments already noted by Ladokhin and coworkers—that the parameters λ_{max} and Γ do not vary linearly with f_{fold} (Ladokhin *et al.*, 2000). In Fig. 6.5A, we show λ_{max} for the spectra from folded and unfolded samples of OmpW and for the spectra generated from Eq. (6.12) when f_{fold} is 0.5. It can be observed that their values are not spaced evenly λ_{max}. In fact, the only spectral parameter from tryptophan fluorescence emission that does vary linearly with f_{fold} is the emission intensity at a specific wavelength ($I(\lambda)$) (Ladokhin *et al.*, 2000). The parameter I (330) is highlighted in Fig. 6.5B, and it is spaced evenly among the three spectra. The intensity at 330 nm is a good choice to monitor for membrane protein folding experiments with liposomes because the wavelength is long enough to avoid direct contribution from light scattering (Figs. 6.1 and 6.2).

Figure 6.6 is a more thorough representation of the relationship between spectral parameters determined from the same generated spectra as depicted in Fig. 6.4B and f_{fold}. The parameter I_{max} (Fig. 6.6A) does not vary linearly with f_{fold}. The nonlinearity of I_{max} with f_{fold} is subtle, but the correlation coefficient of a linear fit to the relationship of the two values in this generated data is still not unity, whereas Fig. 6.6B shows that the correlation coefficient of I (330) versus f_{fold} is in fact unity. The additional spectral parameters shown in Figs. 6.6C–F all significantly deviate from linear relationships with f_{fold}. Figure 6.6C shows the nonlinearity of the λ_{max} parameter that was used by Sanchez and coworkers to measure the thermodynamic stability of folded OmpA (Sanchez *et al.*, 2008). Figures 6.6F shows the relationship of the average emission wavelength ($\langle\lambda\rangle$) with f_{fold}. The magnitude of $\langle\lambda\rangle$ depends

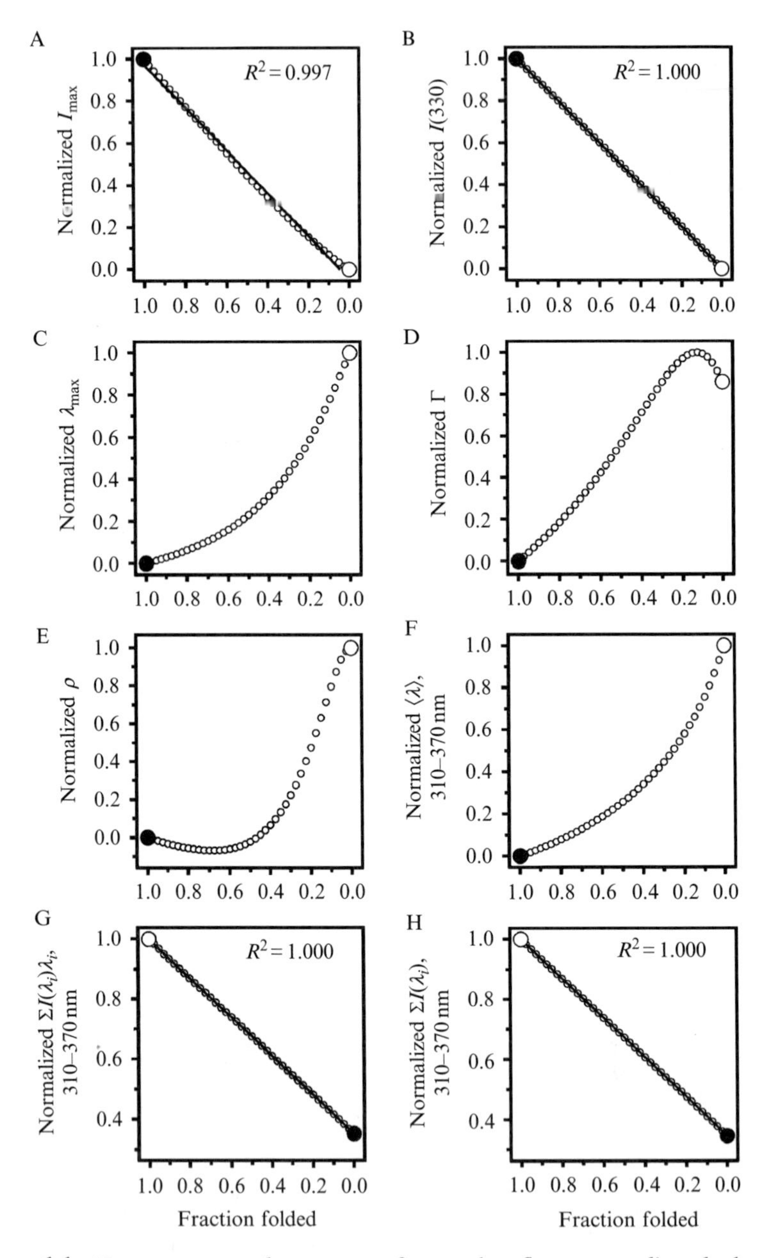

Figure 6.6 Not every spectral parameter of tryptophan fluorescence linearly depends on fraction folded. The endpoints of each plot (large circles) were taken from actual data for folded and unfolded samples of OmpW with LUVs of DLPC in 1.5 *M* and 5.5 *M* guanidine HCl, respectively. Other data points (small circles) come from

on the window of wavelengths over which the parameter is calculated. An expression for calculating $\langle\lambda\rangle$ is given by

$$\langle\lambda\rangle = \frac{\sum I(\lambda_i)\lambda_i}{\sum I(\lambda_i)}. \tag{6.13}$$

where λ_i is the ith emission wavelength sampled and $I(\lambda_i)$ is the emission intensity at that wavelength (Hong and Tamm, 2004). The $\langle\lambda\rangle$ from a window between 300 and 400 nm was used by Hong and Tamm to measure the thermodynamic stability of folded OmpA (Hong and Tamm, 2004). The $\langle\lambda\rangle$ from a window between 310 and 370 nm shown in Fig. 6.6F was used by Huysmans and coworkers to measure the thermodynamic stability of folded PagP (Huysmans *et al.*, 2010).

Both the numerator and denominator of Eq. (6.13) are linear functions of the emission intensity at specific wavelengths. Since intensity at a specific wavelength varies linearly with f_{fold}, both the numerator and denominator also vary linearly with f_{fold}. This conclusion is shown in Figs. 6.6G and 6.6H, which depict the linearity of f_{fold} with the numerator and denominator of Eq. (6.13), respectively. Equation (6.13) is then a ratonal function being that it is a linear function divided by another linear function. Therefore, as Fig. 6.6F shows, $\langle\lambda\rangle$ does not vary linearly with f_{fold}. However, as Hong and Tamm show, $\langle\lambda\rangle$ can be normalized to convert it back to its linear components during further analysis to measure thermodynamic values from data (see below) (Hong and Tamm, 2004).

4. Choosing an Appropriate Tryptophan Spectral Property to Measure the Thermodynamic Stabilities of Folded Membrane Proteins

How does the complex relationship of tryptophan spectral parameters to f_{fold} affect the measurements of stabilities of membrane proteins? In order to measure the stabilities of OmpA and PagP, the three groups we discussed

Eq. (6.12) and represent 2.5% steps of simulated mixtures of the folded and unfolded samples. (A) Peak intensity (I_{max}). Solid line represents a linear fit to the data points. (B) Emission intensity at 330 nm. Solid line represents a linear fit to the data points. (C) Position of maximum emission (λ_{max}). (D) Spectral width (Γ) at half the peak intensity. (E) Log-normal asymmetry parameter (ρ). (F) Average emission wavelength $\langle\lambda\rangle$ using a window of emission from 310–370 nm. (G) Numerator of Eq. (6.13) using a window of emission from 310–370 nm. Solid line represents a linear fit to the data points and has a slope of -0.648. (H) Denominator of Eq. (6.13) using a window of emission from 310–370 nm. Solid line represents a linear fit to the data points and has a slope of -0.655.

above each used a two-state linear extrapolation model that compares the relationship of observed spectral parameters to chemical titrations of denaturant ($Y_{obs}([D])$) to find the standard state free energy of their proteins in the absence of denaturant ($\Delta G^\circ{}_w$). An example of such a model is given by:

$$Y_{obs}([D]) = \frac{(S_{unf}[D] + Y_{unf,w}) + [(S_{fold}[D] + Y_{fold,w})\exp(-(\Delta G^\circ_w + m[D])/RT)]}{1 + \exp(-(\Delta G^\circ_w + m[D])/RT)}. \tag{6.14}$$

where $Y_{unf,w}$ and $Y_{fold,w}$ are the values of the chosen spectral parameter in the absence of denaturant for the unfolded and folded conformations, respectively; S_{unf} and S_{fold} are the slopes of linear baselines in the unfolded and folded regions of the data, respectively; m is a constant that describes how steeply the protein's free energy depends on $[D]$; R is the gas constant; and T is the temperature in Kelvin (Street *et al.*, 2008).

The model in Eq. (6.14) assumes that a protein's equilibrium constant (K_{eq}) in a given denaturant concentration can be determined by (Street *et al.*, 2008):

$$K_{eq} = \frac{Y_{obs}([D]) - (S_{unfold}[D] + Y_{unfold,w})}{(S_{fold}[D] + Y_{fold,w}) - Y_{obs}([D])} \tag{6.15}$$

This assumption can only be true if $Y_{obs}([D])$ varies linearly with f_{fold}. Therefore, the parameter λ_{max} will give an erroneous value for ΔG°_w if used in Eq. (6.14). So will the $\langle\lambda\rangle$ parameter unless it is normalized in Eq. (6.14). The normalization can be done with a ratio of the denominator of Eq. (6.13) calculated for each of the folded and unfolded conformations of the protein being studied (Hong and Tamm, 2004). This ratio (Q_R) is given by:

$$Q_R = \frac{\sum I(\lambda_i)_{fold}}{\sum I(\lambda_i)_{unfold}} \tag{6.16}$$

for the same window of emission intensities as used for calculating $\langle\lambda\rangle$. Equation (6.16) can then be used with Eq. (6.14) to give the following two-state linear extrapolation model to be used when Y_{obs} is $\langle\lambda\rangle$ (Hong and Tamm, 2004):

$$<\lambda>([D]) = \frac{(S_{unf}[D] + <\lambda>_{unf,w}) + \frac{1}{Q_R}\left[(S_{fold}[D] + <\lambda>_{fold,w})\exp\left(-\frac{(\Delta G^\circ_w + m[D])}{RT}\right)\right]}{1 + \frac{1}{Q_R}\exp\left(-\frac{(\Delta G^\circ_w + m[D])}{RT}\right)} \tag{6.17}$$

Of all popular tryptophan spectral parameters, only two will give correct values of $\Delta G_{\mathrm{w}}^{\circ}$: emission intensity and $\langle\lambda\rangle$, as long as it is normalized to a linear function of emission intensity as in Eq. (6.17).

To demonstrate the validity of that conclusion, we simulated a two-state reversible equilibrium folding transition of OmpW that might occur with a denaturant titration. For the simulation, we assumed the protein had an unfolding free energy of $\Delta G_{\mathrm{w}}^{\circ}$ equal to 10.0 kcal mol^{-1} in the absence of denaturant and an m-value of 2.50 kcal mol^{-1} M^{-1}. We also assumed that S_{unf} and S_{fold} from Eq. (6.14) would both be zero for each spectral parameter. In Fig. 6.7A, we show the plot of f_{fold} versus $[D]$ that would be needed to produce our assumed values of $\Delta G_{\mathrm{w}}^{\circ}$, m, S_{unf}, and S_{fold}. The other panels of Fig. 6.7 show plots generated for I_{max}, λ_{max}, I (330), and $\langle\lambda\rangle$ (for the emission window 310–370 nm) that replace the f_{fold} values from Fig. 6.7A with the corresponding values for each spectral parameter as calculated using Eq. (6.12). We then fit Eq. (6.14) to each generated curve in Figs. 6.7B–E and Eq. (6.17) to the generated curve in Fig. 6.7E and determined the values of $\Delta G_{\mathrm{w}}^{\circ}$ and m that would result from the respective spectral parameters. We also fit Eq. (6.17) to the generated curve in Fig. 6.7E and determined the same thermodynamic values. The results of these fits are shown in Table 6.1. The only parameter that reproduced our assumed values for $\Delta G_{\mathrm{w}}^{\circ}$ and m were those that varied linearly with intensity, i.e., I (330) and the normalized $\langle\lambda\rangle$. The other spectral parameters misrepresented the thermodynamic values, primarily because they incorrectly positioned the denaturant midpoint of the conformational transition of the protein (C_m). These incorrect midpoints can be easily seen in Fig. 6.6. For example, when f_{fold} is 0.50, the λ_{max} would be only 23% of the way through its transition (Fig. 6.6C). We also note that the misrepresentations of the thermodynamic parameters shown in Table 6.1 would be even worse if the spectral parameters had nonzero values of S_{unf} and S_{fold}.

5. Conclusions

Measuring the thermodynamic stability of membrane proteins folded in liposomes is approachable by monitoring the tryptophan fluorescence emission by the proteins. We have suggested appropriate procedures to improve the accuracy of these measurements by managing the light scattering from the liposomes and by selecting the appropriate spectral parameter to relate to the fraction of folded protein in experimental samples. Our suggestions are tailored to the use of denaturant titrations. We conclude that the direct contribution of light scattering to tryptophan spectra should be removed by fitting emission intensities to a sum of a normal distribution describing light scattering and a log-normal distribution describing tryptophan emission. We further demonstrate that the emission intensity at a specific wavelength

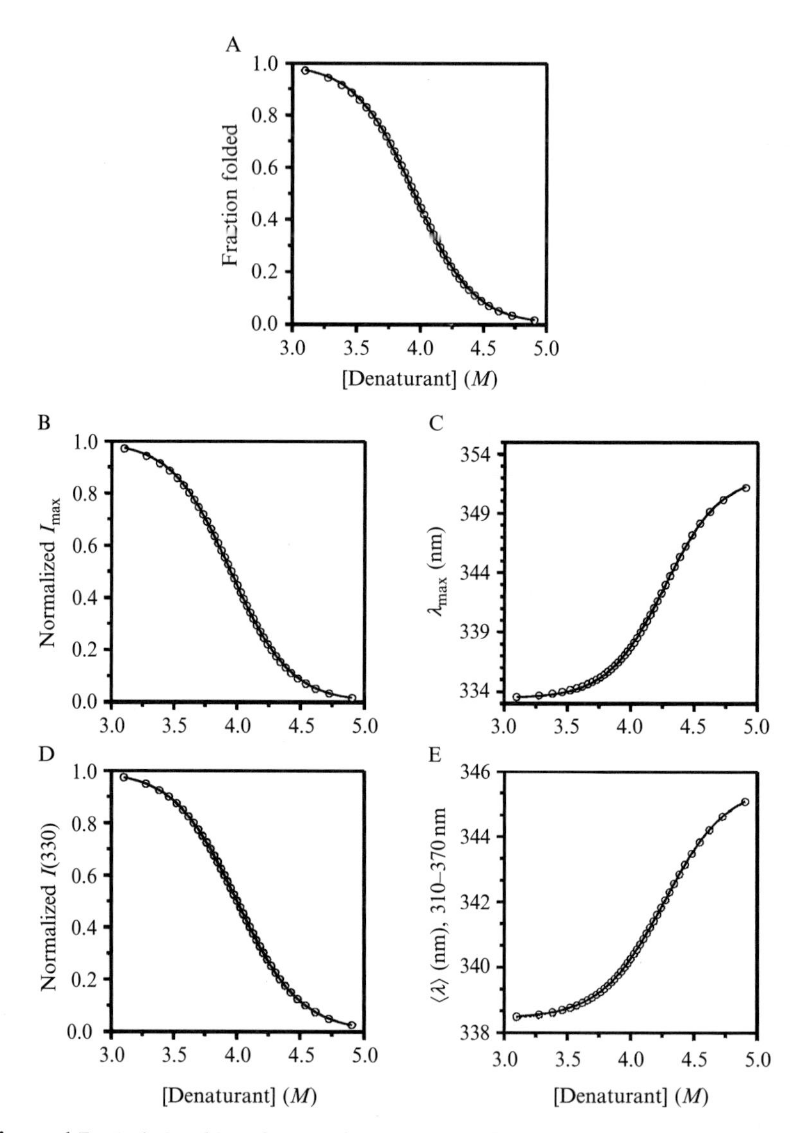

Figure 6.7 Relationship of spectral parameters of tryptophan fluorescence to isotherms from chemical denaturation. The vertical-axis positions of the data points in each plot come from Eq. (6.12) and represent 2.5% steps of simulated mixtures of folded and unfolded samples of OmpW. The horizontal-axis positions of the data points come from Eq. (6.14) and represent a simulated equilibrium, two-state denaturant titration of OmpW at 37 °C. For the simulation, ΔG°_{w} was assumed to be 10 kcal mol^{-1}, the *m*-value was assumed to be 5 kcal mol^{-1} M^{-1}, and the baseline regions were assumed to have zero slope. Solid lines represent a best fit of Eq. (6.14) to the simulated data. (A) Fraction folded. (B) Peak intensity (I_{max}). (C) Position of maximum emission (λ_{max}). (D) Intensity emitted at 330 nm. (E) Average emission wavelength $\langle\lambda\rangle$ using a window of emission from 310–370 nm.

Table 6.1 Thermodynamic parameters recovered from the response of tryptophan spectral parameters to denaturant concentration

	Assumed	I_{max}	I (330)	λ_{max}	$\langle\lambda\rangle$ 310–370 nm	$\langle\lambda\rangle$ normalized
ΔG°_{w} (kcal mol^{-1})	10.0	10.4	10.0	11.9	10.7	10.0
m (kcal mol^{-1} M^{-1})	2.5	2.6	2.5	2.8	2.5	2.5
C_m (M)	4.0	3.9	4.0	4.3	4.3	4.0

(e.g., 330 nm) or the average emission wavelength, normalized to be a linear function of intensity, should be used for the determination of thermodynamic parameters.

6. Materials and Methods

6.1. Protein folding reactions

Both FadL and OmpW were expressed to inclusion bodies and purified in a manner previously described (Burgess *et al.*, 2008). We made fresh unfolded protein stocks prior to each experiment by dissolving inclusion body pellets in a buffer of 8 *M* guanidine HCl. Stock guanidine solutions were prepared from UltraPure powder (Invitrogen). The concentration of guanidine was checked by refractometry. The background buffer for all experiments was 100 m*M* citrate (Sigma) and 2 m*M* EDTA (Sigma), pH 3.8. We used lipid stocks in chloroform (Avanti Polar Lipids) and dried them briefly under nitrogen followed by at least 8 h of dehydration under vacuum. We then wetted the lipids with the background buffer to a lipid concentration of 20 mg mL^{-1} for 1 h with occasional vortexing and then prepared LUVs by extruding the lipid suspensions 21 times through two stacked 0.1 μ*M* filters.

We prepared the protein folding/unfolding reactions in three steps. The first step was a dilution of the unfolded protein stocks to a final guanidine concentration of 3.0 *M* and a final protein concentration of 6.0 μ*M*. Also present was 1.4 m*M* 3-(*N*,*N*-dimethylmyristyl-ammonio)propanesulfonate detergent (Sigma), which is just above its critical micelle concentration (CMC) in 3.0 *M* guanidine. Without detergent the proteins visibly precipitated, even in 3.0 *M* guanidine and the presence of liposomes. The second reaction step was a threefold dilution of the solvated proteins into the presence of LUVs of DLPC at a 2000:1 lipid–protein ratio. For folding reactions, the final guanidine concentration was 1.5 *M*. For unfolding reactions, the final guanidine concentration was 5.5 *M*. In either concentration of guanidine, the threefold dilution was enough to take the detergent below its CMC. If we kept the detergent above its CMC, then the light scattering due to the LUVs disappeared as the LUVs were solubilized by the detergent. Below the detergent CMC, the amount of light scattering by LUVs was the same as for samples of LUVs that had never been exposed to detergent. The third reaction step was a fivefold dilution, which kept the same guanidine concentrations from the second step. The final protein concentration was 400 n*M* and the final lipid concentration was 800 μ*M*. After the dilutions, we incubated all samples with gentle mixing at 37 °C for 40–50 h before fluorescence experiments.

6.2. L-Tryptophan, blank LUVs, and NATA reactions

We prepared samples of L-tryptophan (Sigma) in the same manner as the protein folding experiments, except that we did not include any detergent and the final L-tryptophan concentration was 6.0 μ*M*. The final lipid concentration was 800 μ*M*. We prepared blank samples of LUVs at the different final lipid concentrations shown in Fig. 6.3 in the same manner as the protein folding experiments, except that we did not include any detergent or protein. We prepared samples of NATA by dissolving powder (Sigma) in the following solvents: dichloromethane, acetonitrile, isopropyl alcohol, propanol, water, various percentages of methanol in water, and 10 *M* urea.

6.3. Spectrofluorometry

All protein samples were excited at 295 nm in an ISS PC1 spectrofluorometer (Champaign, IL). We used an excitation polarizer at 90° and an emission polarizer at 0°. Excitation slits were 2.4 mm and emission slits were 2.0 mm. The pathlength of our cuvettes was 10 mm. We collected a minimum of four emission scans from 280 to 400 nm for each sample and then averaged the data before fitting. We used Igor Pro v6.12 (www.wavemetrics.com) for all least-squares fitting routines.

ACKNOWLEDGMENTS

The authors thank the National Science Foundation (MCB 0423807 and 0919868) and the National Institutes of Health (R01 GM079440, T32 GM008403) for financial support.

REFERENCES

Ardhammar, M., Lincoln, P., and Nordén, B. (2002). Invisible liposomes: Refractive index matching with sucrose enables flow dichroism assessment of peptide orientation in lipid vesicle membrane. *Proc. Natl. Acad. Sci. USA* **24,** 15313–15317.

Aroti, A., Leontidis, E., Dubois, M., and Zemb, T. (2007). Effects of monovalent anions of the Hofmeister series on DPPC lipid bilayers Part I: Swelling and in-plane equations of state. *Biophys. J.* **93,** 1580–1590.

Burgess, N. K., Dao, T. P., Stanley, A. M., and Fleming, K. G. (2008). β-Barrel proteins that reside in the *Escherichia coli* outer membrane *in vivo* demonstrate varied folding behavior *in vitro*. *J. Biol. Chem.* **39,** 26748–26758.

Burstein, E. A., Vedenkina, N. S., and Ivkova, M. N. (1973). Fluorescence and the location of tryptophan residues in protein molecules. *Photochem. Photobiol.* **18,** 263–279.

Chong, C. S., and Colbow, K. (1976). Light scattering and turbidity measurements on lipid vesicles. *Biochim. Biophys. Acta* **436,** 260–282.

Feng, Y., Yu, Z., and Quinn, P. J. (2002). Effect of urea, dimethylurea, and tetramethylurea on the phase behavior of dioleoylphosphatidylethanolamine. *Chem. Phys. Lipids* **114,** 149–157.

Hong, H., and Tamm, L. K. (2004). Elastic coupling of integral membrane protein stability to lipid bilayer forces. *Proc. Natl. Acad. Sci. USA* **101,** 4065–4070.

Hong, H., Patel, D. R., Tamm, L. K., and van den Berg, B. (2006). The outer membrane protein OmpW forms an eight-stranded β-barrel with a hydrophobic channel. *J. Biol. Chem.* **281,** 7568–7577.

Huang, H. W., Liu, W., Olah, G. A., and Wu, Y. (1991). Physical techniques of membrane studies—study of membrane active peptides in bilayers. *Prog. Surf. Sci.* **38,** 145–199.

Huysmans, G. H. M., Baldwin, S. A., Brockwell, D. J., and Radford, S. E. (2010). The transition state for folding of an outer membrane protein. *Proc. Natl. Acad. Sci. USA* **107,** 4099–4104.

Ladokhin, A. S., Jayasinghe, S., and White, S. H. (2000). How to measure and analyze tryptophan fluorescence in membranes properly, and why bother? *Anal. Biochem.* **285,** 235–245.

Matsuzaki, K., Murase, O., Sugishita, K., Yoneyama, S., Akada, K., Ueha, M., Nakamura, A., and Kobayashi, S. (2000). Optical characterization of liposomes by right angle light scattering and turbidity measurement. *Biochim. Biophys. Acta* **1467,** 219–226.

Ohki, S. (1968). Dielectric constant and refractive index of lipid bilayers. *J. Theor. Biol.* **19,** 97–115.

Permyakov, E. A. (1993). Luminescent Spectroscopy of Proteins. CRC Press, Boca Raton, pp. 40-42.

Sanchez, K. M., Gable, J. E., Schlamadinger, D. E., and Kim, J. E. (2008). Effects of tryptophan microenvironment, soluble domain, and vesicle size on the thermodynamics of membrane protein folding: Lessons from the transmembrane protein OmpA. *Biochemistry* **47,** 12844–12852.

Street, T. O., Courtemanche, N., and Barrick, D. (2008). Protein folding and stability using denaturants. *Methods Cell Biol.* **84,** 295–325.

van den Berg, B., Black, P. N., Clemons, W. M., and Rapoport, T. A. (2004). Crystal structure of the long-chain fatty acid transporter FadL. *Science* **304,** 1506–1509.

CHAPTER SEVEN

Non-B Conformations of CAG Repeats Using 2-Aminopurine

Natalya N. Degtyareva *and* Jeffrey T. Petty

Contents

Abstract

Repetition of trinucleotide sequences is the molecular basis of ~30 hereditary neurological and neurodegenerative diseases, and alternate structures adopted by these sequences are implicated in the etiology of such diseases. Elucidating these structures is important for advancing mechanistic understanding and ultimately treatment. Studies of (CAG) repeats are motivated by their involvement in a number of these diseases, and the structures favored by $(CAG)_8$ are discussed in this contribution. Utilizing the strong effect of base stacking on fluorescence quantum yield, 2-aminopurine is used in place of adenine to determine the secondary structures adopted by such repeated sequences. Alone, $(CAG)_8$ folds into a hairpin comprised of a duplex stem and a single-stranded loop. Energetic studies indicate that the hairpin is anchored by the interactions in the stem and has a strained loop environment. As a model for

Department of Chemistry, Furman University, Greenville, South Carolina, USA

Methods in Enzymology, Volume 492
ISSN 0076-6879, DOI: 10.1016/B978-0-12-381268-1.00019-7
© 2011 Elsevier Inc.
All rights reserved.

intermediates that form during repeat expansion, $(CAG)_8$ was also incorporated into a duplex to form a three-way junction. In contrast to the isolated $(CAG)_8$, this integrated repeat adopts an open, unfolded loop. Enthalpy and entropy changes associated with denaturation indicate that the stability of the three-way junction is dominated by interactions in the duplex arms and that the repeated sequence tracks global unfolding. Because 2-aminopurine provides both structural and energetic information via fluorescence and also is an innocuous substitution for adenine, significant progress in elucidating the secondary structures of (CAG) repeats will be achieved.

1. Introduction

1.1. Diseases associated with repeated sequences

Modifications in the genetic code can alter cellular function and ultimately lead to disease. Besides exogenous agents, mutagenicity can originate within DNA itself, and such genetic instability is illustrated by a class of ~30 neurological and neurodegenerative diseases that are associated with repeated DNA sequences (Bacolla and Wells, 2009; Gatchel and Zoghbi, 2005; Kovtun and McMurray, 2008; Mirkin, 2007). Fragile X syndrome and other associated diseases are distinguished by their preponderance in humans and by their non Mendelian genetics in which the probability of mutation increases in subsequent generations. A critical insight was that sequence expansion beyond a threshold dictates development and severity of such diseases, and an underlying question is whether alternative secondary structures of these long DNA sequences are significant (Bacolla and Wells, 2009; Cleary and Pearson, 2003; Mirkin, 2007). Central to models for repeat expansion are self-associated forms of single-stranded DNA, such as the hairpins adopted by CAG and CTG repeats (Lenzmeier and Freudenreich, 2003; Sinden *et al.*, 2007). These particular sequences are implicated in a large subset of the repeat-based neurological diseases, and their folding is driven by duplex stem formation (Amrane *et al.*, 2005; Mariappan *et al.*, 1998; Mitas *et al.*, 1995; Zheng *et al.*, 1996). The resulting mismatches of like bases destabilize the hairpin, but this instability is tempered by stacking with neighboring G/C pairs (Amrane *et al.*, 2005; Mitas, 1997; Paiva and Sheardy, 2004).

Expansion of repeated sequences occurs when one strand is preferentially extended relative to its complementary sequence during processes such as replication, so it is important to consider how DNA context influences the secondary structures of repeated sequences. One type of model structure is the slipped intermediate heteroduplex that has a single-stranded repeated sequence that emanates from the duplex (Pearson *et al.*, 2002; Sinden *et al.*, 2007). Using enzymatic probes, the key structural elements that have been identified are strand junctions, hairpin loops, and the core of the repeated sequences. While folding in CTG repeats is not influenced by their DNA context, folding in CAG repeats is disrupted in these heteroduplexes, and this structural difference may be important in the biological processing of these two types of repeats. Beyond enzymatic and chemical probing and relatively low-resolution microscopy techniques, high-resolution structural studies of repeated sequences are significantly advanced by using fluorogenic bases.

1.2. 2-Aminopurine as a structural and energetic probe

Via changes in intensity, spectral position, lifetime, and anisotropy, fluorescence spectroscopy offers a sensitive spectroscopic perspective on both the structure and energetics of nucleic acids (Hawkins *et al.*, 2008; Lakowicz, 2006). Low-fluorescence quantum yields of the natural nucleobases restrict such investigations, thus motivating development of fluorescent base analogs (Sinkeldam *et al.*, 2010). In general, these modified bases have three features that furnish structural information at the base level. First, selective detection of the modified analogs is possible because their emission intensity and spectra are resolved from the natural nucleobases. Second, chemical synthesis allows precise placement of the analogs within the primary sequence of the nucleic acid. Third, spectra can depend on interactions such as base stacking and thus provide a convenient spectroscopic means to assess local structure within the overall structure. An important issue is the extent to which modified bases reflect and possibly influence DNA structure, and in this regard, 2-aminopurine has been extensively investigated (Rist and Marino, 2002). This isomer of adenine has the exocyclic amine group in the second versus the sixth position on adenine, and it maintains base pairing with thymine in B-form DNA (Johnson *et al.*, 2004; Lycksell *et al.*, 1987; Sowers *et al.*, 1986). A key feature of 2-aminopurine is its high fluorescence quantum yield of 50–70% under physiological conditions, and its value for nucleic acid structural studies is that the quantum yield is quenched ~100-fold due to stacking interactions with adjacent bases (Rachofsky *et al.*, 2001; Ward *et al.*, 1969). Because secondary structural elements of nucleic acid structure depend on base stacking, this large variation in fluorescence quantum yield allows fine discrimination of structural features (Rist and Marino, 2002).

The goal of our studies is to determine the structure of repeated (CAG) sequences using selective substitutions of adenine with 2-aminopurine, and two studies provided guidance. First, a constrained hairpin with a loop comprised of three (CAG) repeats showed that the spectral properties of the substituted 2-aminopurines depend on their location in the primary sequence (Lee *et al.*, 2007). Variations in intensity and anisotropy suggest that base stacking and hydrogen bonding are significant factors in stabilizing the repeated sequences in the loop. Second, key structural and energetic features of RNA have been investigated using 2-aminopurine (Ballin *et al.*, 2007, 2008). In these studies, extrahelical/bulged, base paired, and looped bases were identified by the fluorescence intensities and lifetimes of selectively substituted 2-aminopurines. These studies are distinguished by their use of control single-stranded sequences to isolate structural from temperature and base context effects on the fluorescence properties of 2-aminopurine. Our studies use these important contributions to understand the structures and thermodynamics of repeated (CAG) sequences.

2. Materials and Methods

2.1. Materials

Oligonucleotides from Integrated DNA Technologies (Coralville, IA) were synthesized using standard phosphoramidite chemistry and purified using denaturing gel electrophoresis with 7 *M* urea and 8% polyacrylamide (Table 7.1). A 5-base standard was used to identify DNA strands based on their length, and the desired band was excised and then electrocuted from a dialysis bag. Following buffer exchange using gel filtration chromatography (NAP-10, GE Healthcare), the DNA was lyophilized and resuspended in buffer. Three-way junctions were formed by annealing equimolar quantities of 50- and 74-base oligonucleotides, with the longer DNA strand containing a central $(CAG)_8$ with flanking 25-base regions that complement the shorter strand (Fig. 7.1). Measurements were conducted in buffers comprised of 10 m*M* Tris/Tris$-H^+$ (pH $= 7.9$) with 10 m*M* $MgCl_2$ and 50 m*M* NaCl or 10 m*M* $H_2PO_4^-/HPO_4^{2-}$ (pH $= 7$) with 50 m*M* NaCl. The results show no buffer dependence. Nomenclature for modified oligonucleotides identifies the position of 2-aminopurine within the primary sequence (Fig. 7.1). Starting from the 5′ end, individual replacements of adenine with 2-aminopurine in the isolated $(CAG)_8$ hairpin were made in the first, second, third, fourth, fifth, and sixth repeats, and these are designated by 1-AP_{HP}, 2-AP_{HP}, 3-AP_{HP}, 4-AP_{HP}, 5-AP_{HP}, and 6-AP_{HP}, respectively. For the three-way junction, five substitutions notated as 1-AP_J, 3-AP_J, 4-AP_J, 5-AP_J, and 6-AP_J were made in the first, third, fourth, fifth, and sixth repeats of the central $(CAG)_8$, respectively. Modifications labeled α-AP_J and β-AP_J were also made in the two (CAG) repeats in the duplex region that precedes the junction.

2.2. Conformational integrity

To evaluate how 2-aminopurine influences the secondary structure of repeated sequences, circular dichroism, thermal denaturation, gel electrophoresis, and enzyme recognition are used. In the ultraviolet spectral

Table 7.1 Single-stranded oligonucleotides

Sequence	Length[a]	ε[b]
5′-CACCATGCCGGTA TTTAAA CAG CAG CAG CAG CAG CAG CAG CAG CAG CAG CAG CAG TACGTA CTGCAGCTCGAGG-3′	74	720,500
5′-CACCATGCC GGT ATT TAA ACAG CAG CAG CAG CAG CAG CAG CAG CAG CAG CAG CAG TACGTA CTGCAGCTCGAGG-3′[c]	74	708,700
5′-CCTCGAGCTGCAG TACGTA CTGCTG—CTGCTG TTTAAA TACCGGCATGGTG-3′[d]	50	462,500
5′-CACCATGCCGGTA TTTAAA CAGCAG CAGCAG TACGTA CT GCAGCTCGAGG-3′	50	484,500
5′-CAG CAG CAG CAG CAG CAG CAG CAG-3′	24	237,300
5′-CAG CAG CAG CAG CAG CAG CAG CAG-3′[e]	24	225,500
5′-G AAA CAG CAG TTTT CTG CTG TTT C-3′ (DS-CAG)	24	210,900
5′-AAA CAG CAG-3′ (SS_{β}-CAG)	9	86,300
5′-CAG CAG CAG-3′ (SS-CAG)	9	78,000

[a] Length in bases.
[b] Extinction coefficients (M^{-1} cm^{-1}) of the unfolded single strands.
[c] Seven variants of this oligonucleotide were used, each with a single substitution of 2-aminopurine, indicated by the underlined bases.
[d] Connecting line represents the region corresponding to the $(CAG)_8$.
[e] Six different substitutions of 2-aminopurine were made in $(CAG)_8$.

region, base–base interactions dominate the circular dichroism response to provide a sensitive indication of nucleic acid structure (Cantor and Schimmel, 1980). The hairpins adopted by (CAG) repeats have a conformation that is most similar to B-form DNA, which suggests that canonical G/C base pairs drive folding (Paiva and Sheardy, 2004). With respect to the unmodified sequences, similar circular dichroism spectra indicate that single 2-aminopurines throughout $(CAG)_8$ have a minimal impact on conformation (Fig. 7.2). To gain an energetic perspective, melting temperatures, enthalpy changes, and entropy changes associated with global denaturation were derived from absorbance changes at 260 nm, and similar thermodynamic values with and without 2-aminopurine indicate that the stabilities of repeat structures are not perturbed by modification (Tables 7.2 and 7.3). To monitor DNA shape, nondenaturing gel electrophoresis shows that the mobility of $(CAG)_8$, 24-base oligonucleotide is comparable to that of a ~12 base pair duplex, and no changes with 2-aminopurine were observed (Degtyareva *et al.*, 2009). For the three-way junction, only one band with a mobility that is distinct from its constituent single strands was observed (Degtyareva *et al.*, 2010). Using restriction digestion at the two sites adjacent

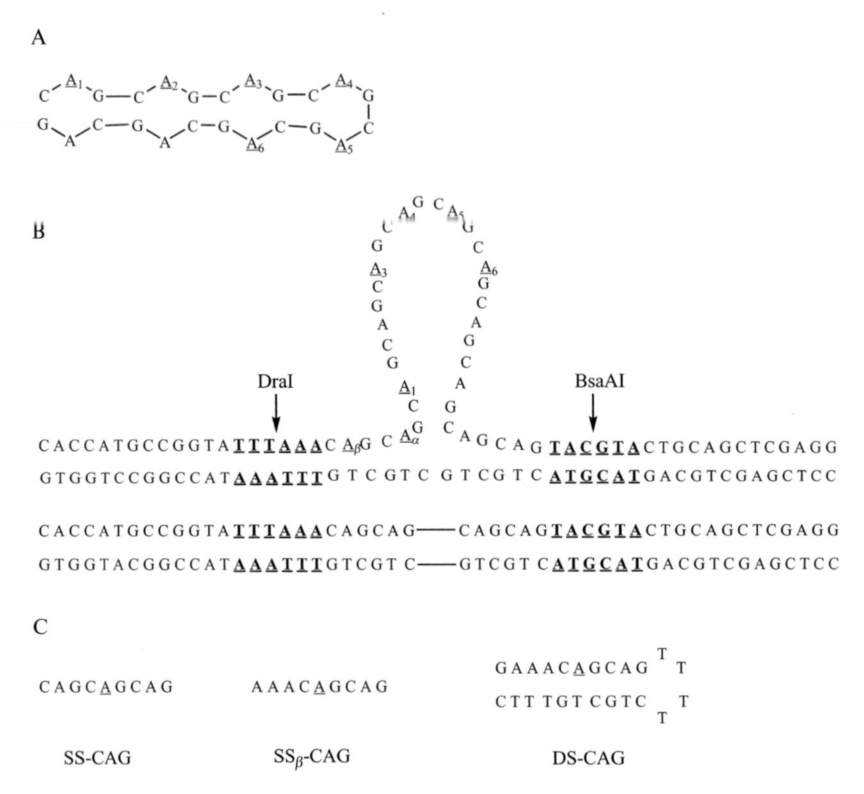

Figure 7.1 (A) Structure for the hairpin formed by $(CAG)_8$. (B) Structure for the three-way junction formed with $(CAG)_8$. Restriction enzyme sites for DraI and BsaAI are indicated in bold and underlined. Below this structure is the duplex that comprises the three-way junction. The connecting lines represent the position of the abstracted repeat sequence. (C) Single- and double-stranded oligonucleotides used as structural references. In all these oligonucleotides, the positions of 2-aminopurine substitution are underlined and/or enumerated.

to the strand junction, gel electrophoresis shows proper strand alignment by liberation of the terminal 16 base pair fragments. Again, 2-aminopurine does not influence these results.

2.3. Calculation of thermodynamic parameters

Because thermal denaturation produces a common single-stranded state, differences in the thermodynamic parameters between the variants can be attributed to the low-temperature folded forms. Changes in the extinction coefficients of the bases and the fluorescence quantum yield of 2-aminopurine permit the folded and unfolded states to be defined, from which the energetic factors that impact global and local structure are deduced.

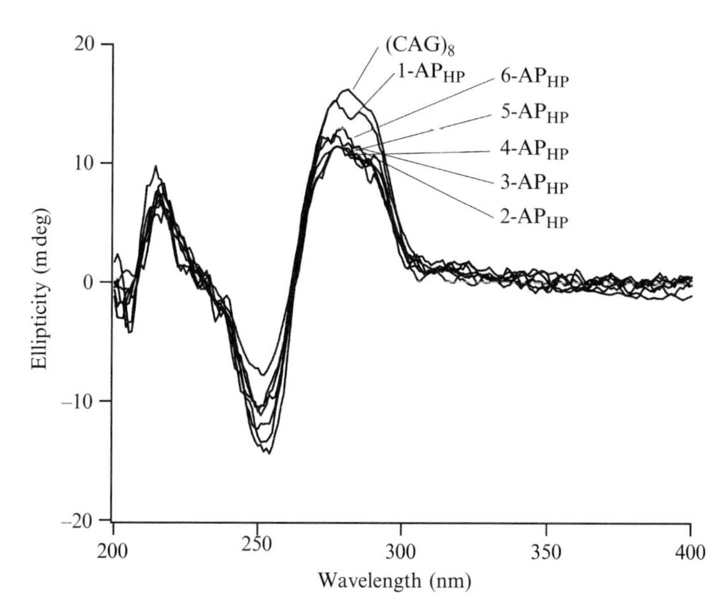

Figure 7.2 Circular dichroism spectra of unmodified and 2-aminopurine modified $(CAG)_8$ sequences. The similarity in the spectra suggest that 2-aminopurine has no significant effect on the conformation of $(CAG)_8$. Spectral shapes are consistent with a B-like duplex stem (Paiva and Sheardy, 2004).

Standard methods of extracting thermodynamic information are used to generate plots of fraction of folded species, θ, versus temperature (Mergny and Lacroix, 2003). Different baselines corresponding to the low- and high-temperature states are used to estimate the uncertainties in the derived quantities. Melting temperatures are determined as the intersection point between the θ dependence on temperature and the median between the two baselines. A more thorough thermodynamic analysis uses a two-state analysis to extract enthalpy and entropy changes after accounting for the molecularity of the transitions. Data was analyzed between $\theta = 0.2$–0.8, where intermediate temperatures give relatively comparable concentrations of both folded and unfolded forms. Changes in entropy and enthalpy were measured for triplicate samples.

2.4. Oligonucleotide concentration

Structural references are used to deduce the secondary structures of repeated sequences, and such comparisons must consider oligonucleotide concentrations. A sensitive method utilizes the high fluorescence quantum yield of unconjugated 2-aminopurine, and the exonuclease BAL-31 cleaves the fluorogenic nucleotide from the oligonucleotide. Concomitant base unstacking in dilute solution results is a significant enhancement in

Table 7.2 Thermodynamic data for $(CAG)_8$ hairpin

Modification	T_m (°C)[a]	ΔH (kcal/mol)[a]	ΔS (cal/mol/K)[a]	T_m (°C)[b]	ΔH (kcal/mol)[b]	ΔS (cal/mol/K)[b]	K_q (M^{-1})
1-AP_{HP}	58.8 (1.0)[d]	32.0 (1.9)	96 (6)	ND[c]	ND	ND	9.8 (0.5)
2-AP_{HP}	60.1 (1.4)	28.5 (2.4)	86 (8)	52.8 (0.3)	31.2 (0.9)	96 (3)	5.0 (0.5)
3-AP_{HP}	60.3 (1.0)	29.9 (4.7)	90 (14)	54.4 (0.9)	32.7 (1.3)	100 (4)	9.1 (0.8)
4-AP_{HP}	61.1 (1.0)	32.1 (1.4)	96 (4)	50.4 (0.6)	32.9 (2.9)	95 (9)	11.8 (1.0)
5-AP_{HP}	59.2 (1.4)	28.9 (1.5)	87 (5)	47.0 (1.0)	20.0 (2.0)	61 (6)	14.8 (1.2)
6-AP_{HP}	60.2 (1.0)	29.5 (3.7)	88 (11)	51.8 (0.3)	32.7 (1.3)	100 (4)	10.6 (0.4)
Unmodified	57.4 (0.6)	29.0 (1.5)	88 (5)				
SS-CAG							9.6 (1.0)
DS-CAG							8.2 (0.6)

[a] Data derived from absorbance changes at 260 nm.
[b] Data derived from fluorescence changes of 2-aminopurine modified forms of $(CAG)_8$.
[c] Values not determined.
[d] Standard deviations in parentheses derived from three measurements.

Table 7.3 Thermodynamic data for $(CAG)_8$ three-way junction

Modification	T_m (°C)[a]	ΔH (kcal/mol)[a]	ΔS (cal/mol/K)[a]	T_m (°C)[b]	ΔH (kcal/mol)[b]	ΔS (cal/mol/K)[b]	K_q (M^{-1})
β-AP_J	75.9 (0.4)[d]	220 (22)	606 (61)	77.0 (0.6)	238 (24)	630 (63)	7.5 (0.2)
α-AP_J	77.2 (0.1)	199 (20)	543 (54)	78.7 (0.3)	237 (32)	650 (65)	9.8 (0.5)
1-AP_J	76.6 (0.1)	ND[c]	ND	ND	ND	ND	11.1 (0.2)
3-AP_J	76.8 (0.8)	ND	ND	77.0 (1.7)	224 (22)	606 (61)	15.0 (0.2)
4-AP_J	76.4 (0.1)	194 (19)	522 (53)	77.1 (0.6)	190 (20)	516 (52)	13.5 (0.7)
5-AP_J	75.9 (0.3)	ND	ND	77.6 (1.6)	ND	ND	15.2 (0.7)
6-AP_J	76.4 (0.6)	217 (22)	600 (60)	76.5 (0.3)	250 (25)	685 (69)	14.9 (0.4)
Unmodified	77.2 (0.2)	199 (20)	540 (54)				
Duplex	81.4 (0.3)	243 (25)	665 (67)				

[a] Data derived from absorbance changes at 260 nm.
[b] Data derived from fluorescence changes of 2-aminopurine modified forms of the three-way junctions.
[c] Values not determined.
[d] Standard deviations in parentheses derived from three measurements.

fluorescence intensity, which is then related to concentration using a standard curve. Typically, 5–20 nmol of DNA (in base pairs) were mixed with 0.5 units of BAL-31 in digestion buffer (20 m*M* Tris–HCl, pH 8.0, 600 m*M* NaCl, 12 m*M* $CaCl_2$, 12 m*M* $MgCl_2$, 1 m*M* EDTA) in a total volume of 50 μL and incubated at 30 °C for 20–30 min. The reaction mixture was then diluted with buffer, and fluorescence intensities of the standard and digested samples were measured. Alternatively, because the sequences of the oligonucleotides are known, concentrations were also calculated using extinction coefficients for the cleaved nucleotides (Cavaluzzi and Borer, 2004). Finally, concentrations of intact oligonucleotides were determined using extinction coefficients at 260 nm derived from the nearest-neighbor approximation, with the small contribution from 2-aminopurine (1000 M^{-1} cm^{-1} at 260 nm) included in the calculation (Fox *et al.*, 1958). To account for the effect of temperature, high-temperature posttransition baselines were extrapolated back to 25 °C. Variations in DNA concentration determined by all three methods were less than 10%.

2.5. Structural references

To relate the fluorescence from the 2-aminopurine labeled repeated sequences to particular secondary structural elements of DNA, single- and double-stranded references are used (Ballin *et al.*, 2007, 2008) (Fig. 7.1). Nine-base oligonucleotides provide the environments for 2-aminopurine in solvent-exposed, single-stranded DNA, and a 2-aminopurine/thymine base pair in a duplex stem provides a reference for the fluorophore in a base-stacked and solvent-sequestered state. Although structurally analogous to the adenine/thymine pair, the 2-aminopurine/thymine pair has faster proton exchange rates with the solvent and lower thermal stability, indicative of less effective stacking within the duplex (Lycksell *et al.*, 1987; Patel *et al.*, 1992; Xu *et al.*, 1994). In both reference oligonucleotides, the immediate base environment of the 2-aminopurine is identical to that in the repeated sequence. In addition to secondary structure and neighboring bases, temperature also influences the fluorescence quantum yield of 2-aminopurine (Fig. 7.3). To dissect these contributions during thermal denaturation, the ratios of fluorescence intensities from the single-stranded reference and repeated sequences are used, as the former maintain an unfolded state over the entire temperature range. At high temperatures, similar fluorescence intensities indicate that both types of DNA achieve the same unfolded state (Figs. 7.3 and 7.4).

2.6. Acrylamide quenching

To complement information derived from the inherent fluorescence of 2-aminopurine, extrinsic fluorescence quenching by acrylamide is also used to assess solvent exposure (Lakowicz, 2006). By relating quenching

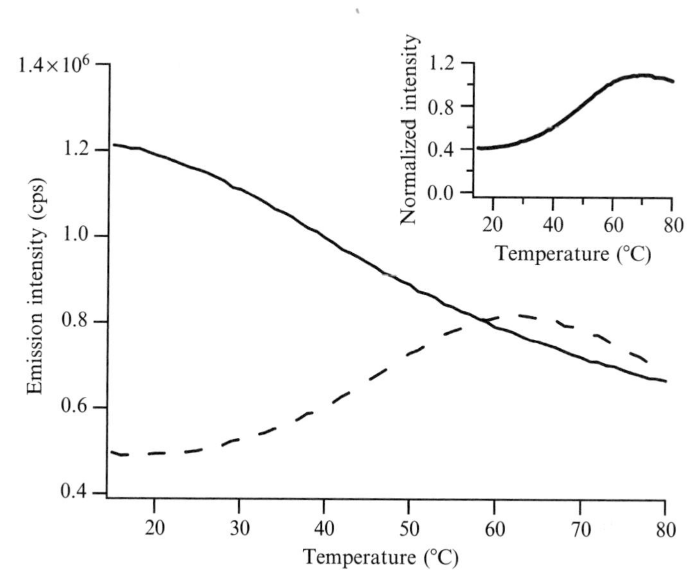

Figure 7.3 Emission intensities of 3-AP_{HP} (dashed) and SS-CAG (solid) as a function of temperature. The intensity change associated with the single-stranded reference is due solely to temperature, as this oligonucleotide is unfolded over the entire temperature range. The inset shows the normalized intensity for 3-AP_{HP} using the reference to account for temperature effects on the emission intensity. Because of denaturation of the $(CAG)_8$ hairpin at high temperature, both oligonucleotides achieve comparable intensities.

efficiency to secondary structural elements of DNA, structural variations within the oligonucleotides are inferred (Ballin *et al.*, 2007, 2008). Standard Stern–Volmer analysis is used to extract quenching constants, K_q, for the 2-aminopurines (Fig. 7.5, Tables 7.2 and 7.3). The smallest quenching constants are associated with bases that are most protected from the solvent, as in a duplex. As the fluorogenic base becomes more exposed to the aqueous solution, as for single-stranded DNA, the quenching constant increases.

3. Structure and Thermodynamics of Isolated and Integrated $(CAG)_8$

3.1. $(CAG)_8$ hairpin

The overall conclusion from 2-aminopurine studies of $(CAG)_8$ alone is that this repeated sequence favors a hairpin displaying suppressed fluorescence in the stem and enhanced fluorescence in the loop (Figs. 7.1 and 7.4). Folding driven by strand association is indicated by the similar behaviors of the distant modifications 2-AP_{HP}/3-AP_{HP} and 6-AP_{HP}. These three variants exhibit relatively lower emission intensities when compared to single-stranded

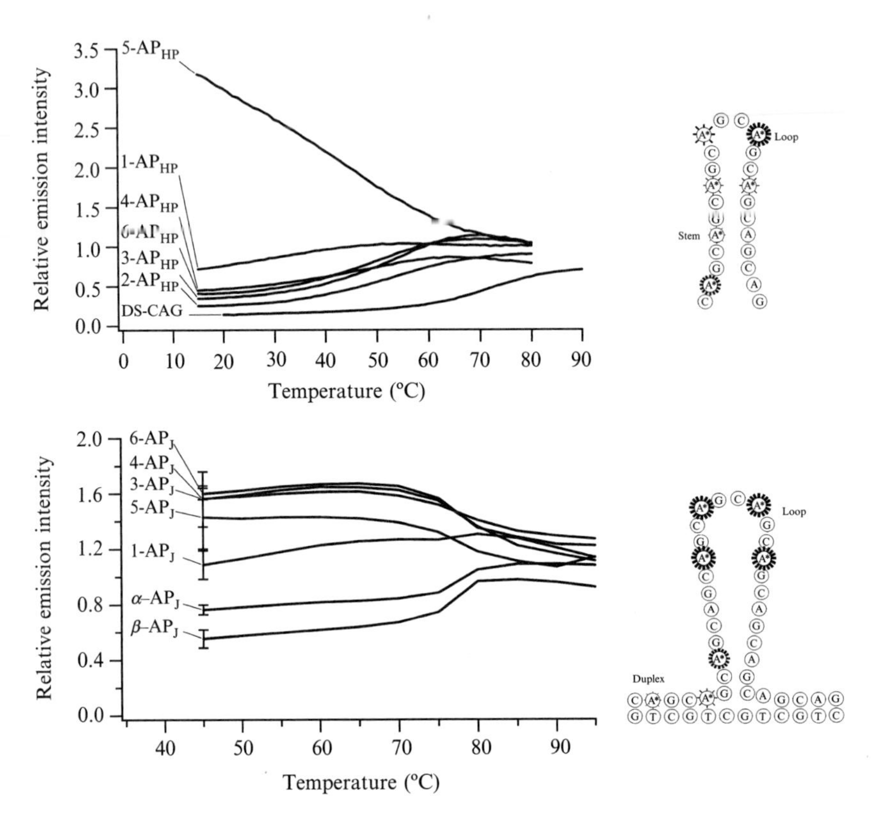

Figure 7.4 (Top) Emission intensities of 2-aminopurine-modified $(CAG)_8$ as a function of temperature normalized using the emission from SS-CAG ($\lambda_{ex} = 307$ nm, $\lambda_{em} = 370$ nm). (Bottom) Normalized emission intensities of the modified three-way junctions as a function of temperature. To the right of each graph are the models for the secondary structure based on fluorescence intensities. The graphic emphasis at each position indicates its level of solvent exposure.

DNA, and these intensities are comparable at high temperature. These two observation support base stacking in the folded, low-temperature form of $(CAG)_8$ and are consistent with the propensity of repeated sequences to form into hairpins with B-like conformations in the stem (Paiva and Sheardy, 2004). Incorporation of the resulting purine–purine mismatches in the stem of $(CAG)_8$ is gauged using a duplex with a 2-aminopurine/thymine base pair. Comparable intensities for matched and mismatched pairs involving 2-aminopurine indicate their comparable degrees of stacking with neighboring bases, a conclusion that is supported by NMR studies of duplexes containing adenine/adenine mismatches (Mariappan *et al.*, 1998). The other major structural element identified by 2-aminopurine is a loop between the duplex. Relative to the single-stranded reference, 5-AP_{HP} has a higher

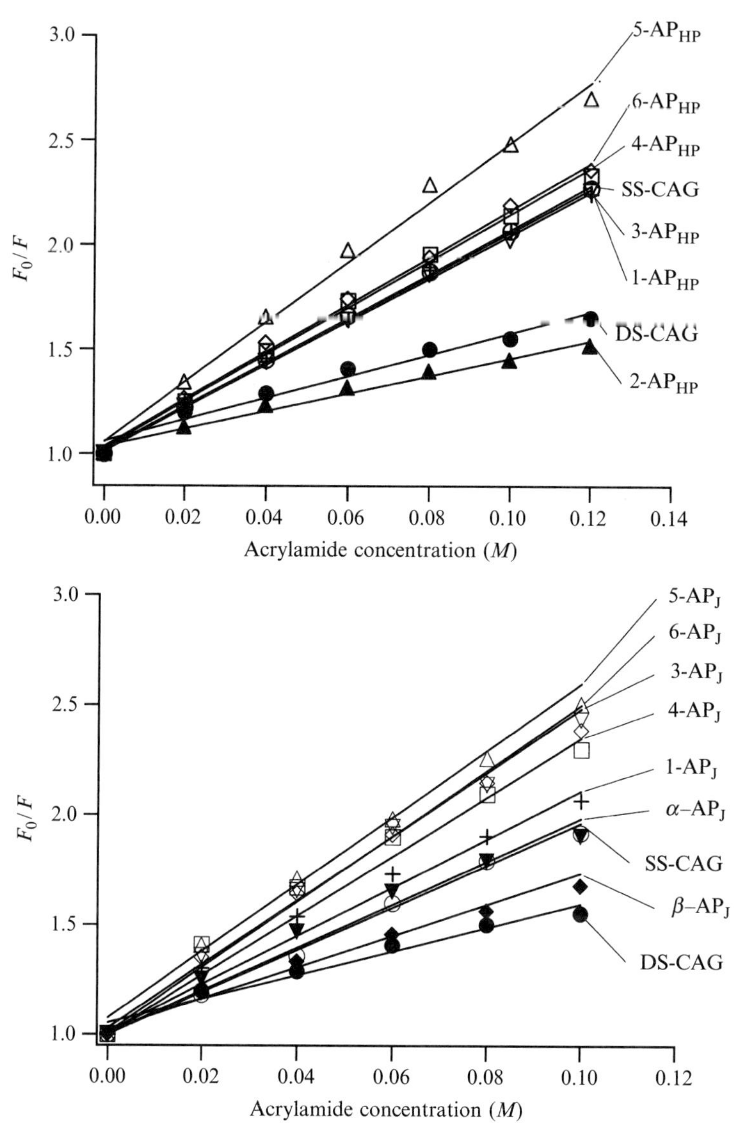

Figure 7.5 Stern–Volmer plots of the acrylamide quenching of 2-aminopurine variants of $(CAG)_8$ (top) and the three-way junction (bottom). F_0 and F are the fluorescence intensities in the absence and presence of acrylamide, respectively. Linear fits were used to derive quenching constants using the standard Stern–Volmer model.

intensity that diminishes with temperature. This is consistent with a constrained loop that relaxes to enhance base stacking and the associated solvent sequestration in the single-stranded form. An important advantage of

2-aminopurine is the strong effect of base stacking on fluorescence quantum yield, thereby resolving finer details of secondary structure. For $(CAG)_8$, transitional behavior between the stem and loop motifs is evident. For 1-AP_{HP}, its intensity is comparable to the single-stranded reference over the entire temperature range, consistent with frayed, unpaired terminal repeats. Features of both the stem and loop are exhibited by 4-AP_{HP}, which has an intermediate fluorescence intensity and smaller fluorescence enhancement. This behavior is considered in the context of models of repeat folding, which indicate that stacking interactions in the stem are maximized by the tetraloop AGCA (Fig. 7.1; Hartenstine *et al.*, 2000). In $(CAG)_8$, these bases comprise the fourth and fifth repeats, yet variations in the fluorescence properties of 2-aminopurine substitutions in these repeats suggest an asymmetric loop environment. Local variations in the interactions with neighboring bases could be dictating the structure of the loop (Mariappan *et al.*, 1996; Senior *et al.*, 1988; Varani, 1995).

Structural patterns deduced from fluorescence are supported by acrylamide quenching studies, which show clustering into three groups (Fig. 7.5, Table 7.2). The least efficient quenching is observed for 2-AP_{HP} and the duplex reference, which supports protection for 2-aminopurine within a duplex stem from this external quenching agent. Higher solvent exposure comparable to the single-stranded reference is observed for 1-AP_{HP}, 3-AP_{HP}, 4-AP_{HP}, and 6-AP_{HP}. While the frayed ends of 1-AP_{HP} and the single-stranded reference are expected to be structurally similar, the fluorescence properties of 3-AP_{HP}, 4-AP_{HP}, and 6-AP_{HP} indicate that these 2-aminopurine substitutions are more solvent sequestered. Thus, these similarities suggest that acrylamide quenching constants should serve as general guides to the structures of repeated sequences. The largest quenching constant is observed for 5-AP_{HP}, consistent with the high fluorescence intensity from this region of the loop. Quenching is as efficient as observed for 2-aminopurine at the exposed bulge of a RNA duplex, but is smaller than that of the free nucleobase (Ballin *et al.*, 2007; Degtyareva *et al.*, 2009).

As utilized in structural studies, denaturation of $(CAG)_8$ produces a single-stranded oligonucleotide, so high temperature fluorescence spectra provide an important structural reference. Furthermore, temperature-dependent fluorescence changes allow the fractional conversion from folded to unfolded states to be derived, from which enthalpy and entropy changes are determined using Arrhenius analysis (Table 7.2). A context for interpreting local energetic changes is provided by global unfolding using hyperchromic absorbance changes at 260 nm. These enthalpy and entropy changes are similar to values obtained from calorimetric analysis, which suggests that neglecting heat capacity changes does not significantly compromise interpretation of the results (Amrane *et al.*, 2005; Paiva and Sheardy, 2004). For 2-AP_{HP}, 3-AP_{HP}, and 6-AP_{HP}, local enthalpy and entropy changes derived from fluorescence are similar to the global values.

This consistency suggests that base pairing and stacking the stem dictate the overall stability of the folded hairpin. In contrast, 5-AP$_{HP}$ has both smaller enthalpy and entropy changes for denaturation relative to the global values. These energetic features complement the relatively high fluorescence from this region of the hairpin and support limited base interactions in the loop of the hairpin. Thermodynamic measurements also provide insight into how the mismatch influences the stability of the stem. Global melting temperatures derived from absorbance are 6–8 °C higher than those derived from fluorescence. This behavior suggests that neighboring base interactions with 2-aminopurine weaken and thereby enhance solvent exposure prior to global denaturation (Lycksell *et al.*, 1987; Nordlund *et al.*, 1993).

3.2. (CAG)$_8$ three-way junction

Fluorescence from 2-aminopurine shows that integration of (CAG)$_8$ into a duplex disrupts intrastrand association within the repeat sequence to produce an open and solvent-exposed loop (Figs. 7.1 and 7.4). To understand this context-dependent structural change for (CAG)$_8$, seven modifications were used to map solvent exposure throughout the three-way junction. Within the duplex arms, (CAG) repeats base pair with opposing (GTC) repeats in the complementary strands, as shown by β-AP$_J$ and α-AP$_J$. Emission intensities from these variants are lower than the single-stranded reference and increase with denaturation at high temperature, and both observations support base stacking of 2-aminopurine that is relieved with thermal denaturation. However, the strand junction perturbs base pairing and stacking within the duplex, as indicated by the higher intensities of β-AP$_J$ and α-AP$_J$ relative to the duplex reference with a 2-aminopurine/thymine base pair. Furthermore, α-AP$_J$ has a higher intensity than β-AP$_J$, suggesting that the level of solvent exposure increases with proximity to the junction. This trend in solvent exposure continues into the repeated (CAG)$_8$ sequence, with 1-AP$_J$ acting as single-stranded DNA over the entire temperature range. Collectively, the fluorescence properties of these three modifications indicate that the transition from duplex to repeated sequence is accompanied by disrupted base interactions that increase solvent exposure of the bases. This disturbance appears to be transmitted into the repeated sequence, with 3-AP$_J$, 4-AP$_J$, 5-AP$_J$, and 6-AP$_J$ exhibiting similarly high fluorescence intensities when compared with the single-stranded reference. In addition, mirroring the loop region of the (CAG)$_8$ hairpin, these four variants of the three-way junction display solvent exposure that diminishes at higher temperatures, suggesting that constraints induced by the junction relax when the three-way junction denatures. This openness of the repeated sequence strikingly contrasts with the folded (CAG)$_8$, as most clearly demonstrated by the third and sixth repeats. Hairpin formation in (CAG)$_8$ alone is shown by comparable and relatively low levels of solvent exposure in 3-AP_{HP} and 6-AP_{HP}. In contrast, similarly placed 2-aminopurines in 3-AP_J and 6-AP_J are more solvent exposed when compared with the single-stranded reference.

Acrylamide quenching studies support a model in which solvent exposure becomes comparable to single-stranded DNA in the vicinity of the strand junction and is consistently high in the core of the repeated sequence (Fig. 7.5, Table 7.3). Quenching for β-AP_J and the duplex reference are comparable, which suggests that the β modification is protected from extrinsic quenching by base stacking and pairing. Quenching becomes increasingly efficient proceeding toward the junction from α-AP_J and onto 1-AP_J. Quenching constants that are consistently higher than single-stranded DNA are observed for 3-AP_J, 4-AP_J, 5-AP_J, and 6-AP_J. Additional support for the open conformation throughout the integrated $(CAG)_8$ is provided by the similarity of these quenching constants when compared with 5-AP_{HP} in the loop of the $(CAG)_8$ hairpin.

To understand the thermodynamic factors that contribute to the stability of the three-way junction, thermal denaturation was followed using absorbance and fluorescence spectral changes (Table 7.3). After dissecting the repeated sequence, the resulting duplex is shown to dominate the stability of the three-way junction. Specifically, enthalpy and entropy changes for the three-way junction are slightly (~20%) lower than the values for the duplex. Furthermore, the melting profile of the three-way junction is monophasic with no evidence of a transition associated with the $(CAG)_8$ hairpin. Local unfolding was monitored using the fluorescent variants, and the similarities of the melting temperatures, enthalpy changes, and entropy changes relative to the values derived from absorbance indicate that local structural changes track global unfolding. Collectively, these observations suggest that the stability of the three-way junction is dictated by its duplex arms with slight instability induced by the repeated sequence.

The overall conclusion is that DNA context critically impacts the structure of $(CAG)_8$. Enzymatic probes also highlight the distinctive features of (CAG) repeats when compared with (CTG) repeats (Pearson *et al.*, 2002). In slipped intermediates, enzymatic recognition of the latter repeats is consistent with their folded, hairpin structures that also form in the analogous isolated repeats. In contrast, (CAG) repeats preferentially interact with single-strand specific enzymes and proteins, indicative of their random coil conformation. Such a conformation differs from the folded and largely protected hairpins favored by isolated (CAG) hairpins. This behavior again supports the dominant effect of the duplex arms and the strand junction on the structure of such repeated sequences. The significance of these earlier results and our studies using 2-aminopurine lies in the importance of secondary structure in the biological recognition and processing of repeated sequences via enzyme recognition (Kovtun and McMurray, 2008; Pearson *et al.*, 1997).

4. Conclusions

The adenine isomer 2-aminopurine provides an avenue to understand the secondary structures of repeated DNA sequences, and a profound difference is noted when $(CAG)_8$ is isolated from versus integrated into a duplex of canonical base pairs. Alone, $(CAG)_8$ forms a hairpin in which folding is driven by duplex formation in the stem. In contrast, such distant intrastrand interactions are absent when $(CAG)_8$ is incorporated into a duplex. This structural distinction suggests that the duplex has a significant effect on the structures of repeated sequences. These results provide the foundation for understanding the secondary structures favored by larger repeated sequences that are implicated in debilitating neurological diseases. Such structures will be elucidated from the repeat level resolution of structural and energetic information that is gleaned from 2-aminopurine. By identifying key structural elements favored by repeated sequences, the path to understanding the mechanisms of repeat expansion will have a stronger molecular basis.

ACKNOWLEDGMENTS

We are grateful to the National Science Foundation (CHE-0718588) and the Henry Dreyfus Teacher–Scholar Awards Program for the primary support for this work. We also greatly appreciate partial support from the National Science Foundation (CBET-0853692) and the National Institutes of Health (R15GM071370 and P20 RR-016461 (from the National Center for Research Resource)).

REFERENCES

Amrane, S., Sacca, B., Mills, M., Chauhan, M., Klump, H. H., and Mergny, J.-L. (2005). Length-dependent energetics of $(CTG)_n$ and $(CAG)_n$ trinucleotide repeats. *Nucleic Acids Res.* **33,** 4065–4077.

Bacolla, A., and Wells, R. D. (2009). Non-B DNA conformations as determinants of mutagenesis and human disease. *Mol. Carcinog.* **48,** 273–285.

Ballin, J. D., Bharill, S., Fialcowitz-White, E. J., Gryczynski, I., Gryczynski, Z., and Wilson, G. M. (2007). Site-specific variations in RNA folding thermodynamics visualized by 2-aminopurine fluorescence. *Biochemistry* **46,** 13948–13960.

Ballin, J. D., Prevas, J. P., Bharill, S., Gryczynski, I., Gryczynski, Z., and Wilson, G. M. (2008). Local RNA conformational dynamics revealed by 2-aminopurine solvent accessibility. *Biochemistry* **47,** 7043–7052.

Cantor, C. R., and Schimmel, P. R. (1980). Biophysical Chemistry. First ed., W. H. Freeman and Company, New York.

Cavaluzzi, M. J., and Borer, P. N. (2004). Revised UV extinction coefficients for nucleoside-5'-monophosphates and unpaired DNA and RNA. *Nucleic Acids Res.* **32,** e13.

Cleary, J. D., and Pearson, C. E. (2003). The contribution of cis-elements to disease-associated repeat instability: Clinical and experimental evidence. *Cytogenet. Genome Res.* **100,** 25–55.

Degtyareva, N. N., Reddish, M. J., Sengupta, B., and Petty, J. T. (2009). Structural studies of a trinucleotide repeat sequence using 2-aminopurine. *Biochemistry* **48,** 2340–2346.

Degtyareva, N. N., Barber, C. A., Sengupta, B., and Petty, J. T. (2010). Context dependence of trinucleotide repeat structures. *Biochemistry* **49,** 3024–3030.

Fox, J. J., Wempen, I., Hampton, A., and Doerr, I. L. (1958). Thiation of nucleosides. I. Synthesis of 2-amino-6-mercapto-9-β-D-ribofuranosylpurine ("Thioguanosine") and related purine nucleosides1. *J. Am. Chem. Soc.* **80,** 1669–1675.

Gatchel, J. R., and Zoghbi, H. Y. (2005). Diseases of unstable repeat expansion: Mechanisms and common principles. *Nat. Rev. Genet.* **6,** 743–755.

Hartenstine, M. J., Goodman, M. F., and Petruska, J. (2000). Base stacking and even/odd behavior of hairpin loops in DNA triplet repeat slippage and expansion with DNA polymerase. *J. Biol. Chem.* **275,** 18382–18390.

Hawkins, M. E., Ludwig, B., and Michael, L. J. (2008). Fluorescent pteridine probes for nucleic acid analysis. *Methods in Enzymology* Vol. 450, pp. 201–231. Academic Press, San Diego, CA, Chapter 10.

Johnson, N. P., Baase, W. A., and von Hippel, P. H. (2004). Low-energy circular dichroism of 2-aminopurine dinucleotide as a probe of local conformation of DNA and RNA. *Proc. Natl. Acad. Sci. USA* **101,** 3426–3431.

Kovtun, I. V., and McMurray, C. T. (2008). Features of trinucleotide repeat instability *in vivo*. *Cell Res.* **18,** 198–213.

Lakowicz, J. R. (2006). Principles of Fluorescence Spectroscopy, Third ed., Springer, New York.

Lee, B. J., Barch, M., Castner, E. W., Volker, J., and Breslauer, K. J. (2007). Structure and dynamics in DNA looped domains: CAG triplet repeat sequence dynamics probed by 2-aminopurine fluorescence. *Biochemistry* **46,** 10756–10766.

Lenzmeier, B. A., and Freudenreich, C. H. (2003). Trinucleotide repeat instability: A hairpin curve at the crossroads of replication, recombination, and repair. *Cytogenet. Genome Res.* **100,** 7–24.

Lycksell, P. O., Graslund, A., Claesens, F., McLaughlin, L. W., Larsson, U., and Rigler, R. (1987). Base pair opening dynamics of a 2-aminopurine substituted Eco RI restriction sequence and its unsubstituted counterpart in oligonucleotides. *Nucleic Acids Res.* **15,** 9011–9025.

Mariappan, S. V., Garcoa, A. E., and Gupta, G. (1996). Structure and dynamics of the DNA hairpins formed by tandemly repeated CTG triplets associated with myotonic dystrophy. *Nucleic Acids Res.* **24,** 775–783.

Mariappan, S. V., Silks, L. A., 3rd, Chen, X., Springer, P. A., Wu, R., Moyzis, R. K., Bradbury, E. M., Garcia, A. E., and Gupta, G. (1998). Solution structures of the Huntington's disease DNA triplets, (CAG)n. *J. Biomol. Struct. Dyn.* **15,** 723–744.

Mergny, J.-L., and Lacroix, L. (2003). Analysis of thermal melting curves. *Oligonucleotides* **13,** 515–537.

Mirkin, S. M. (2007). Expandable DNA repeats and human disease. *Nature* **447,** 932–940.

Mitas, M. (1997). Trinucleotide repeats associated with human disease. *Nucleic Acids Res.* **25,** 2245–2254.

Mitas, M., Yu, A., Dill, J., Kamp, T. J., Chambers, E. J., and Haworth, I. S. (1995). Hairpin properties of single-stranded DNA containing a GC-rich triplet repeat: $(CTG)_{15}$. *Nucleic Acids Res.* **23,** 1050–1059.

Nordlund, T. M., Xu, D., and Evans, K. O. (1993). Excitation energy transfer in DNA: Duplex melting and transfer from normal bases to 2-aminopurine. *Biochemistry* **32,** 12090–12095.

Paiva, A. M., and Sheardy, R. D. (2004). Influence of sequence context and length on the structure and stability of triplet repeat DNA oligomers. *Biochemistry* **43,** 14218–14227.

Patel, N., Berglund, H., Nilsson, L., Rigler, R., McLaughlin, L. W., and Gräslund, A. (1992). Thermodynamics of interaction of a fluorescent DNA oligomer with the anti-tumour drug netropsin. *Eur. J. Biochem.* **203,** 361–367.

Pearson, C., Ewel, A., Acharya, S., Fishel, R., and Sinden, R. (1997). Human MSH2 binds to trinucleotide repeat DNA structures associated with neurodegenerative diseases. *Hum. Mol. Genet.* **6,** 1117–1123.

Pearson, C. E., Tam, M., Wang, Y.-H., Montgomery, S. E., Dar, A. C., Cleary, J. D., and Nichol, K. (2002). Slipped-strand DNAs formed by long (CAG){middle dot}(CTG) repeats: Slipped-out repeats and slip-out junctions. *Nucleic Acids Res.* **30,** 4534–4547.

Rachofsky, E. L., Osman, R., and Ross, J. B. (2001). Probing structure and dynamics of DNA with 2-aminopurine: Effects of local environment on fluorescence. *Biochemistry* **40,** 946–956.

Rist, M. J., and Marino, J. P. (2002). Fluorescent nucleotide base analogs as probes of nucleic acid structure, dynamics and interactions. *Curr. Org. Chem.* **6,** 775.

Senior, M. M., Jones, R. A., and Breslauer, K. J. (1988). Influence of loop residues on the relative stabilities of DNA hairpin structures. *Proc. Natl. Acad. Sci. USA* **85,** 6242–6246.

Sinden, R. R., Pytlos-Sinden, M. J., and Potaman, V. N. (2007). Slipped strand DNA structures. *Front. Biosci.* **12,** 4788–4799.

Sinkeldam, R. W., Greco, N. J., and Tor, Y. (2010). Fluorescent analogs of biomolecular building blocks: Design, properties, and applications. *Chem. Rev.* **110,** 2579–2619.

Sowers, L. C., Fazakerley, G. V., Eritja, R., Kaplan, B. E., and Goodman, M. F. (1986). Base pairing and mutagenesis: Observation of a protonated base pair between 2-aminopurine and cytosine in an oligonucleotide by proton NMR. *Proc. Natl. Acad. Sci. USA* **83,** 5434–5438.

Varani, G. (1995). Exceptionally stable nucleic acid hairpins. *Annu. Rev. Biophys. Biomol. Struct.* **24,** 379–404.

Ward, D. C., Reich, E., and Stryer, L. (1969). Fluorescence studies of nucleotides and polynucleotides. I. Formycin, 2-aminopurine riboside, 2, 6-diaminopurine riboside, and their derivatives. *J. Biol. Chem.* **244,** 1228–1237.

Xu, D., Evans, K. O., and Nordlund, T. M. (1994). Melting and premelting transitions of an oligomer measured by DNA base fluorescence and absorption. *Biochemistry* **33,** 9592–9599.

Zheng, M., Huang, X., Smith, G. K., Yang, X., and Gao, X. (1996). Genetically unstable CXG repeats are structurally dynamic and have a high propensity for folding NMR UV spectroscopic study. *J. Mol. Biol.* **264,** 323–336.

CHAPTER EIGHT

Disulfide Bond-Mediated Passenger Domain Stalling as a Structural Probe of Autotransporter Outer Membrane Secretion *In Vivo*

Jonathan P. Renn[1] *and* Patricia L. Clark

Contents

Abstract

Autotransporters (ATs) are the largest class of extracellular virulence proteins secreted by Gram-negative pathogenic bacteria, but the details of their outer membrane (OM) secretion mechanism remain unclear. Recently, a novel strategy has been developed to study OM secretion of AT proteins by introducing

Department of Chemistry and Biochemistry, University of Notre Dame, Notre Dame, Indiana, USA
[1] Current address: Department of Molecular Biosciences, Northwestern University, Evanston, Illinois, USA

Methods in Enzymology, Volume 492
ISSN 0076-6879, DOI: 10.1016/B978-0-12-381268-1.00030-6
© 2011 Elsevier Inc.
All rights reserved.

pairs of cysteine (Cys) residues into the central passenger domain sequence. Upon oxidation in the periplasm, these Cys residues form a long loop that stalls AT OM secretion. This Cys-loop stalling technique has been used to investigate such questions as the directionality of AT OM secretion and the extent of AT passenger domain folding during secretion. Here, we will describe how to use the Cys-loop approach to produce disulfide-bonded, stalled AT OM secretion intermediates, and how these stalled "snapshots" can be used to investigate structural aspects of the AT OM secretion mechanism.

1. Protein Secretion: An Essential Component of Bacterial Virulence

A healthy human body includes billions of bacterial cells, which perform beneficial, synergistic functions primarily related to food digestion and nutrient adsorption (Savage, 1977). In contrast, colonization of the human body by pathogenic strains of the same bacteria can lead to devastating and potentially fatal diseases. What features distinguish these pathogenic bacterial strains from the "helpful" bacteria of our gut flora? One striking difference is the large number of proteins secreted to the surface of pathogenic bacteria. These secreted proteins perform diverse functions, including those related to cell adhesion, subversion of the host immune response, and host cell invasion. In all cases, secreted proteins are synthesized in the cell cytoplasm and transported across the bacterial cell membrane(s) to reach its outer surface. Extracellular secretion of virulence proteins is particularly challenging in Gram-negative pathogens, because these proteins must cross three distinct compartments—the inner membrane (IM), the periplasmic space, and the outer membrane (OM)—in order to reach the outer surface of the cell (Fig. 8.1). Transport of all water-soluble molecules across the IM, including proteins and small molecules such as ATP, ions, and water itself, is tightly regulated. For protein secretion across the IM, the hydrolysis of ATP provides an external energetic driving force. In contrast, the OM is porous to ions and small molecules. The permeability of the OM means that the periplasmic space lacks a significant concentration of ATP and that there is no proton gradient across the OM (Mansell *et al.*, 2008). In the absence of ATP or an ion gradient, it is much less clear what provides the driving force for efficient secretion of virulence proteins across the OM (Thanassi *et al.*, 2005).

2. The Autotransporter Secretion Pathway

Gram-negative bacteria have evolved several distinct mechanisms to secrete proteins across the OM (Henderson *et al.*, 1998, 2004; Rego *et al.*, 2010; Saier, 2006). Some of these mechanisms couple together IM and OM

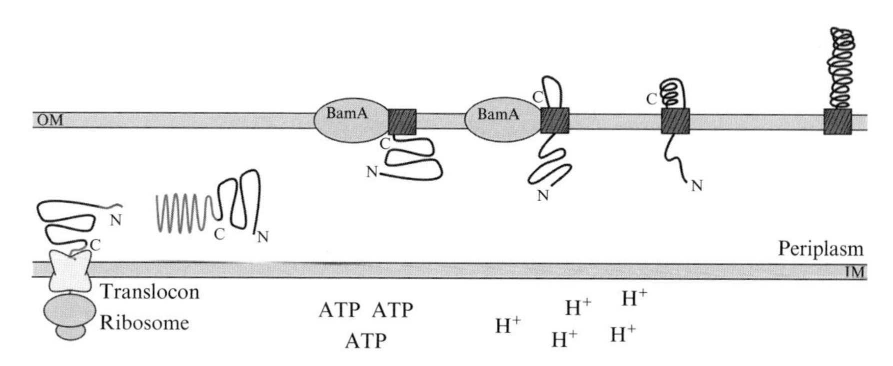

Figure 8.1 The autotransporter secretion pathway. The N-terminal signal sequence (pink) directs transport across the OM in an ATP dependent manner. The passenger domain (black) represents the mature folded virulence protein. The C-terminal porin domain (blue) is required for OM secretion. (See Color Insert.)

secretions in order to use ATP stored in the cytoplasm to drive transport across both membranes (Galan and Collmer, 1999). However, the most common OM secretion pathway is the autotransporter (AT; also called Type Va) secretion pathway, in which protein transport across the OM is energetically uncoupled from IM secretion (Remaut *et al.*, 2008). Each monomeric AT protein sequence consists of three functional parts: an N-terminal signal sequence that facilitates translocation across the IM via *sec*, a central passenger domain that will become the mature extracellular virulence protein, and a C-terminal OM porin domain essential for transport of the passenger domain across the OM (Henderson *et al.*, 1998, 2004; Pohlner *et al.*, 1987). There are three crystal structures of AT porin domains available; each consists of a 12-stranded transmembrane β-barrel fold (Barnard *et al.*, 2007; Oomen *et al.*, 2004; van den Berg, 2010).

The precise mechanism of AT OM secretion is currently the subject of much debate. Originally, it was proposed that AT passenger domains are transported across the OM through their own C-terminal porin domains (Loveless and Saier, 1997). But the inner diameter of this porin is too small (1–2 nm) to permit transport of a folded passenger domain (Barnard *et al.*, 2007; Oomen *et al.*, 2004; van den Berg, 2010). This suggests that if the passenger domain uses its own porin to reach the extracellular environment, it must travel through the porin in an unfolded conformation, folding only after it exits the cell. Yet in this model, it is not clear what keeps the passenger domain in an unfolded conformation as it transits the periplasm. Alternatively, others have suggested that the passenger domain could cross the OM through another, larger porin such as BamA, the essential and conserved OM porin known to facilitate OM insertion of other porins, including AT porin domains (Bernstein, 2007; Ieva and Bernstein, 2009;

Ieva *et al.*, 2008). Yet the inner diameter of a similar-sized OM porin, the 24-strand transmembrane β-barrel of the chaperone/usher system, is still too narrow to accommodate the folded structures of most AT passengers (Remaut *et al.*, 2008). Crucially, correct processing of most AT proteins, including their OM secretion, occurs in heterologous hosts, including nonpathogenic laboratory strains of *Escherichia coli* (Junker *et al.*, 2009; Renn and Clark, 2008), indicating that AT secretion is not dependent on other, strain-specific factors.

3. Overview of Cys-Loop Stalling

Disulfide bond-mediated stalling (Cys-loop stalling) of an AT passenger domain during its secretion across the OM has emerged as a powerful technique with which to investigate aspects of the AT OM secretion mechanism *in vivo*. The premise is as follows (Fig. 8.2): Two widely spaced (>40 aa) cysteine residues are introduced within the AT passenger domain sequence, at the position desired for OM secretion stalling. Upon entry into the periplasm, these Cys residues will be oxidized to form a disulfide bond, creating a long loop. Once this loop structure forms, OM secretion can begin, but stalls at a point corresponding to the location of the loop (Junker *et al.*, 2009). The long lifetime of these stalled intermediates in the OM enables diverse biochemical and biophysical assays of OM secretion, *in vivo*. For example, we and others have used this Cys-loop stalling approach to confirm that passenger domain structure formation can block the OM secretion process (Jong *et al.*, 2007; Junker *et al.*, 2009), determine the directionality of passenger domain secretion across the OM (Junker *et al.*, 2009), measure the extent of folding of the passenger domain during OM secretion (Junker *et al.*, 2009), and identify interactions between AT proteins and other components of the OM (Sauri *et al.*, 2009). Here we explain in detail how Cys-loop stalling of an AT passenger domain during its transport across the OM can be used to facilitate these and other structural studies of the OM secretion mechanism *in vivo*.

4. Architecture and Processing of AT Passenger Domains

To date, five AT passenger domain crystal structures have been published (Fig. 8.3; Emsley *et al.*, 1996; Gangwer *et al.*, 2007; Kenjale *et al.*, 2009; Otto *et al.*, 2005; van den Berg, 2010). Four include a long right-handed parallel β-helix, in agreement with the prediction that 97% of all AT passenger domains (now >1600 total) include β-helical structure

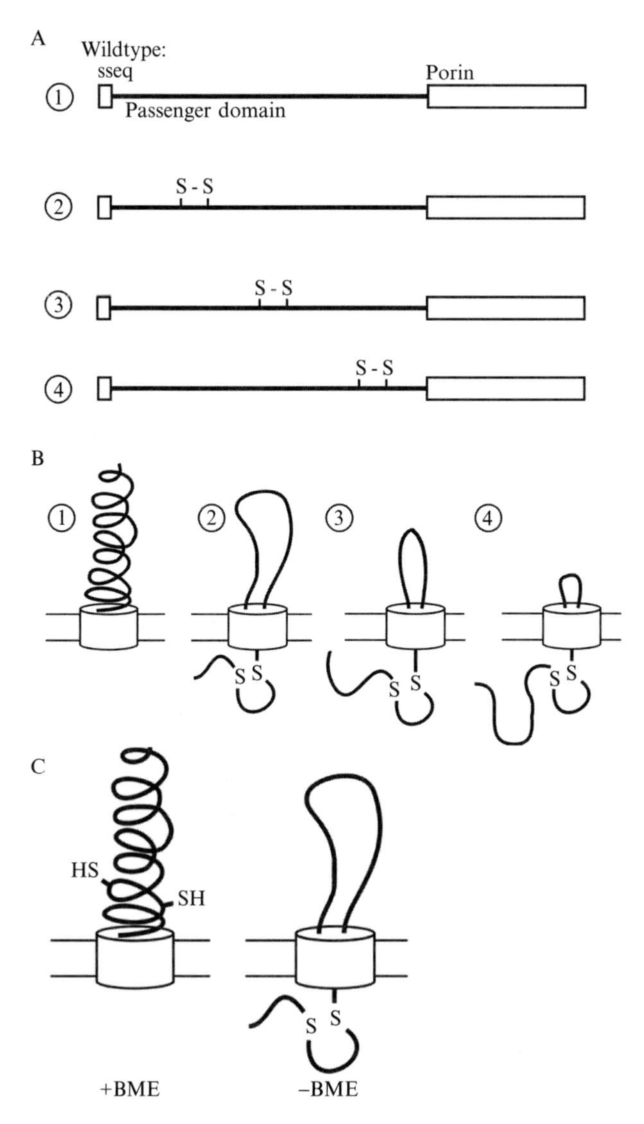

Figure 8.2 Schematic of Cys-loop constructs for disulfide bond-mediated passenger domain stalling. (A) Pairs were placed in either the N-terminus (2), middle (3), or C-terminus (4) of the passenger domain. (B) The predicted surface-exposed portions of the passenger domain as a result of the different placements of disulfide bond pairs from (A). (C) Reduction of the disulfide bond by BME releases the stalled construct and allows for secretion of the passenger domain. Adapted from (Junker *et al.*, 2009).

(Junker *et al.*, 2006). Combined, these results suggest that the β-helix is a common structural feature of AT passenger domains, particularly for AT passengers of average or greater length (> 500 aa). By contrast, the unusually

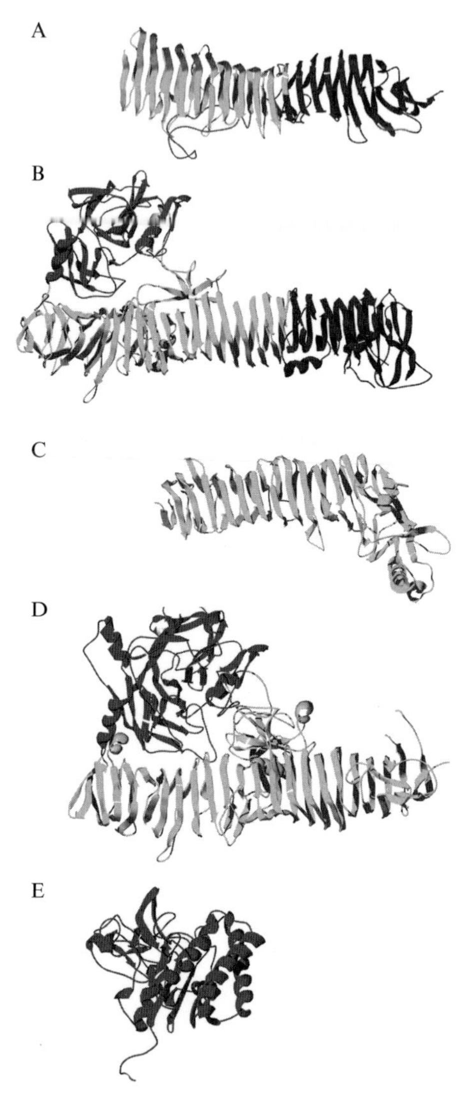

Figure 8.3 Crystal structure of AT passenger domains. There are five crystal structures available for AT passenger domains: (A) pertactin (PDB ID:1dab; Emsley *et al.*, 1996), (B) hemoglobin protease (1wxr; Otto *et al.*, 2005), (C) VacA (2qv3; Gangwer *et al.*, 2007), (D) IgA protease (3h09; Johnson *et al.*, 2009), and (E) EstA (3kvn; van den Berg, 2010) are shown. β-Helical structure is shown in green, although the C-terminal stable core structures of pertactin (Junker *et al.*, 2006) and the Hbp homolog Pet (Renn and Clark, 2008) are shown in blue. Non-β-helical domains containing AT virulence function are shown in red, while other non-β-helical structure is shown in orange. (See Color Insert.)

short passenger domain of EstA (306 aa) consists of only a globular domain (Fig. 8.3E). Once an AT passenger domain is secreted across the IM, it undergoes a series of processing steps (Fig. 8.1): the signal sequence is cleaved from the N-terminus, the C-terminal porin domain is inserted into the OM, and the passenger domain is transported to the outside surface of the cell. After OM secretion, the passenger domain is typically cleaved from the porin domain. Cleavage can occur either via an autocatalytic mechanism within the porin domain (such as for pertactin and the serine protease (SPATE) subfamily of ATs; Dautin *et al.*, 2007), or via an intrinsic protease activity within the passenger domain (such as IgA protease; Pohlner *et al.*, 1987), or via a separate OM protease such as NalP (which cleaves IgA protease and App; van Ulsen *et al.*, 2003). For some ATs (including pertactin, BrkA, and AIDA-I), the cleaved passenger domain remains non-covalently attached to the cell surface (Benz and Schmidt, 1992; Leininger *et al.*, 1991; Oliver *et al.*, 2003), while for others (including VacA and the SPATES), the cleaved passenger is released into the extracellular milieu (Cover and Blaser, 1992; Eslava *et al.*, 1998; Otto *et al.*, 1998). A small fraction of AT passenger domains (including Hia, EstA, and VirG/IcsA) do not require cleavage from their porin domain (St Geme and Cutter, 2000; Steinhauer *et al.*, 1999; van den Berg, 2010).

5. Heterologous Passenger Domain Secretion

Since its discovery, there have been attempts to exploit the AT secretion pathway as a novel mechanism to deliver heterologous passenger domains to the cell surface. Such an approach could be used for *in vitro* evolution, employing a random mutagenesis and selection strategy analogous to phage display. The first example of extracellular display of a heterologous passenger domain used the IgA protease AT from *Neisseria gonorrhoeae* (Klauser *et al.*, 1990, 1992). The 1094 aa IgA protease passenger domain was replaced by the 112 aa β-subunit of cholera toxin. Interestingly, this heterologous passenger domain sequence contained two Cys residues, at positions 9 and 86, and was secreted across the OM only when the reducing agent β-mercaptoethanol (BME) was added to the culture medium during protein expression. This result provided the first evidence that the formation of bulky, stable structure in the periplasm such as a disulfide-bonded loop could block OM secretion of the AT passenger domain. In another study, the non-β-helical domain 2 of Hbp (Fig. 8.3B) was replaced with calmodulin (Jong *et al.*, 2007). For this construct, addition of calcium to the growth medium stabilized the folded structure of calmodulin in the periplasm, and was sufficient to block OM secretion of the chimera; this effect was reversed by addition of the chelating agent

EGTA to the growth medium. Taken together, these results suggest that periplasmic formation of either a long disulfide-bonded loop or a tightly folded structure can be incompatible with OM secretion of the AT passenger domain. These results have key implications for the development of the Cys-loop stalling procedure, as described below.

6. Selecting a Model Autotransporter for Cys-Loop Stalling

The selection of an AT protein for Cys-loop stalling should be carefully considered, as some AT passenger domain sequences and structures are much better suited to this approach than others. Below we describe some general considerations for AT protein selection, and additional considerations that will depend on the specific experimental goal(s) (i.e., determining the directionality of OM secretion, measuring extracellular folding of the passenger domain during OM secretion, or identifying interactions between AT proteins and other components of the OM).

One important consideration is the fate of the wild-type passenger domain after OM secretion (see above). If the passenger domain is cleaved from the porin domain after its secretion across the OM, the change in molecular weight upon cleavage of the porin from the passenger can be used as a convenient readout for the completion of OM secretion, regardless of whether the passenger stays associated with the OM, or not (Junker *et al.*, 2009). If the passenger domain is released from the cell surface after OM secretion, the accumulation of processed passenger domain can be measured in the spent culture medium. Conversely, if the passenger domain remains noncovalently attached to the cell surface after OM secretion and cleavage from the porin domain, the accumulation of the processed passenger domain can be assayed in whole cell lysates or purified OM fractions. It is important to consider, however, that due to spatial and steric constraints, the accumulation of processed AT passenger domains retained on the cell surface might be lower than that for passengers released into the culture medium.

Another important consideration is the availability of a high-resolution structural model, as this will facilitate rational design of the locations of Cys mutations for Cys-loop stalling. Rational design guided by structure might, for example, prevent undesirable effects of the Cys residues on passenger domain folding, stability, or aggregation. Ideally, the Cys residues should be introduced at surface-exposed sites, substituting for similar-sized, polar amino acids such as serine.

An AT passenger domain structure will also reveal what types of structural subdomains comprise the passenger domain, if any. For example, while pertactin, Hbp, and IgA protease passenger domains each includes a long β-helix, each also includes other, non-β-helical structure, ranging from a long 34 aa loop in pertactin to entire globular domains in Hbp and IgA protease (Fig. 8.3; Johnson *et al.*, 2009; Otto *et al.*, 2005). A multidomain AT passenger might be an ideal choice for studies of heterologous passenger domain secretion. For numerous AT proteins, deletion or mutation of the C-terminal portion of the passenger domain, which is typically β-helical, can disrupt OM secretion (Dutta *et al.*, 2003; Oliver *et al.*, 2003). Conversely, many multidomain AT passengers have a globular N-terminal domain, and if this domain can be deleted without adversely affecting OM secretion (Renn and Clark, 2008), the non-β-helical domain could simply be replaced by the heterologous protein domain.

The overall length of the passenger domain is another point to consider. AT passenger domains range in size from ~200 to >4000 aa (Junker *et al.*, 2006). All other factors being equal, longer passenger domains will be secreted less efficiently than shorter passenger domains. If the desired experimental goal is to understand the OM secretion pathway, one approach is to introduce mutations designed to alter OM secretion efficiency. But if the wild-type passenger domain is not efficiently secreted, it might be difficult to analyze the effects of mutations that further reduce OM secretion efficiency. Also, longer passenger domains can reduce the time resolution for pulse–chase analyses of OM secretion (Sijbrandi *et al.*, 2003; Skillman *et al.*, 2005).

Many AT passenger domains have an enzymatic or binding activity, and if this activity can be conveniently assayed in liquid bacterial cultures (i.e., substrates or binding partners are commercially available or otherwise accessible, and AT activity is easily measured), an activity assay can provide a convenient mechanism to measure the amount of active AT protein secreted to the cell surface. A convenient activity assay can also facilitate both *in vitro* and *in vivo* measurements of the effects of point mutations, deletions, and the replacement of portions of the passenger domain with heterologous domains on the folding and secretion of the remainder of the passenger domain, assuming that the active site residues are still present.

Finally, some passenger domain sequences include endogenous Cys residues, which could affect the suitability for Cys-loop stalling. Potential complications arising from endogenous Cys residues are discussed below.

7. Disulfide Mediated Passenger Domain Stalling

7.1. Disulfide bond formation in the periplasm

DsbA is the primary enzyme responsible for disulfide bond formation in the periplasm (Tan and Bardwell, 2004). DsbA catalyzes the oxidation of the two cysteine sulfhydryls to form a covalent S–S bond linking the two amino acid side chains. Upon translocation of a secreted protein across the IM and into the periplasm, DsbA will form disulfide bonds between sequential Cys residues in the primary sequence, regardless of the number of amino acids separating the Cys residues or the proximity of the Cys residues to one another in the final folded structure (Rietsch *et al.*, 1996). Other proteins in the Dsb family play supporting roles: DsbB is involved in regenerating the disulfide bond in DsbA, and DsbC functions primarily as a disulfide bond isomerase, to rearrange these initially sequential but possibly nonnative disulfide bond pairings, for example, in proteins that form multiple nonsequential disulfide bonds (Bardwell, 1994).

7.2. Does the wild-type passenger domain sequence include cysteine residues?

The position of stalling during AT OM secretion can be controlled by introducing a single pair of widely spaced Cys residues at distinct points along the passenger domain sequence (Junker *et al.*, 2009). Yet, a significant number of wild-type AT passenger domains contain endogenous Cys residues, typically separated by up to 11 aa (Letley *et al.*, 2006). These endogenous Cys residues can compete with the Cys residues introduced for Cys-loop stalling, and cause stalling at undesired locations. In some cases, these endogenous Cys residues can be substituted with Ser with no measurable effect on OM secretion or folding or stability of the passenger domain (Renn and Clark, unpublished results). However, when the two Cys residues in the C-terminus of the *Helicobacter pylori* VacA AT passenger domain were substituted with Ser, lower levels of OM secretion were observed (Letley *et al.*, 2006). Hence, another important consideration when selecting a model AT protein for Cys-loop stalling is whether the wild-type AT passenger sequence includes Cys residues, and if so, whether these can be substituted with another amino acid without affecting OM secretion and/or folding and stability of the AT passenger.

7.3. Designing cysteine pairs for Cys-loop stalling

In a breakthrough study of Hbp by Luirink and coworkers, two sets of Cys pairs were introduced into the passenger domain (Jong *et al.*, 2007). One pair of Cys residues was separated by only 4 aa, while the Cys residues in the second pair were separated by 238 aa. While the pair of closely spaced Cys residues did not significantly affect OM secretion, the more widely spaced pair resulted in significantly reduced levels (~60%) of OM secretion. An extension of this approach was employed for Cys-loop mediated stalling of pertactin OM secretion: Three separate pairs of Cys residues were introduced into various positions along the passenger domain; in each pair, the individual Cys residues were separated by >40 aa, and resulted in 100% stalling of OM secretion (Junker *et al.*, 2009).

Junker *et al.* applied the following criteria when designing the Cys pairs for pertactin stalling: the amino acid residues selected for Cys pairs were >40 aa apart from one another, and surface exposed based on the passenger domain crystal structure (Junker *et al.*, 2009). But, as described above for Hbp, even widely spaced Cys pairs can be less than 100% effective at stalling AT OM secretion. Incomplete Cys-loop stalling can make it challenging to interpret subsequent analyses. For example, if Cys-loop formation produces only a partial blockade of OM secretion, fluorescence microscopy (see Protocol) cannot be used to unambiguously determine whether stalling has occurred at the OM, as the unstalled fraction will also produce a fluorescence signal at the cell surface. For this and other reasons, complete stalling is advantageous: it creates a homologous pool of stalled intermediates, which should simplify subsequent analyses of stalled intermediates. Hence, the effects of the Cys-loop on OM secretion efficiency should be quantified, and if complete stalling is not achieved, alternative Cys-loop locations should be investigated.

7.4. β-Mercaptoethanol-dependent stalling

It is important to remember that introducing a pair of Cys residues into the AT passenger domain could affect OM secretion levels for reasons other than those related to disulfide bond stalling. For this reason, it is important to perform control experiments to compare OM secretion levels of the wild-type AT with the Cys-loop construct under conditions where the disulfide bond cannot form, for example by expressing the wild type and Cys-containing constructs in culture media that includes 0.1 m*M* BME. BME increases the total reduction potential of the periplasm, preventing disulfide bond formation and permitting a direct comparison of the secretion efficiencies of the wild-type AT and the construct bearing Cys residues (Junker *et al.*, 2009).

Another important consideration is the effects of Cys-loop stalling on the OM secretion mechanism. Unstalled OM secretion is a dynamic process, whereas stalling for a prolonged period of time might enable the AT protein to reach an equilibrium or near-equilibrium conformation that would be inaccessible to the AT protein during unstalled OM secretion. Such an "off-pathway" conformation might not faithfully represent the conformation of the AT protein during unstalled OM secretion. Determining whether OM secretion proceeds normally when the Cys-loop mediated stalling is reversed can test whether the stalled intermediate remains on-pathway (Junker *et al.*, 2009). After stalling, BME can be added to the culture medium to reduce the AT passenger disulfide bond. If upon reduction of the disulfide bond the previously stalled AT protein is able to continue on with OM secretion (characterized by cleavage of the passenger from the porin and/or release of mature passenger into the culture media), this indicates that the Cys-loop stalled species represents a *bone fide*, on-pathway intermediate of the OM secretion process. Transferring the cells to fresh media lacking the AT expression inducer (IPTG, etc.) at the same time that BME is added can be helpful to insure that detected mature passenger arises from the formerly stalled preprotein, rather than from newly synthesized preprotein translated after the addition of BME (Junker *et al.*, 2009).

8. Methods to Measure OM Secretion and Folding of the Stalled AT Passenger

Once complete stalling of the passenger domain has been achieved, structural questions can be addressed, including the directionality of OM secretion, the extent of passenger domain folding that occurs during OM secretion, and interactions formed between the AT protein and other components of the OM and periplasm. Here, we present two convenient methods to interrogate the structure of the stalled passenger domain, and confirm its location on the outer surface of the cell: protease digestion and fluorescence microscopy (Fig. 8.4).

8.1. Procedure: Whole cell protease digestion

Protease digestion of whole cells expressing a stalled AT Cys-loop construct can be used to confirm the localization of the stalled AT passenger domain at the surface of the cell, and determine the extent of folding of the extracellular, exposed portion of the passenger (Junker *et al.*, 2009). Ideally, a nonspecific protease such as proteinase K (proK) or thermolysin should be used, to increase the likelihood that cleavage sites in the passenger domain will be accessible to the protease. ProK-resistant passenger domain fragments can be identified using in-gel trypsin digestion followed by MALDI–TOF mass spectrometry.

8.1.1. Preliminary considerations

The length of time that the *E. coli* cells are digested with protease must be kept as short as possible, to ensure that cell membranes remain intact during the digestion and subsequent manipulations. Prolonged digestion with high concentrations of proK could disrupt the integrity of the IM and/or OM. To test the integrity of the IM after digestion, Junker *et al.* used a pertactin AT passenger-only construct, lacking the signal sequence and porin domain: this AT construct accumulates only in the *E. coli* cytoplasm (Junker *et al.*, 2006). Cell cultures expressing this construct were digested with proK. No pertactin fragments were found in the digestion media, suggesting the proK digestion did not adversely affect IM integrity (Junker

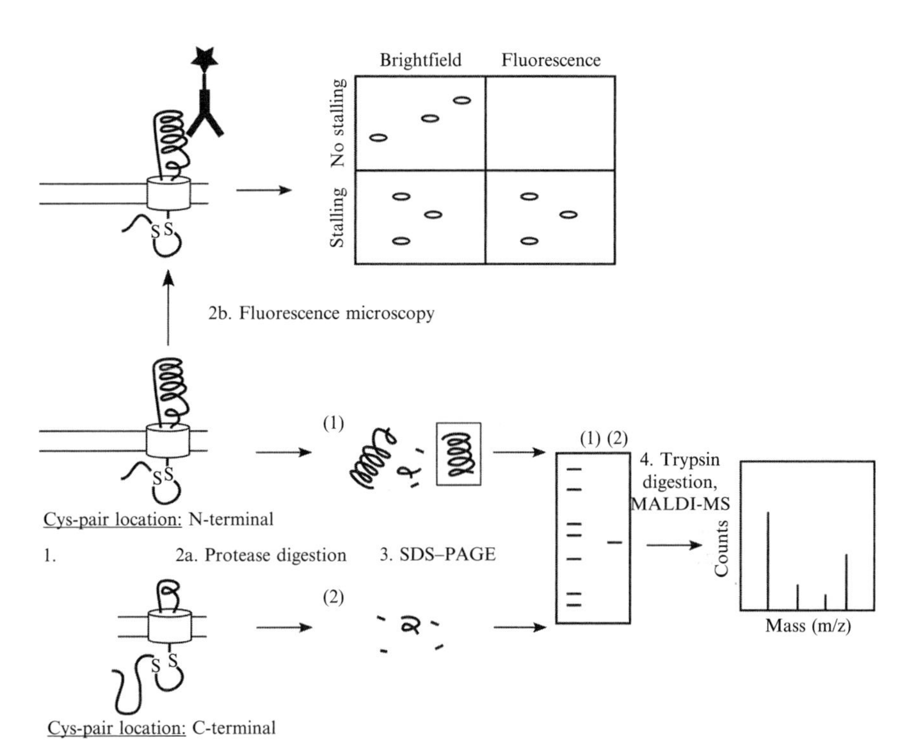

Figure 8.4 Analysis and results of stalled passenger domain. Stalling can be analyzed in two ways: whole cell protease digestion (2a) and fluorescence microscopy (2b). C-terminal stalling does not create a protease resistant fragment, however, N-terminal stalling does, which can be detected with SDS–PAGE (3). The location of the protease resistant fragment can be determined by MALDI–TOF MS (4). Alternatively, fluorescence microscopy can be used to detect stalled passenger domain (2b). Stalled passenger domain is detected with a fluorescent antibody. Cells will be fluorescent if stalled passenger is present.

et al., 2009); any abundant cytoplasmic marker protein could be used in a similar fashion. To test the integrity of the OM, release of a periplasmic marker protein such as maltose binding protein (MBP) into the digestion media can be used. MBP levels can be compared between digested and undigested cells to determine whether protease digestion has disrupted the OM, allowing periplasmic proteins to leak out (Junker *et al.*, 2009)—and correspondingly, allowing proK to enter the periplasm to digest the periplasmic face of the stalled AT protein, complicating the interpretation of digestion results. In addition, aliquots of cells digested with proK can be plated and the number of colony forming units (CFU) can be counted and compared to undigested cells to ensure that protease digestion has not affected cell viability. It is important to realize, however, that OM permeability might be compromised without any measurable effect on viability.

If the experimental objective is to identify stably folded passenger domain structures, the protease concentration must be high enough to distinguish between folded and unfolded conformations, yet still low enough to preserve membrane integrity (as measured by the controls described above).

Detailed protocol

1. Inoculate 1000 mL growth medium with 20 mL overnight culture of *E. coli* bearing an inducible plasmid encoding the AT model protein, with or without the Cys-loop mutations.
2. Culture with shaking at 37 °C until the culture reaches an optical density of 0.45–0.5 at 600 nm.
3. Induce with 0.5 m*M* IPTG (or other inducer) for 45 min. This short induction time allows for the accumulation of detectable quantities of the Cys-loop stalled preprotein, while avoiding prolonged induction-related cell stress.
4. Centrifuge culture at 5000×*g* for 10 min, then resuspend cells in 1 mL 50 m*M* Tris, pH 8.0, supplemented with 7.5 m*M* $CaCl_2$. Calcium is required for optimal proK activity.
5. Initiate digestion by adding freshly prepared proK stock solution to a final concentration of 0.1 mg/mL.
6. Digest cells for 10 min and pellet cells by centrifugation for 10 min at 21,000×*g*.
7. Remove the digestion supernatant and boil the cell pellet for 10 min in SDS loading buffer to inactivate proK.
8. Separate the digestion fragments by SDS–PAGE and visualize by western blotting, or Coomassie or silver staining. Laboratory strains of *E. coli* release few proteins into the culture medium, enabling detection of released fragments (Junker *et al.*, 2009).
9. For mass spectral analysis, use coomassie staining. Cut the band(s) of interest from the gel, dice each band into small pieces, and vortex for 10 min in 300 mL 25 m*M* ammonium carbonate with 50% acetonitrile.
10. Remove the liquid phase and repeat step 9 until all Coomassie stain is removed (as determined by visual inspection).
11. Dry the gel pieces in a vacuum centrifuge at 30 °C, then rehydrate the gel slices in a solution of trypsin (100 ng/mL) in 25 m*M* ammonium carbonate. Incubate overnight at 37 °C.
12. Remove the gel pieces from the digestion supernatant and extract the peptides from the gel pieces by vortexing and/or sonicating in 50% acetonitrile containing 5% formic acid.
13. Determine the mass of the extracted peptides by MALDI–TOF mass spectrometry.
14. Using a theoretical tryptic digestion of the passenger domain, map the identified tryptic peptide masses to the passenger domain sequence.

8.2. Procedure: Fluorescence microscopy

Confocal fluorescence microscopy can be used as an orthogonal approach to confirm that the passenger domain is located at the surface of the cell.

8.2.1. Preliminary conditions

The *E. coli* cells must not be fixed or permeabilized for the microscopy experiment. To reduce the possibility of cell lysis, cells must be centrifuged at slow speeds. Likewise, cells should be resuspended by gentle shaking, rather than pipetting up and down. All washing steps are essential, both to remove the autofluorescent LB growth medium and to wash away unbound antibodies. Thorough washing will reduce background fluorescence.

The proper amount of cells and antibodies should be used in each experiment. This is not an exact science: initially, serial dilutions of the cell cultures and primary and secondary antibodies should be used to determine appropriate dilution factors. Ideally, the microscopy fields will include numerous but distinctly individual cells, rather than clumps of cells.

In addition to the wild-type AT and stalled AT constructs, a positive control (such as an AT protein with a mutated cleavage site that stays covalently attached to its porin domain) and negative control (such as an empty vector or an AT passenger domain lacking the porin that cannot be secreted across the OM) should also be analyzed to calibrate the fluorescence intensity and cellular distribution measured for the stalled AT construct. In addition to the fluorescence image, a brightfield image of each field should also be recorded, to confirm the location and morphology of each cell. Identical microscope settings should be used for each sample, to facilitate comparisons.

Detailed protocol

1. Inoculate 25 mL of fresh LB media supplemented with 0.5 μg/mL ampicillin with 1 mL overnight culture.
2. Grow cells until the culture reaches an optical density of 0.45–0.5 at 600 nm.
3. Induce expression of the AT by addition of 0.5 m*M* IPTG to the culture and induce for 30 min.
4. Pellet the cells by centrifuge spinning at 1500×*g* for 5 min.
5. Remove the supernatant and resuspend the cells gently in 25 mL phosphate buffered saline (PBS).
6. Pellet cells and wash by resuspending in 25 mL PBS two times.
7. Place cover slips in a sterile petri dish.
8. Apply 125 μL of 0.1% (1 mg/mL) poly-L-lysine to each cover slip and incubate for 30–60 min at room temperature.
9. Remove excess poly-L-lysine by aspiration. Add 150 μL of PBS to each cover slip to wash. Aspirate to remove.

10. Apply 125 μL of diluted cell culture onto a poly-L-lysine-coated cover slip and incubate for 30 min.
11. Remove the culture by aspiration. Wash the cover slip three times with PBS as described in step 9.
12. Apply 100 μL of primary antibody (~1 μg/mL) to the cover slip and incubate for 10 min.
13. Remove the unbound primary antibody by aspiration and wash three times with 150 μL of PBS, as described in step 9.
14. Apply 100 μL of secondary antibody (e.g., a 1:100 dilution of Cy3-labeled antirabbit IgG) and incubate for 10 min.
15. Remove the secondary antibody and wash five times as in step 13.
16. Invert cover slip and place on the microscope slide.
17. Analyze cells with a fluorescence microscope (e.g., Applied Precision DeltaVision Core, 1.42 numerical aperture, 100× objective, 40 ms exposure time), and deconvolve using constrained iterative deconvolution (Junker *et al.*, 2009).

9. Applications: Using Cys-Loop Stalling to Define the Mechanism of AT OM Secretion

There are three proposed models for secretion of AT passengers across the OM. The "hairpin" model proposed that the passenger domain crosses the OM from C → N-terminus through its own C-terminal porin domain (Dautin and Bernstein, 2007; Junker *et al.*, 2009; Kostakioti and Stathopoulos, 2004, 2006; Pohlner *et al.*, 1987). The "threading" model also proposes that the passenger crosses the OM through its own porin domain, but from N → C-terminus (Dautin and Bernstein, 2007; Kostakioti and Stathopoulos, 2006). The N → C-terminus model is unlikely, as Cys-loop stalling has revealed that the C-terminus of the pertactin AT passenger domain appears first during secretion across the OM (Junker *et al.*, 2009). In addition, N → C-terminal OM secretion would require a mechanism to target the N-terminus of the AT passenger to the porin domain, and no such targeting sequence or motif has been identified. Alternatively, the "concerted" model suggests that the AT passenger domain does not cross the OM through its own pore, but is instead transported across the OM through the pore of the essential OM protein BamA (Ieva *et al.*, 2008; Nishimura *et al.*, 2010). It has been shown that BamA is required for AT OM transport (Ieva *et al.*, 2008; Jain and Goldberg, 2007), but this may be because BamA is required for insertion of the transmembrane OM porin domain (Voulhoux *et al.*, 2003), rather than playing a direct role in transporting the passenger domain across the OM. Other evidence supporting the concerted model includes indications that some folded structures, including long Cys-loop constructs can be secreted to various extents by a monomeric AT (Jong *et al.*, 2007; Skillman *et al.*, 2005), a recent study showing that Cys-loop stalled and slowly secreted mutants of the EspP and Hbp ATs can be cross-linked to BamA (Ieva and Bernstein, 2009; Sauri *et al.*, 2009). These recent results demonstrate Cys-loop stalling as a valuable approach to address structural questions regarding the AT OM secretion mechanism, one that is likely to contribute to the resolution of current controversies and ambiguities in the field.

REFERENCES

Bardwell, J. C. (1994). Building bridges: Disulphide bond formation in the cell. *Mol. Microbiol.* **14,** 199–205.

Barnard, T. J., *et al.* (2007). Autotransporter structure reveals intra-barrel cleavage followed by conformational changes. *Nat. Struct. Mol. Biol.* **14,** 1214–1220.

Benz, I., and Schmidt, M. A. (1992). AIDA-I, the adhesin involved in diffuse adherence of the diarrhoeagenic *Escherichia coli* strain 2787 (O126:H27), is synthesized via a precursor molecule. *Mol. Microbiol.* **6,** 1539–1546.

Bernstein, H. D. (2007). Are bacterial 'autotransporters' really transporters? *Trends Microbiol.* **15,** 441–447.

Cover, T. L., and Blaser, M. J. (1992). Purification and characterization of the vacuolating toxin from *Helicobacter pylori*. *J. Biol. Chem.* **267,** 10570–10575.

Dautin, N., and Bernstein, H. D. (2007). Protein secretion in Gram-negative bacteria via the autotransporter pathway. *Annu. Rev. Microbiol.* **61,** 89–112.

Dautin, N., *et al.* (2007). Cleavage of a bacterial autotransporter by an evolutionarily convergent autocatalytic mechanism. *EMBO J.* **26,** 1942–1952.

Dutta, P. R., *et al.* (2003). Structure–function analysis of the enteroaggregative *Escherichia coli* plasmid-encoded toxin autotransporter using scanning linker mutagenesis. *J. Biol. Chem.* **278,** 39912–39920.

Emsley, P., *et al.* (1996). Structure of *Bordetella pertussis* virulence factor P.69 pertactin. *Nature* **381,** 90–92.

Eslava, C., *et al.* (1998). Pet, an autotransporter enterotoxin from enteroaggregative *Escherichia coli*. *Infect. Immun.* **66,** 3155–3163.

Galan, J. E., and Collmer, A. (1999). Type III secretion machines: Bacterial devices for protein delivery into host cells. *Science* **284,** 1322–1328.

Gangwer, K. A., *et al.* (2007). Crystal structure of the *Helicobacter pylori* vacuolating toxin p55 domain. *Proc. Natl. Acad. Sci. USA* **104,** 16293–16298.

Henderson, I. R., *et al.* (1998). The great escape: Structure and function of the autotransporter proteins. *Trends Microbiol.* **6,** 370–378.

Henderson, I. R., *et al.* (2004). Type V protein secretion pathway: The autotransporter story. *Microbiol. Mol. Biol. Rev.* **68,** 692–744.

Ieva, R., and Bernstein, H. D. (2009). Interaction of an autotransporter passenger domain with BamA during its translocation across the bacterial outer membrane. *Proc. Natl. Acad. Sci. USA* **106,** 19120–19125.

Ieva, R., *et al.* (2008). Incorporation of a polypeptide segment into the beta-domain pore during the assembly of a bacterial autotransporter. *Mol. Microbiol.* **67,** 188–201.

Jain, S., and Goldberg, M. B. (2007). Requirement for YaeT in the outer membrane assembly of autotransporter proteins. *J. Bacteriol.* **189,** 5393–5398.

Johnson, T. A., *et al.* (2009). Active-site gating regulates substrate selectivity in a chymotrypsin-like serine protease: The structure of *Haemophilus influenzae* immunoglobulin A1 protease. *J. Mol. Biol.* **389,** 559–574.

Jong, W. S., *et al.* (2007). Limited tolerance towards folded elements during secretion of the autotransporter Hbp. *Mol. Microbiol.* **63,** 1524–1536.

Junker, M., *et al.* (2006). Pertactin beta-helix folding mechanism suggests common themes for the secretion and folding of autotransporter proteins. *Proc. Natl. Acad. Sci. USA* **103,** 4918–4923.

Junker, M., *et al.* (2009). Vectorial transport and folding of an autotransporter virulence protein during outer membrane secretion. *Mol. Microbiol.* **71,** 1323–1332.

Kenjale, R., *et al.* (2009). Structural determinants of autoproteolysis of the *Haemophilus influenzae* Hap autotransporter. *Infect. Immun.* **77,** 4704–4713.

Klauser, T., *et al.* (1990). Extracellular transport of cholera toxin B subunit using Neisseria IgA protease beta-domain: Conformation-dependent outer membrane translocation. *EMBO J.* **9,** 1991–1999.

Klauser, T., *et al.* (1992). Selective extracellular release of cholera toxin B subunit by *Escherichia coli*: Dissection of Neisseria Iga beta-mediated outer membrane transport. *EMBO J.* **11,** 2327–2335.

Kostakioti, M., and Stathopoulos, C. (2004). Functional analysis of the Tsh autotransporter from an avian pathogenic *Escherichia coli* strain. *Infect. Immun.* **72,** 5548–5554.

Kostakioti, M., and Stathopoulos, C. (2006). Role of the alpha-helical linker of the C-terminal translocator in the biogenesis of the serine protease subfamily of autotransporters. *Infect. Immun.* **74,** 4961–4969.

Leininger, E., *et al.* (1991). Pertactin, an Arg-Gly-Asp-containing *Bordetella pertussis* surface protein that promotes adherence of mammalian cells. *Proc. Natl. Acad. Sci. USA* **88,** 345–349.

Letley, D. P., *et al.* (2006). Paired cysteine residues are required for high levels of the *Helicobacter pylori* autotransporter VacA. *Microbiology* **152,** 1319–1325.

Loveless, B. J., and Saier, M. H., Jr. (1997). A novel family of channel-forming, autotransporting, bacterial virulence factors. *Mol. Membr. Biol.* **14,** 113–123.

Mansell, T. J., *et al.* (2008). Engineering the protein folding landscape in gram-negative bacteria. *Curr. Protein Pept. Sci.* **9,** 138–149.

Nishimura, K., *et al.* (2010). Autotransporter passenger proteins: Virulence factors with common structural themes. *J. Mol. Med.* **88,** 451–458.

Oliver, D. C., *et al.* (2003). A conserved region within the *Bordetella pertussis* autotransporter BrkA is necessary for folding of its passenger domain. *Mol. Microbiol.* **47,** 1367–1383.

Oomen, C. J., *et al.* (2004). Structure of the translocator domain of a bacterial autotransporter. *EMBO J.* **23,** 1257–1266.

Otto, B. R., *et al.* (1998). Characterization of a hemoglobin protease secreted by the pathogenic *Escherichia coli* strain EB1. *J. Exp. Med.* **188,** 1091–1103.

Otto, B. R., *et al.* (2005). Crystal structure of hemoglobin protease, a heme binding autotransporter protein from pathogenic *Escherichia coli*. *J. Biol. Chem.* **280,** 17339–17345.

Pohlner, J., *et al.* (1987). Gene structure and extracellular secretion of *Neisseria gonorrhoeae* IgA protease. *Nature* **325,** 458–462.

Rego, A. T., *et al.* (2010). Two-step and one-step secretion mechanisms in gram-negative bacteria: Contrasting the type IV secretion system and the chaperone–usher pathway of pilus biogenesis. *Biochem. J.* **425,** 475–488.

Remaut, H., *et al.* (2008). Fiber formation across the bacterial outer membrane by the chaperone/usher pathway. *Cell* **133,** 640–652.

Renn, J. P., and Clark, P. L. (2008). A conserved stable core structure in the passenger domain beta-helix of autotransporter virulence proteins. *Biopolymers* **89,** 420–427.

Rietsch, A., *et al.* (1996). An in vivo pathway for disulfide bond isomerization in *Escherichia coli*. *Proc. Natl. Acad. Sci. USA* **93,** 13048–13053.

Saier, M. H., Jr. (2006). Protein secretion and membrane insertion systems in gram-negative bacteria. *J. Membr. Biol.* **214,** 75–90.

Sauri, A., *et al.* (2009). The Bam (Omp85) complex is involved in secretion of the autotransporter haemoglobin protease. *Microbiology* **155,** 3982–3991.

Savage, D. C. (1977). Microbial ecology of the gastrointestinal tract. *Annu. Rev. Microbiol.* **31,** 107–133.

Sijbrandi, R., *et al.* (2003). Signal recognition particle (SRP)-mediated targeting and Sec-dependent translocation of an extracellular *Escherichia coli* protein. *J. Biol. Chem.* **278,** 4654–4659.

Skillman, K. M., *et al.* (2005). Efficient secretion of a folded protein domain by a monomeric bacterial autotransporter. *Mol. Microbiol.* **58,** 945–958.

St Geme, J. W., 3rd, and Cutter, D. (2000). The *Haemophilus influenzae* Hia adhesin is an autotransporter protein that remains uncleaved at the C terminus and fully cell associated. *J. Bacteriol.* **182,** 6005–6013.

Steinhauer, J., *et al.* (1999). The unipolar *Shigella* surface protein IcsA is targeted directly to the bacterial old pole: IcsP cleavage of IcsA occurs over the entire bacterial surface. *Mol. Microbiol.* **32,** 367–377.

Tan, J. T., and Bardwell, J. C. (2004). Key players involved in bacterial disulfide-bond formation. *Chembiochem* **5,** 1479–1487.

Thanassi, D. G., *et al.* (2005). Protein secretion in the absence of ATP: The autotransporter, two-partner secretion and chaperone/usher pathways of Gram-negative bacteria (review). *Mol. Membr. Biol.* **22,** 63–72.

van den Berg, B. (2010). Crystal structure of a full-length autotransporter. *J. Mol. Biol.* **396,** 627–633.

van Ulsen, P., *et al.* (2003). A *Neisserial* autotransporter NalP modulating the processing of other autotransporters. *Mol. Microbiol.* **50,** 1017–1030.

Voulhoux, R., *et al.* (2003). Role of a highly conserved bacterial protein in outer membrane protein assembly. *Science* **299,** 262–265.

CHAPTER NINE

Strategies for the Thermodynamic Characterization of Linked Binding/Local Folding Reactions Within the Native State: Application to the LID Domain of Adenylate Kinase from *Escherichia coli*

Travis P. Schrank,* W. Austin Elam,† Jing Li,† *and* Vincent J. Hilser†,‡

Contents

Abstract

Conformational fluctuations in proteins have emerged as an important aspect of biological function, having been linked to processes ranging from molecular recognition and catalysis to allostery and signal transduction. In spite of the

* Department of Biochemistry and Molecular Biology, University of Texas Medical Branch, Galveston, Texas, USA
† T.C. Jenkins Department of Biophysics, Johns Hopkins University, Baltimore, Maryland, USA
‡ Department of Biology, Johns Hopkins University, Baltimore, Maryland, USA

Methods in Enzymology, Volume 492
ISSN 0076-6879, DOI: 10.1016/B978-0-12-381268-1.00020-3

© 2011 Elsevier Inc.
All rights reserved.

realization of their importance, however, the connections between fluctuations and function have largely been empirical, even when they have been quantitative. Part of the problem in understanding the role of fluctuations in function is the fact that the mere existence of fluctuations complicates the interpretation of classic mutagenesis approaches. Namely, mutagenesis, which is typically targeted to an internal position (to elicit an effect), will change the fluctuations as well as the structure of the native state. Decoupling these effects is essential to an unambiguous understanding of the role of fluctuations in function. Here, we use a mutation strategy that targets surface-exposed sites in flexible parts of the molecule for mutation to glycine. Such mutations leave the ground-state structure unaffected. As a result, we can assess the nature of the fluctuations, develop a quantitative model relating fluctuations to function (in this case, molecular recognition), and unambiguously resolve the probabilities of the fluctuating states. We show that when this approach is applied to *Escherichia coli* adenylate kinase (AK), unique thermodynamic and structural insights are obtained, even when classic mutagenesis approaches targeted to the same region yield ambiguous results.

1. Introduction

It is now widely accepted that the native state of proteins is a dynamic and heterogeneous collection of conformations (i.e., an ensemble) that surround the average structure of the molecule, usually typified by the crystallographic structure. Dynamic excursions of proteins to excited (minor) states have been associated with biochemically important processes such as protein folding, catalysis, and binding (Korzhnev *et al.*, 2006; Manson *et al.*, 2009; Wolf-Watz *et al.*, 2004). However, relatively few studies give insight into the physical/conformational character of the states that make up these functionally important dynamics (Vallurupalli *et al.*, 2007). One particularly important experimental approach that has emerged over the past two decades has been the use of hydrogen–deuterium exchange to investigate exchange-competent states in the native state ensembles of proteins (Bai, 2006; Krishna *et al.*, 2004; Rogero *et al.*, 1986). Although several models for exchange have been forwarded (Miller and Dill, 1995), with the precise origin of exchange for all positions in the protein unresolved (LeMaster *et al.*, 2009), denaturant-dependent exchange experiments strongly suggest that exchange occurs as a result of local unfolding reactions that expose buried or hydrogen bonded amides to solvent (Englander, 1975).

Consistent with this view, a computational model of native state fluctuations that is based on local unfolding (i.e., the COREX algorithm) has also had success reconciling many other aspects of protein behavior, including allosteric effects of ligand binding, noncooperative protein unfolding at low temperatures, and the interaction of the effects of pH and denaturants in determining the apparent *m*-value of unfolding (Babu *et al.*, 2004; Liu *et al.*, 2007; Pan *et al.*, 2000; Whitten *et al.*, 2001). These results, taken in total, suggest that local unfolding to highly disordered states may be a ubiquitous property of the native state ensembles of most, if not all, proteins. Nevertheless, the agreement between COREX and experiment, although highly suggestive, cannot be taken as unambiguous proof of local unfolding. Indeed, there can be numerous models that are consistent with the data (Miller and Dill, 1995). To establish that local unfolding is present in the ensembles of proteins and plays a functionally relevant role, experimental strategies are needed that will directly challenge local unfolding relative to other types of conformational excursions.

The low probability of excited states is a primary difficulty in the experimental investigation of native state fluctuations (Baldwin and Kay, 2009). Many spectroscopic techniques (such as fluorescence or circular dichroism) commonly used to investigate proteins report average properties of the molecule in solution, and therefore cannot easily detect minor populations. Significant progress has been made toward this end through the development of modern NMR relaxation–dispersion methods, although these methods too have a lower population limit of detectability ($\sim$0.5%; Baldwin and Kay, 2009). Given the experimental difficulties in detecting states with low probability, it would be useful to develop strategies that increase the population of such states, but doing so in a way that does not complicate the interpretation of experimental results.

To address these challenges, we have implemented a strategy that specifically targets mutational effects so that they are manifested as changes in the fluctuations around the ground-state structure and not as changes to the ground-state structure itself. Specifically, we utilize an entropy-enhancing mutation strategy that promotes conformational excursions to locally unfolded states, and we monitor the effects of the mutations to the thermodynamic properties of measurable binding reactions. Numerous studies have demonstrated that positions on the periphery of the binding sites are often among the most dynamic and unstable regions within the protein (Ferreon and Hilser, 2003a; Henzler-Wildman *et al.*, 2007; Rundqvist *et al.*, 2009). Not surprisingly, the dynamics of these unstable regions is also often found to be modulated by binding (Boehr *et al.*, 2010; Ferreon and Hilser, 2003a). Because the approach we describe below specifically addresses the coupling between the fluctuations and binding, it provides a potential tool for probing the coupling between binding and fluctuations in many protein systems. Indeed, our laboratory has previously applied this strategy to successfully study the binding and fluctuations of the RT-loop of the SEM-5 SH3 domain, as well as the LID domain of adenylate kinase (AK) from *Escherichia coli* (Ferreon *et al.*, 2004; Manson *et al.*, 2009; Schrank *et al.*, 2009).

2. A Mutation Strategy to Amplify Locally Unfolded States

Although local unfolding appears to be a ubiquitous property of folded proteins, the probabilities of individual locally unfolded states are, in general, relatively small. As discussed above, mutational strategies that will selectively target such states are of great experimental utility. However, classic mutagenesis studies are targeted toward disruption of protein structure, a strategy that carries the risk of spurious changes that can greatly compromise the interpretation of results, especially when the point of the study is to clarify the structure and energetics of fluctuations. This point is highlighted in Fig. 9.1, which shows how typical mutation studies change the ground-state structure of a protein. In the example, a mutation which removes an interior group in the protein necessarily changes the ground-state structure and energy relative to the wild type (Fig. 9.1A). Such mutation may also change the number and energy of excited states, but deconvolution of the relative effects is clearly not straight-forward. An alternative ideal case would involve making mutations that do not change the ground-state structure of the molecule, but instead change the probability of fluctuations (Fig. 9.1B). Such an approach would greatly simplify the interpretation of results and would allow quantitative access to the energetics associated with the fluctuations. Although this ideal experiment

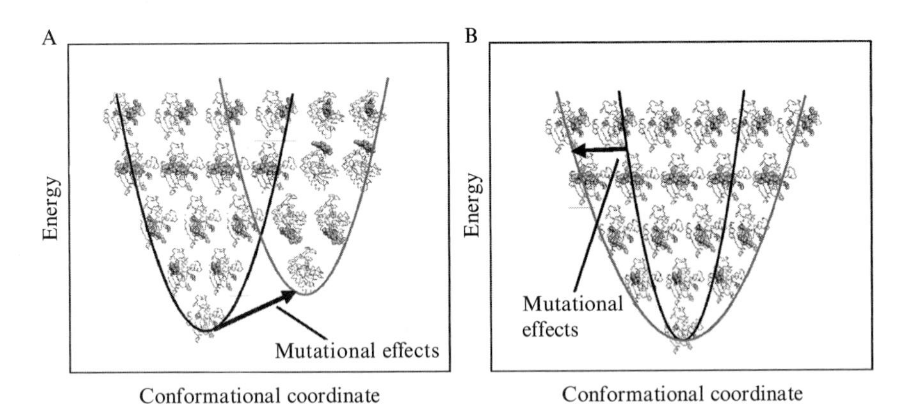

Figure 9.1 Expected effects of contact-based (classical) versus entropy based mutations on the native state conformational energy landscape. *Black*: The conformational space available to the wild-type protein. *Gray*: The conformational space available to the mutant protein. (A) The expected changes in a mutation designed to perturb contacts within the native fold of the molecule. (B) The idealized effect of the surface-exposed Gly mutation, which only perturbs the conformational entropy of excited conformational states and leaves the ground-state fold unaffected.

is not possible, we utilize an approach that most closely approximates these conditions—the introduction of conformational entropy-enhancing mutations through substitution of surface-exposed residues with glycine (Gly).

To see how such a mutation strategy achieves this objective, we consider the following. In unfolded states, proteins and peptides have multiple conformational degrees of freedom due to the presence of the numerous rotatable bonds. For the majority of amino acids, which contain a β-carbon, the available conformational space is somewhat limited relative to Gly, which contains only an hydrogen (H). This limited conformational space arises because many (in fact most) conformations accessible to Gly are sterically occluded when the H of Gly is substituted with a more bulky side chain (D'Aquino *et al.*, 1996). This principle is demonstrated in Fig. 9.2. A simulation was performed by sampling a uniform statistical distribution of φ and ψ angles for all residues in an AAAAA or AAGAA peptide, and constructing an ensemble of peptide conformations based on these angles and other well-known geometric constraints of polypeptides (Manson *et al.*, 2009; Whitten *et al.*, 2008). All conformers producing steric collisions were then discarded. The φ and ψ angles of the central residue in the sterically allowed space are displayed as a histogram in Fig. 9.2. Large differences between Ala and Gly can be appreciated, as previously demonstrated from the classical Ramachandran map (Ramakrishnan and Ramachandran, 1965). Using this simulation, the ratio of available conformational space for the Ala or Gly containing peptide can be calculated as (Leach *et al.*, 1966),

$$\Omega_{\text{Ala-Gly}} = \frac{n_{\text{allowed,Ala}} / n_{\text{test,Ala}}}{n_{\text{allowed,Gly}} / n_{\text{test,Gly}}}, \tag{9.1}$$

where n_{allowed} is the number of conformers passing the steric check, and n_{test} is the total number of randomly sampled conformations. Accordingly, the expected change in entropy of an unfolded state affected by the mutation can then be estimated by using the Boltzmann entropy equation (Leach *et al.*, 1966),

$$\Delta\Delta S_{\text{Ala-Gly}} = -R\ln(\Omega_{\text{Ala-Gly}}). \tag{9.2}$$

According to this simulation, the entropy difference of unfolding an Ala versus a Gly residue is 2.2 cal/mol K (Table 9.1), similar to values measured from calorimetric protein unfolding experiments using Ala/Gly mutant pairs (2.3–2.7 cal/mol; D'Aquino *et al.*, 1996) and other computational methods (D'Aquino *et al.*, 1996; Leach *et al.*, 1966). Therefore, based on the constraints imposed by molecular geometry, Ala to Gly mutations should invariably decrease the stability through enhancement of the entropy of the

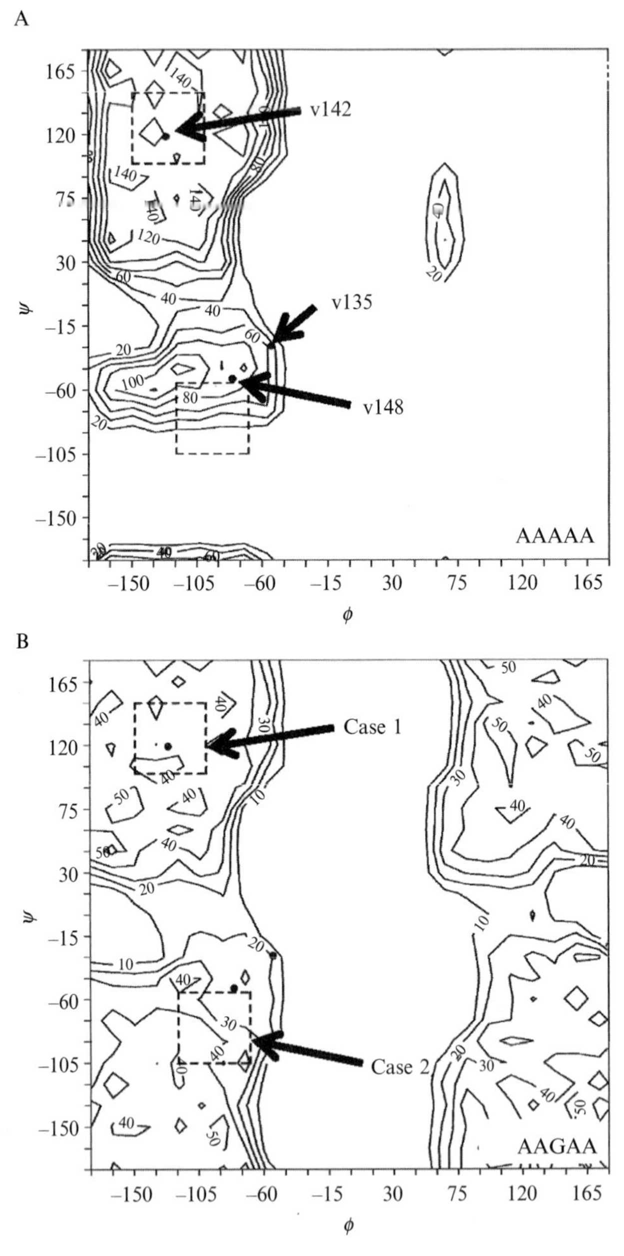

Figure 9.2 Hard sphere collision simulation of the relative conformational space available to Ala and Gly in a disordered peptide. All simulations were performed using the MPMOD program (Manson *et al.*, 2009; Whitten *et al.*, 2008). Conformers were generated until 15,000 sterically allowed (Ponder and Richards, 1987; Ramachandran and Sasisekharan, 1968; Ramakrishnan and Ramachandran, 1965)

Table 9.1 Hard sphere collision approximation of the relative conformational space available to Ala and Gly in a disordered peptide

Test space (Φ, Ψ)	$\Omega_{\text{Ala-Gly}}$, $\Delta\Delta S_{\text{a,g}}$ (cal/mol K)	$\Omega_{\text{Val-Gly}}$, $\Delta\Delta S_{\text{v,g}}$ (cal/mol K)
All available ($0 \leftrightarrow 360$, $0 \leftrightarrow 360$)	3.03, 2.20	4.28, 2.89
Case-1 ($-150 \leftrightarrow -100$, $100 \leftrightarrow 150$)	1.08, 0.14	1.16, 0.29
Case-2 ($-120 \leftrightarrow -70$, $-75 \leftrightarrow -25$)	3.09, 2.24	7.27, 3.94

All simulations were performed using the MPMOD algorithm (Manson *et al.*, 2009; Whitten *et al.*, 2008). Equal probabilities were assumed for all valine rotamers. $\Omega_{\text{Ala-Gly}}$ is as defined in Eq. (9.1). Van der Waals radii were taken from published values (Ponder and Richards, 1987; Ramachandran and Sasisekharan, 1968; Ramakrishnan and Ramachandran, 1965).

unfolded state. Indeed, several authors have used this robust feature of protein folding as a tool for protein destabilization (Huyghues-Despointes *et al.*, 1999; Maity *et al.*, 2003).

Thus, the mutation strategy is expected to promote conformational excursions to highly dissimilar, that is disordered conformational states. To illustrate this point, the above simulation was again used to test the relative availability of conformational space within a given widow of φ/ψ space. Interestingly, such a calculation reveals that the density of available (steric) conformations is similar for an Ala or Gly residing in a φ/ψ space accessible to both amino acid types (e.g., *Case-1*; Fig. 9.2; Table 9.1). If the conformation of a mutated position is accessible to both Ala and Gly, the entropy of the folded ground state (a range of available φ–ψ space compatible with the compact fold) should be minimally affected (see *Case-1*; Table 9.1). A mutation of a residue in a β-sheet (see position of Valine 142, Fig. 9.2), for instance, is expected to provide a relatively isolated entropic stabilization to states that can access distant regions of φ/ψ space (in this case, greater than 90 ° in rotation around one or both dihedral angles). Others have noted that the entropy of an Ala or Gly confined to an α-helix should be similar in conformational entropy (D'Aquino *et al.*, 1996). However, mutation of residues that reside near an overlapping boundary of

conformations were found for each peptide (AAXAA) in question. The number of attempts required to achieve this goal was recorded for each peptide, as well as the φ and ψ angles of each acceptable conformer. Slightly reduced atomic radii were applied to the peptide backbone to increase sampling efficiency (Ponder and Richards, 1987). (A, B) Contour histogram of acceptable conformations of Gly and Ala. *Case-1*: A space accessible to a folded residue that will have similar conformational freedom if mutated from Val or Ala to Gly. *Case-2*: A space accessible to a folded residue that may have different conformational freedom if mutated from Val or Ala to Gly (an example of a nonideal mutation site). *Points*: Favorable folded conformations of mutation sites selected for AK, PDB ID 1AKE (Müller and Schulz, 1992).

the available Ala and Gly spaces (*Case-2*; Fig. 9.2, Table 9.1) might allow access to new states similar in conformation to the folded (i.e., crystallographic) state. The magnitude of this effect in terms of entropy will be highly dependent on the conformational envelope of the state in question. In short, considerable data support the notion that Ala (or Val) to Gly mutations will robustly effect an increase in the conformational entropy of unfolded or highly disordered states, thus destabilizing the conformationally constrained folded state. Since the relevant changes are intrinsic to the unfolded state itself, the mutational effects should be realized for any mutation site, independent of structural context.

Based on such considerations, we have designed (Val to Gly) mutations to promote disordered states in highly dynamic structures, in an effort to increase the population of highly disordered or unfolded states involving these structures. To ensure that the mutations closely resembled the ideal strategy described in Fig. 9.1, the selected Val residues were surface exposed, with the side chain involved in few to no molecular contacts. This strategy allows for destabilization without perturbation to the conformation of the folded state, a condition that minimizes undesirable/confounding changes in structure.

To illustrate the extensive thermodynamic information that can be obtained from such an approach, we will summarize the results of a recent ITC analysis of three mutations of AK from *E. coli* (v142g, v135g, v148g). The highly exposed location of the mutation sites, as well as the structural relationship to the binding cleft, is depicted in Fig. 9.3. The positions of the mutated residues in φ/ψ space are also indicated in Fig. 9.2. As a counter example, we also present data gathered for a mutation of the same region that is designed to destabilize the *folded* conformation by eliminating a native state hydrogen bond (s129a), a common strategy used to probe mutational effects.

3. Thermodynamic Properties of Linked Folding and Binding Reactions

Thermodynamically linked folding (local or global) and binding has been found to be a property of many proteins (Cliff *et al.*, 2005; Wright and Dyson, 2009). In such cases, ITC binding experiments become a powerful tool to investigate the folding reaction. The simplest conformational model that can be used to describe any linked folding and binding reaction, is the model where the (locally) unfolded state has little or no affinity for the ligand, or is *binding incompetent* (BI), and is in equilibrium with a folded, high affinity or *binding competent* (BC) state. Eftink *et al.* term this model a "mandatory coupling model," to emphasize that binding occurs with only

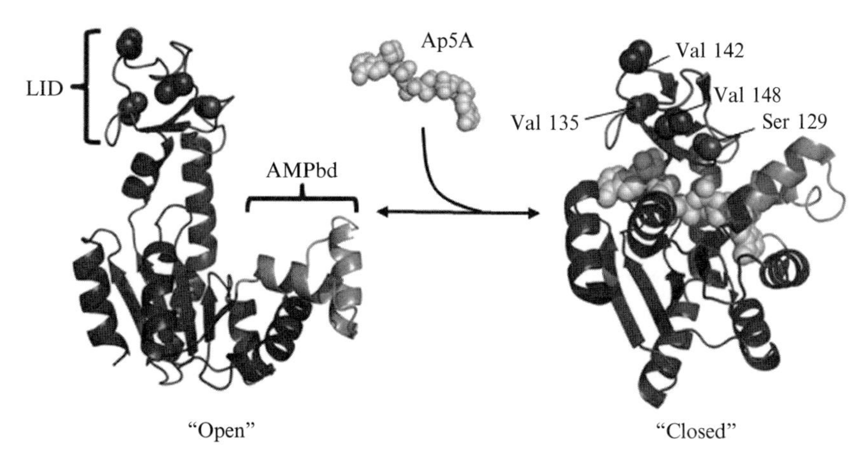

Figure 9.3 Mutational strategy and binding associated conformational changes in AK. AK from *E. coli* is a much studied enzyme that catalyzes the reaction 2ADP ↔ ATP + AMP. *Open*: Graphic of Apo-AK, PDB ID 4AKE (Müller *et al.*, 1996). *Ap5A*: P^1,P^5-di (adenosine-5′) pentaphosphate, a nonhydrolysable bisubstrate analog inhibitor of AK. *Closed*: Graphic of the Ap5A/AK complex, PDB ID 1AKE (Müller and Schulz, 1992). *LID*: The "LID domain" is a highly dynamic structural appendage associated with the ATP binding site. *AMPbd*: The "AMP binding domain" is a highly dynamic structural appendage associated with the AMP binding site. Definitions of these regions have been previously published (Shapiro *et al.*, 2000). *Spheres*: Indicate the location of selected Val (or serine) to Gly mutations sites. Adapted from previously published results (Schrank *et al.*, 2009).

one of the two possible states (Eftink *et al.*, 1983). In this case, the addition of ligand will promote the population of the BC state through mass action. The impact of such a model on direct (ΔH and ΔG) and indirect (ΔC_p) calorimetric observables of binding has been published (Eftink *et al.*, 1983).

According to this model, the apparent (measured) binding constant will have the form

$$K_{app}(T) = \frac{[BCX]}{[BC + BI][X]}. \tag{9.3}$$

Since the equilibrium between the BC and BI states can be written as $K_{conf}(T) = [BI]/[BC]$, Eq. (9.3) reduces to

$$K_{app}(T) = \frac{K_0(T)}{(1 + K_{conf}(T))}, \tag{9.4}$$

where $K_0(T)$ is the intrinsic association constant between the BC state and the ligand.

$$K_0(T) = \frac{[\mathrm{BCX}]}{[\mathrm{BC}][\mathrm{X}]}. \tag{9.5}$$

Considering Eqs. (9.3)–(9.5), it is clear that an increase in the probability of the BI should decrease the measured binding affinity of the system. The free energy of binding will consist of two terms,

$$\Delta G_{\mathrm{app}}(T) = -RT\ln K_0(T) + RT\ln(1 + K_{\mathrm{conf}}(T)), \tag{9.6}$$

$$\Delta G_{\mathrm{app}}(T) = \Delta G_0(T) - \Delta G_{\mathrm{conf,app}}(T), \tag{9.7}$$

where the first term is the intrinsic free energy of interaction between the BC state and the ligand, and the second term is the apparent conformational free energy contributed by the equilibrium between the BI and BC states (Eftink *et al.*, 1983). We note that in Eq. (9.6), the term

$$K_{\mathrm{conf}}(T) = \exp(\Delta G_{\mathrm{conf}}/-RT), \tag{9.8}$$

where

$$\Delta G_{\mathrm{conf}}(T) = \Delta H_{\mathrm{conf}}\left(\frac{T_{\mathrm{m,conf}} - T}{T_{\mathrm{m,conf}}}\right) - (T_{\mathrm{m,conf}} - T)\Delta C_{\mathrm{p,conf}} + T\Delta C_{\mathrm{p,conf}}\ln\left(\frac{T_{\mathrm{m,conf}}}{T}\right), \tag{9.9}$$

and ΔH_{conf}, $T_{\mathrm{m,conf}}$, and $\Delta C_{\mathrm{p,conf}}$ are the enthalpy, transition midpoint temperature, and heat capacity difference between the BI and BC states.

The apparent enthalpy for the binding process can be obtained through the derivative of Eq. (9.4) with respect to $1/T$,

$$\Delta H_{\mathrm{app}}(T) = -R\frac{\mathrm{d}\ln K_{\mathrm{app}}(T)}{\mathrm{d}(1/T)}, \tag{9.10}$$

to yield

$$\Delta H_{\mathrm{app}}(T) = \Delta H_0(T) + \frac{K_{\mathrm{conf}}}{(1 + K_{\mathrm{conf}})}\Delta H_{\mathrm{conf}}(T) \tag{9.11}$$

where, the first term corresponds to the enthalpy of the intrinsic association process and the second term corresponds to the enthalpy difference between the BI and BC state, weighted according to the population of molecules in the BI state.

$$\Delta H_{\text{app}}(T) = \Delta H_0(T) + P_{\text{BI}}\Delta H_{\text{conf}}(T) \tag{9.12}$$

$$\Delta H_{\text{app}}(T) = \Delta H_0(T) + \Delta H_{\text{conf,app}}(T) \tag{9.13}$$

Similarly, two terms are expected to contribute to the apparent ΔC_{p} of binding, one corresponding to the intrinsic and one to the conformational reaction. Formally,

$$\Delta C_{\text{p,app}} = \Delta C_{\text{p,int}} + \Delta C_{\text{p,conf,app}}, \tag{9.14}$$

where $\Delta C_{\text{p,int}}$ is the change in heat capacity for the BC/ligand association reaction, and $\Delta C_{\text{p,conf,eff}}$ is the equilibrium dependent contribution to the observed heat capacity. This conformational term can be written as (Eftink *et al.*, 1983; Prabhu and Sharp, 2005),

$$\Delta C_{\text{p,conf,eff}} = \Delta C_{\text{p,conf}} P_{\text{BI}} + \frac{\Delta H_{\text{conf}}{}^2}{RT^2} P_{\text{BI}} P_{\text{BC}}, \tag{9.15}$$

as the sum of the probability weighted change in heat capacity associated with the conformational reaction itself ($\Delta C_{\text{p,conf}}$), as well as the mean squared fluctuation in the enthalpy (ΔH_{conf}) of the available conformational states.

Using the above expressions (Eftink *et al.*, 1983), it is possible to simulate the expected effect of a linked folding reaction on the experimental observables provided by ITC (Fig. 9.4A). ITC data measuring the apparent free energy and enthalpy of the binding between AK and its nonhydrolyzable bisubstrtate analog P^1,P^5-di(adenosine) pentaphosphate (Ap5A) are also presented for comparison (Fig. 9.4B and C). The qualitative agreement between the simulation and the experiment is clear and can be used to highlight key features that can serve as a potential diagnostic of linked conformational/binding reactions.

First, we note that the enthalpy of many protein–ligand binding reactions is linear with temperature (as in the case with the low temperature data, Fig. 9.4C). However, the onset of a curvature of ΔH with increasing population of the BI state is expected based on the linkage equations. This phenomenon represents an apparent nonconstant $\Delta C_{\text{p,app}}$ (the derivative of ΔH_{app} with respect to temperature). Although not directly observed by ITC, the $\Delta C_{\text{p,app}}$ at a given temperature range can be estimated from the slope of the line connecting two measurements of ΔH_{app}, which differ (slightly) in temperature. Such nonconstant $\Delta C_{\text{p,app}}$ values may be indicative of a linked process in a binding reaction of interest. The pronounced curvature of the ΔH_{app} of AK binding provides a clear example of this effect.

As stated above, increasing the population of a BI (nonbinding) state in the unbound ensemble will decrease the apparent binding affinity of the system, regardless of the free energy of the intrinsic association reaction. The

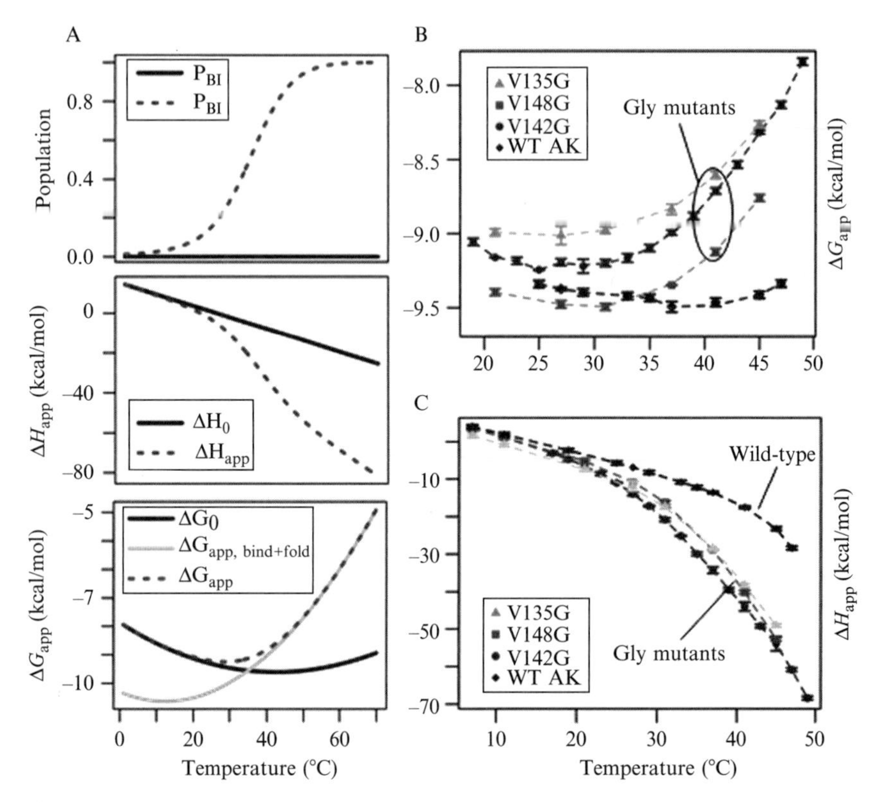

Figure 9.4 Agreement of expected thermodynamic observables of linked folding and binding with those of from AK Val to Gly Mutants. (A) Simulation of expected calorimetric observables according to Eqs. (9.3)–(9.15) and populations of the relevant BI state. Simulation parameters defining the linked folding reaction were: $\Delta H_{conf}(35\ ^{\circ}C) = 33$ kcal/mol, $\Delta C_{p,conf} = 660$ cal/mol K, $T_{m,conf} = 35\ ^{\circ}C$. Parameters defining the intrinsic binding (and conformational) reaction were chosen to approximate the AK data. (B, C) ITC measured (apparent) thermodynamic parameters of the AK/P^1,P^5-Di(adenosine-5′) pentaphosphate (Ap5A) binding reaction. Ap5A is a non-hydrolysable bisubstrate inhibitor of AK. Solution conditions were 60 m*M* PIPES acid, 2 m*M* EDTA, and pH 7.85. Adapted from previously published results (Schrank *et al.*, 2009).

decreased affinity of all AK mutants (above 37 °C) is consistent with the expected effects of mutations that increase the probability of a BI state. The temperature dependence of the free energy for such reactions can be more completely understood by considering three population regimes; (1) the *nonpopulated* regime, where the BI state is not significantly populated; (2) the *partially populated* regime, where the BI state is partially populated, and (3) the *fully populated* regime, where BI is populated >99%. As expected, because the BI state never becomes significantly populated in the *nonpopulated* regime, binding reactions that fall within this regime are easily described by a two-state equilibrium, which provides direct access to the intrinsic association reaction. On the other extreme, the *fully populated* regime can also be described by a two-state equilibrium. Although in this regime, the apparent ΔH, ΔS, and ΔC_p represent the sums of the contributions from the intrinsic binding reaction and the conformational transition (see the high temperature data in Fig. 9.4A). Within this regime, it can be difficult to decouple the intrinsic energies and those arising from conformational changes.

Only within the partially populated regime does the energetics provide unambiguous evidence of linked conformational and binding reactions. Owing to the evolving difference in the amount of BI at different temperatures, within this regime, the free energy function within this regime cannot be adequately described by a two-state model. However, in practice such a determination can be difficult to make, owing in large part to the wide temperature range over which small (i.e., low enthalpy) conformational transitions occur, as well as the corresponding level of precision that would be required to make such a determination.

As discussed previously (Eftink *et al.*, 1983), another hallmark feature of coupled thermodynamic processes, which may also serve as an indicator of a linked equilibrium, is the observation of entropy/enthalpy compensation, whether such effects are manifested is highly dependent on the thermodynamic parameters describing the conformational reaction. In general, protein folding is enthalpically favorable and entropically unfavorable. As a result, induced folding via ligand association should produce a considerable degree of compensation. Indeed, we note that for AK the relatively small change in ΔG_{app} is accompanied by relatively large changes in ΔH_{app} (see Fig. 9.4B and C).

In summary, it appears that mutants and WT AK proteins demonstrate several key thermodynamic features of linked folding and binding. However, it should be clear that other models could fit the data equally well. For this reason, qualitative agreement between the model and the data should not be considered unequivocal evidence in support of the model. Instead, we use this result as the starting point and take steps to validate the model through structural and dynamic analyses.

4. Strategies for Quantitative Interpretation of Measured Enthalpies for a Linked Folding and Binding System

A significant benefit of investigating ligand binding reactions with ITC is that in addition to binding affinities, which can be obtained from numerous techniques, ITC provides direct access to the enthalpy of the reaction. Access to the ΔH of a binding reaction renders ITC unique in its ability to determine the cooperativity of the reaction.

To demonstrate this unique potential, we rewrite the conformational contribution to the observed enthalpy of binding (ΔH_{app}, as in Eq. (9.11)) as,

$$\Delta H_{\text{conf,app}}(T) = \left(\frac{e^{\Delta G_{\text{conf}}(T)/-RT}}{1+e^{\Delta G_{\text{conf}}(T)/-RT}}\right)\left(\Delta H_{\text{m,conf}} - \Delta C_{\text{p,conf}}\left(T - T_{\text{m,conf}}\right)\right), \tag{9.16}$$

where the temperature dependent free energy ($\Delta G_{\text{conf}}(T)$) is defined by the Gibbs–Helmholtz equation (Eq. (9.9)). This form illustrates that this observable quantity is determined not only by the enthalpy but also all other parameters determining ΔG_{conf}, namely the changes in heat capacity ($\Delta C_{\text{p,conf}}$) and conformational entropy (ΔS_{conf}). Moreover, two distinct copies of ΔH_{conf} (and $\Delta C_{\text{p,conf}}$) are retained in the expression, both within and outside of the exponential term. The exponential term represents the rate of change of the probability of the BI state with temperature (also known as the van't Hoff enthalpy, ΔH_{vH}), while the preexponential copy represents the magnitude of the calorimetrically observable heat evolved (also known as the calorimetric enthalpy, ΔH_{cal}). As a consequence of having both the calorimetric and the van't Hoff enthalpies in the expression, use of Eq. (9.16) as a fitting function carries the added advantage of rigorously imposing two-state thermodynamic behavior by requiring that $\Delta H_{\text{vH}} = \Delta H_{\text{cal}}$. Indeed, if Eq. (9.16) produces a satisfactory fit, a thermodynamically two-state process is highly suggested.

Although the benefits of using Eq. (9.16) for nonlinear least squares (NLS) fitting are clear, the calorimetrically observed apparent enthalpy, ΔH_{app}, contains at least one more term, representing the intrinsic heat of ligand association to the BC state (ΔH_0). In principle, this term could be retained in the fitting equation, although at the cost of two more free parameters (i.e., a line). However, to simplify the analysis of AK, we subtracted the extrapolated low temperature baseline enthalpy prior to fitting.

Finally, all other possible linked contributions to the observed enthalpy should be considered. For example, protonation or deprotonation events that are coupled to the BI to BC transition will impact the results and possibly alter the analysis. Likewise, the globally unfolded state (i.e., the state where all residues are unfolded) can contribute enthalpy to the measured heat of binding at high temperatures. For such a case, the heat corresponding to the unfolding of additional residues can be estimated from CD unfolding experiments, and subtracted from the data prior to fitting, as was done for AK (Schrank *et al.*, 2009).

NLS fitting of Eq. (9.16) to the conformational contribution to the enthalpy of binding ($\Delta H_{\text{conf,eff}}$) for the WT and three Val to Gly mutants of AK is shown in Fig. 9.5, and the parameter estimates are summarized in Table 9.2. Interestingly, all four AK variants could be described by a single

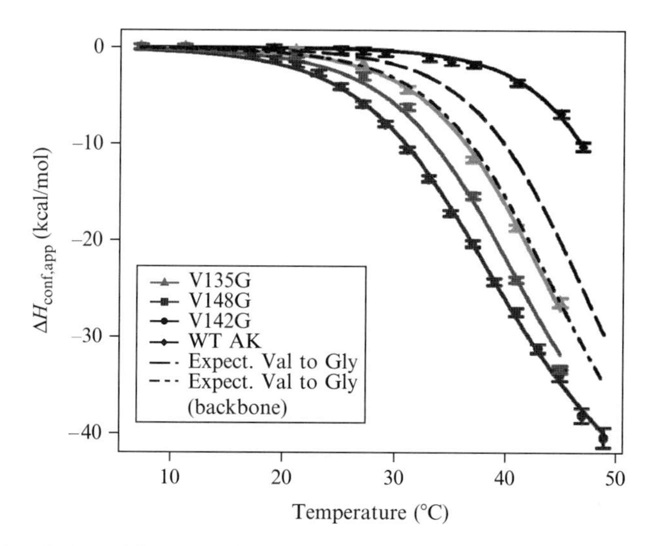

Figure 9.5 Fitting of $\Delta H_{\mathrm{conf,app}}$ for thermodynamic parameters of the binding associated conformational reaction. Corrected data (see text) and fitting functions representing $\Delta H_{\mathrm{conf,app}}$. Error bars are two times standard experimental error or the average thereof. *Solid lines*: represent the fitting functions. Dashed line is a simulation of the data for the expected entropy gain Val to Gly mutation based published entropy estimates (D'Aquino *et al.*, 1996; Lee *et al.*, 1994), and the fitted conformational ΔH and ΔC_{p}. *Dot-dashed line*: represents a simulation of the expected data for a Val to Gly mutation where folding places no conformational restriction on the side chain, that is, only the backbone contribution is taken into account (D'Aquino *et al.*, 1996). Adapted from previously published results (Schrank *et al.*, 2009).

Table 9.2 Thermodynamic parameter estimates for local unfolding of the LID region of AK

	$\Delta H_{\mathrm{m,conf}}$ [a]	$\Delta C_{\mathrm{p,conf}}$ [b]	$T_{\mathrm{m,conf}}$ [c]
V142G	33 ± 1.0	660 ± 60	35.1±0.3
V148G	nv	nv	37.9±0.3
V135G	nv	nv	41.1±0.2
Wild type	nv	nv	52.3±0.1

±,95% Confidence interval (estimated from the fitted model, by profile likelihoods); nv,value not varied in the fitting routine (v142g value applied). Adapted from previously published results (Schrank *et al.*, 2009).
[a] Transition enthalpy at $T_{\mathrm{m,conf}}$ of v142g (kcal/mol).
[b] Change in heat capacity (cal/mol).
[c] Transition midpoint temperature (°C).

value of ΔH_{conf} and $\Delta C_{\mathrm{p,conf}}$ (determined for the most complete data set, v142g), indicating that the conformational process that is being monitored is the same for all of the mutants and the wild type. This agreement suggests that the mutation strategy was successful in perturbing only the entropy of the local folding reaction in question. Furthermore, the conservation of the enthalpy of local folding suggests that the same cooperative unit (subset of residues) of local folding is shared between all mutants and the wild-type protein. As noted above, the high quality of the fits provides strong evidence supporting a two-state thermodynamic model of the local folding reaction.

In summary, NLS fitting of Eq. (9.16) is a powerful method for the extraction of thermodynamic information for linked folding and binding reactions. In addition, the entropy-enhancing mutation strategy has both allowed us to amplify the native state local unfolding to an extent that it can be rigorously analyzed, while the same time preserving both the temperature dependence of the reaction (ΔH_{conf} and $\Delta C_{\mathrm{p,conf}}$), as well as the cooperative unit of the local unfolding event found in the wild-type protein.

5. Interplay of Local Mutational Effects, Global Stability, and Binding Affinity

In general, destabilizing mutations are expected to affect the thermodynamic stability of the native state with respect to the globally unfolded state. However, when mutations are placed in a locally unstable region, the effect on the global stability will depend not only on the destabilizing impact of the mutation itself, but also the relative stability of the region where the mutation is placed (Ferreon *et al.*, 2004). This slightly counterintuitive result can be understood by considering a three-state unfolding reaction. The relevant partition function can be written as

$$Q = K_{\mathrm{conf}} + K_{\mathrm{conf}} K_{\mathrm{unfold}}, \tag{9.17}$$

where

$$K_{\mathrm{conf}} = e^{\Delta G_{\mathrm{conf}}/-RT} = \frac{P_{\mathrm{BI}}}{P_{\mathrm{BC}}} = \frac{P_{\mathrm{LU}}}{P_{\mathrm{F}}}; \quad K_{\mathrm{unfold}} = e^{\Delta G_{\mathrm{unfold}}/-RT} = \frac{P_{\mathrm{U}}}{P_{\mathrm{LU}}}, \tag{9.18}$$

and F represents the fully folded state, LU the BI and locally unfolded state discussed above, and U the state where all residues are unfolded. The probability of all states in the system is therefore given by

$$P_{\mathrm{F}} = \frac{1}{Q}; \quad P_{\mathrm{LU}} = \frac{K_{\mathrm{conf}}}{Q}; \quad P_{\mathrm{U}} = \frac{K_{\mathrm{conf}} K_{\mathrm{unfold}}}{Q}. \tag{9.19}$$

As written, $\Delta G_{\mathrm{unfold}}$ represents the free energy difference between a locally unfolded state and the state where all residues are unfolded. For a mutation perturbing K_{conf} only, the impact on the stability of the globally unfolded state will depend on which terms dominate the partition function. For instance, in the extreme case that $K_{conf} \gg 1$, increasing K_{conf} will not appreciably change the probability of the unfolded state (P_U), as all dominant terms in both numerator and denominator are multiplied by this value. In the opposite case, where the locally unfolded state is highly improbable, mutational effects are fully expressed in the globally unfolded state. These trends are visualized in Fig. 9.6A, where the effect of an identical thermodynamic mutation is simulated in the context of several different local stabilities (according to Eqs. (9.17)–(9.19)). To summarize, the apparent effect of the mutation on global stability is inversely correlated with the local stability of the region surrounding the mutation site (Ferreon *et al.*, 2004).

To assess the effects of the Val to Gly mutations in AK, global unfolding experiments were performed by monitoring CD as a function of temperature (see Fig. 9.6B). The observed destabilization was relatively minimal (a decrease in the unfolding transition temperature (T_m) < 1.5 °C for all mutants), suggesting as noted above that the region in question is relatively unstable to local unfolding (see Fig. 9.6B). In terms of the three-state unfolding system presented in Eqs. (9.17)–(9.19), the observed CD signal for AK can be described by,

$$\langle \theta_{228} \rangle (T) = \theta_{228,F}(T)[P_F(T) + P_{LU}(T)] + \theta_{228,U}[P_U(T)], \qquad (9.20)$$

where there is no (or little) difference in the observed value for the F and LU states. This inference is reasonable as the LU state of the mutant AKs is known to be highly populated before the global unfolding transition ($T_{m,local}$ = 35–41 °C); the CD signal is minimally affected in this range. In the case of v142g, the LU state is populated to a level of ~85% before the initiation of the observed global unfolding transition. Therefore, the two-state fit to the CD data should provide a suitable estimate of $K_{unfold}(T)$, that is, the thermodynamics of the LU to U transition. Having estimates of $K_{unfold}(T)$, $K_{conf,wt}(T)$, and $K_{conf,v142g}(T)$, the probabilities of all three states can be calculated from Eq. (9.19; see Fig. 9.7A). Interestingly, this calculation reproduces the measured (CD) temperature dependence of the globally unfolded state of WT AK (Fig. 9.7A, *dot-dashed line*), although no WT unfolding data is included. Such a prediction, using an independent observable, strongly validates the accuracy of the parameter estimates describing the local unfolding reaction, as well as the thermodynamic analysis applied to both the ITC and CD data. We note that a significant population of the locally unfolded state is revealed for the WT AK at biological temperatures (~5% at 37 °C).

One other thermodynamic relationship can be developed, which can provide additional means of validating the analysis. As described above (Eq. (9.6)), the observed free energy of binding will also depend on the position of the relevant linked conformational equilibrium, $K_{conf}(T)$.

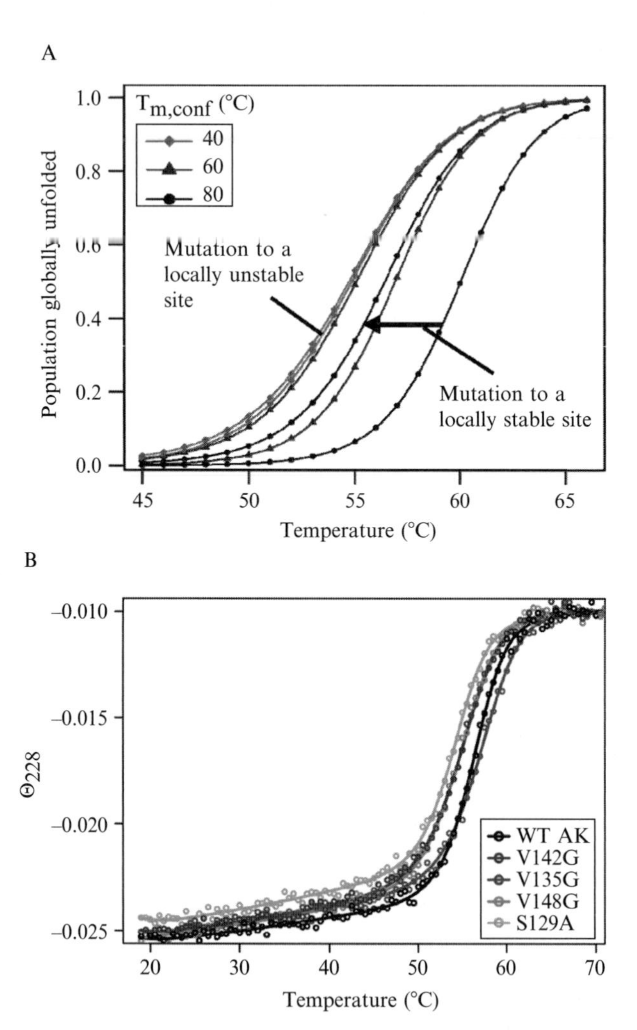

Figure 9.6 Simulation of regional destabilization on global protein stability and representative circular dichrosim unfolding experiments of AK mutants. (A) Simulation based on Eqs. (9.17)–(9.19). Parameters defining $K_{\mathrm{unfold}}(T)$ were $T_{\mathrm{m,unfold}} = 55$ °C, $\Delta H_{\mathrm{unfold}}(T_{\mathrm{m}}) = 89$ kcal/mol, $\Delta C_{\mathrm{p,unfold}} = 2.8$ kcal/mol K. Parameters defining $K_{\mathrm{conf}}(T)$ were: $T_{\mathrm{m,conf}}$ as indicated, $\Delta H_{\mathrm{conf}}(55$ °C$) = 33$ kcal/mol, $\Delta C_{\mathrm{p,unfold}} = 0.66$ kcal/mol K. *Lines*: Simulated WT/mutant protein pairs. The applied thermodynamic mutation entropically destabilizes the locally unfolding region, $\Delta\Delta S_{\mathrm{mutation}} = 5$ cal/mol K. (B) Experimental observation of global of AK mutants. Molar ellipticity at 288 nm, θ_{288} (deg cm^2/dmol res), recorded by circular dichroism spectroscopy. WT (*black*), v142g (*dark gray*), v148g (*gray*), v135g (*light gray*), and s129a (*very light gray*). Conditions were identical to ITC experiments. Lines are fitting functions, representing a two-state thermodynamic model. Adapted from previously published results (Schrank *et al.*, 2009).

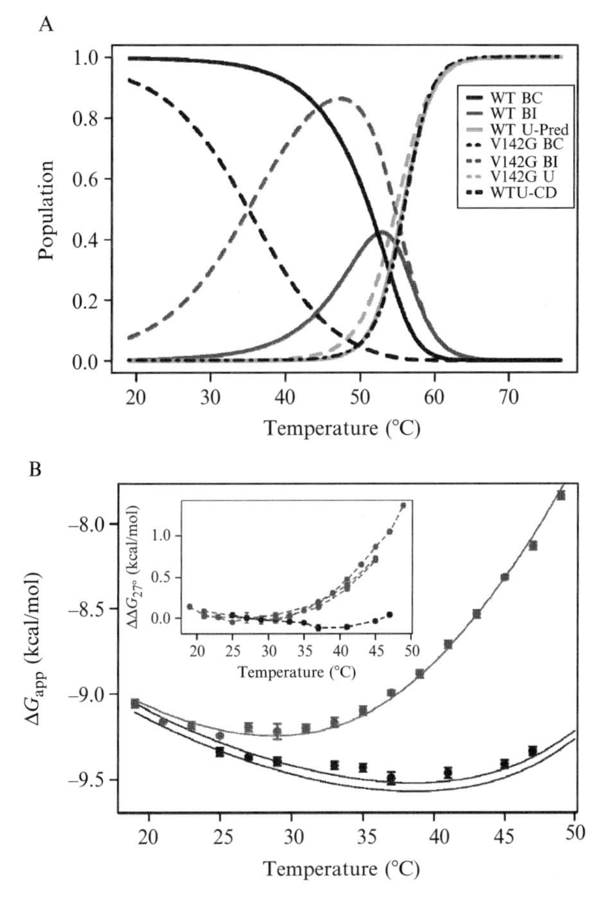

Figure 9.7 Agreement of local unfolding parameter estimates with external experimental observables. (A) Three-state model of unbound adenylate kinase. Populations are calculated from the BC/BI equilibrium, K_{conf}, estimated from the ITC data. The *locally unfolded* (LU)/*unfolded* (U) equilibrium, K_{unfold}, is estimated from the CD unfolding experiments on v142g ($T_{\mathrm{m,unfold}} = 54.7\ ^{\circ}\mathrm{C}$, $\Delta H_{\mathrm{unfold}}(T_{\mathrm{m}}) = 89.4$ kcal/mol, $\Delta C_{\mathrm{p,unfold}} = 2.8$ kcal/mol K). The reasonability of this approximation is seen in this plot, from the high probability of the BI state at 47 °C, prior to the global unfolding transition. Calculations are based on Eqs. (9.17)–(9.19). *Dot-dashed line*: Population of the U state for WT as calculated directly from the CD unfolding parameter estimates. (B) Prediction of WT affinity data based on the fitting of $\Delta H_{\mathrm{conf,app}}$, Eq. (9.22). Lines represent the prediction, plus and minus the average standard error of determining ΔG_{app}. *Inset*: change in ΔG_{app} for each protein, with temperature, referenced from 27 °C. $\Delta\Delta G_{27\ ^{\circ}\mathrm{C}}(\mathrm{Temp}) = \Delta G(\mathrm{Temp}) - \Delta G(27\ ^{\circ}\mathrm{C})$. Adapted from previously published results (Schrank *et al.*, 2009).

If mutational effects are manifested exclusively in the conformational equilibrium, leaving the intrinsic binding interaction unchanged, the difference in the apparent binding affinity between the two proteins will be

$$\Delta\Delta G_{\mathrm{app,wt-mut}} = RT\ln\left(\frac{1+K_{\mathrm{conf,wt}}}{1+K_{\mathrm{conf,mut}}}\right), \tag{9.21}$$

where the two identical terms representing $\Delta G_0(T)$ cancel. For clarity, this can be rewritten as

$$\Delta\Delta G_{\mathrm{app,wt-mut}} = RT\ln\left(\frac{1+K_{\mathrm{conf,wt}}}{1+K_{\mathrm{conf,wt}}\Omega_{\mathrm{mut}}}\right), \tag{9.22}$$

where Ω_{mut} represents the effect of the mutation (i.e., the change in ratio of accessible conformational space to Val and Gly in the disordered state, Eq. (9.1)). Considering these relationships, it becomes clear that when the mutated region is highly unstable ($K_{\mathrm{conf,wt}} \gg 1$), then the effect of the mutation on $\Delta\Delta G$ will be maximized. Conversely, low probabilities of the BI state diminish the mutational effects. In the case of AK, low temperatures suppress local unfolding ($1 > K_{\mathrm{conf,wt}}$). Accordingly, the difference in ΔG_{obs} between mutant and wild-type AKs decreases with decreasing temperature. For v142g AK, the difference in affinity as compared to WT approaches zero, indicating that the ΔG_0 of these proteins are equal. Therefore, we can use the above estimated thermodynamic parameters describing the BI state for the wild type and v142g proteins together with Eq. (9.21), to directly predict the difference in the observed binding affinity of these two proteins.

This calculation was performed by first fitting Eq. (9.6) to the v142g data (the most extensive data set), fixing all parameters given ($\Delta C_{\mathrm{p,intrinsic}}$) or estimated ($\Delta H_{\mathrm{conf}}$, $\Delta C_{\mathrm{p,conf}}$, and $T_{\mathrm{m,conf}}$) in the fitting of the observed enthalpy. $\Delta\Delta G$ was then calculated as a distance from this fitted curve. The agreement of the experimental data and this prediction validates both the proposed binding model and the quality of the fitted parameters obtained from the enthalpy data (Fig. 9.7B). We note that the assumption of low to no binding affinity for the BI state is strongly validated by this simulation. Also, although this prediction cannot be quantitatively applied to v148g and v135g because of apparent small perturbations to the intrinsic free energy of binding (Fig. 9.4B, low temperature), the fact that each of the Val to Gly mutations shows similar temperature dependence of the binding affinity (see Fig. 9.7B, *inset*) further supports the validity of the model for all mutants investigated.

In previous work from our group, Ferreon *et al.* proposed an alternate experimental approach for determining local stability, using the identical mutation strategy under discussion. In this work, they showed that by combining HX and ITC data, the principle of inversely related effects on stability and affinity can be leveraged to determine local stability (Ferreon *et al.*, 2004). Such an approach only requires measurements at a single temperature or solution condition. By this method, they demonstrated that the RT-loop of SEM-5 SH3 domain (a flexible binding loop) behaves as it was 25% in an unfolded conformation (Ferreon *et al.*, 2004). We note that many local unfolding reactions may have small or poorly determined enthalpy. In this case, this affinity/stability method may be more applicable. The enthalpy-based approach demonstrated here is preferred when possible, because of the relatively complete thermodynamic characterization available from this method.

6. Success of the Strategy in Preserving Structure

One important goal of our mutation strategy was to selectively perturb the properties of local unfolded-like conformational excursions, while leaving the ground-state structure unperturbed. For this reason, we have chosen highly surface-exposed residues, involved in few molecular contacts. If our mutation design is successful at achieving the goal of structural preservation, then we can be much more certain that any observed changes in binding are a consequence of modulating the stability of highly disordered or unfolded states. This inference becomes very important when experimental isolation of the intrinsic and conformational contributions to measured binding affinity or enthalpy is not possible.

As a reference, it is well known that the fold shared by homologous proteins is robust to sequence changes. For example, AKs from *Bacillus* sp. share the fold of *E. coli* AK. Furthermore, an interesting series of *Bacillus* AKs, sharing ~70% sequence identity, have been examined by crystallography (Bae and Phillips, 2004). Structural alignment of the LID region (residues 128–159) of the molecules revealed alpha carbon RMSD values of 0.43–0.46 Å (Bae and Phillips, 2004). By extension, the structural effects of a single point mutation, as in the strategy under discussion, are expected to be subtle and not propagate by largely perturbing the conformation of the protein backbone. Of course, local structural consequences of disrupted interactions are expected. To evaluate the success of the proposed (surface-exposed Val to Gly) mutation strategy on preserving the structure of AK, we performed crystallography on two AK mutants (v142g, v148g) and the wild-type molecule.

Only the v148g mutant crystallizes in the identical space group ($P2_12_12$) and asymmetric unit (ASU) as the wild type, and thus merits detailed comparison. When comparison of small structural changes is the goal, one must carefully consider the effects of crystal packing on the resultant structure. In this case, the ASU happens to contain two distinct copies of AK. This provides an excellent opportunity to examine the magnitude and distribution of crystal packing effects. Interestingly, a distinct conformation of the outer region of the LID domain is associated with each crystallographic environment investigated (also observed in the distinct ASU of v142g, data not presented). We note that this region (residues ~130–155) has been demonstrated to be the most dynamic region of the protein on the nano- to picosecond time scale (Shapiro *et al.*, 2000). Therefore, structural plasticity observed in this region may be indicative of a relatively heterogeneous conformational manifold that can be dynamically explored in solution (Shapiro *et al.*, 2000). Alignment demonstrates that the influences of crystal packing are largely identical for both WT and v148g AK (Fig. 9.8A).

However, we can also examine the structural changes effected by our mutations (where crystallographic influences should be minimal) by comparing identically oriented copies of WT and v148g from the shared ASU. The similar backbone structures of analogous chains of v148g and WT can be seen in Fig. 9.8A. Two copies within the ASU also provide another benefit when using medium resolution data (as here), in that true observable structural perturbations effected by mutation should be reproducible in the data from chain A to chain B. Using a very small threshold of perturbation (>0.3 Å), we identified all atoms that are potentially perturbed in both copies of the ASU, comparing WT versus v148g structures. Such atoms are highlighted in Fig. 9.8B. Notably, the structures surrounding the mutation sites do not show a significant increase in the number of perturbed atoms. Furthermore, most of the identified atoms belong to surface-exposed side chains where motility of the side chain results in poor electron density. The positions of such atoms are simply poorly determined. Therefore, within the available resolution of measurement, it appears that our mutation strategy is highly successful at preserving the fully folded structure of the protein. These data have been corroborated by the preservation of ^{1}H–^{15}N chemical shifts (HSQC experiment) of WT and mutant AKs, as well as similar NMR studies of the SH3 SOS-Y binding reaction (Ferreon and Hilser, 2003b; Schrank *et al.*, 2009).

7. Comparison of Interaction Versus Entropy Based Mutation Strategy

The s129a mutation represents an interesting control study, as this mutation should destabilize the LID by a completely distinct mechanism. According to the crystal structure (Schrank *et al.*, 2009), the hydroxyl group

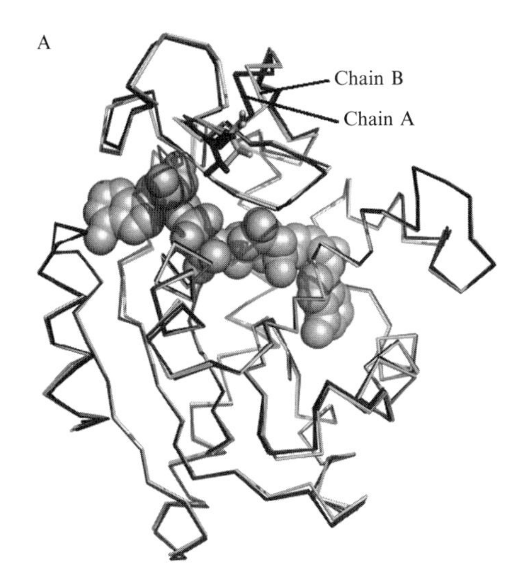

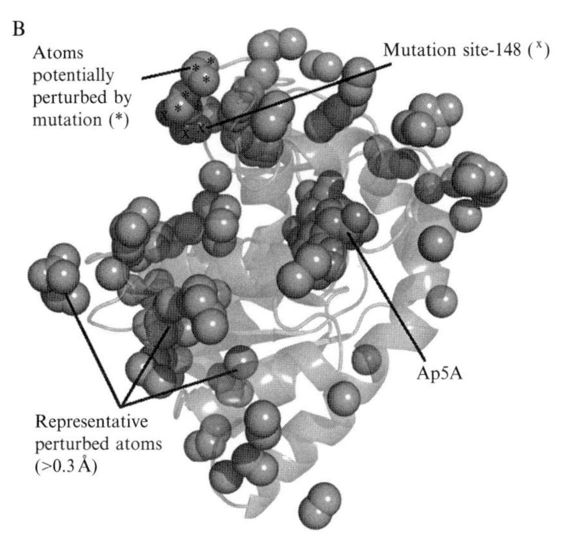

Figure 9.8 Structural conservation of v148g and s129a AK. (A) Alignment of the crystal structures of WT and v148g AK (Schrank *et al.*, 2009). *Black*: Chains (WT and v148g) from position A within the ASU. *Gray*: Chains from position B within the ASU. (B) Analysis of structural possible structural perturbations effected by mutation. *Gray spheres*: Represent all atoms that move > 0.3 Å from the WT to mutant structure in both copies within the ASU. *Asterisk*: The mutation site (148). *Cross*: All perturbed atoms (spheres) that can be connected to the mutation site by a continuous chain (< 6 Å per step) of other perturbed atoms. Adapted from previously published results (Schrank *et al.*, 2009).

of serine 129 is a hydrogen bond donor to a nearby histidine side chain nitrogen. Mutation to Ala disrupts this interaction, while preserving the β-carbon at position 129, and therefore should make very minimal perturbations to the sterically allowed conformational freedom of this position.

Identical ITC experiments were carried out for this mutant. As expected, the LID region appears to be highly destabilized, as a very large increase in the favorable enthalpy of binding is observed. However, it is interesting to note that the gross shape of the change in $\Delta H_{\mathrm{conf,app}}$ with temperature is very different from the shared behavior of the Val to Gly series, see Fig. 9.9A. Furthermore, this data cannot be fitted by the shared conformational scheme (with or without shared ΔH_{conf} and $\Delta C_{\mathrm{p,conf}}$) that so nicely describes Val to Gly mutants (see Fig. 9.9A). Also, the magnitude in the observed differences is inconsistent with what would be expected for local mutational effects, $\sim$10 kcal/mol. Therefore, the thermodynamic model describing the local unfolding reaction must be fundamentally different. We again note the utility of Eq. (9.16) as a test of a two-state system, which in this case is clearly excluded.

Therefore, the case of s129a mutation demonstrates the desirable properties of the investigated Val to Gly mutation strategy. It appears that preservation of energetic relationships (cooperativity) between the many possible states of unfolding is a unique property of the noninvasive, entropy-enhancing strategy that we utilize. Furthermore, we note that crystallographic investigation of the s129a mutation reveals minimal changes, similar to those reported for v148g (unpublished results). Therefore, the presented data support the greater utility of entropy-based, surface-exposed mutations as compared to even structurally conservative interaction-directed mutations.

8. How Similar Are Local and Global Unfolding?

One might expect that a locally unfolded region of an otherwise folded protein may have important differences from an isolated unfolded peptide free in solution. Both the spatially constrained peptide ends and the relatively high likelihood of interactions with the surrounding structures should in principle alter the conformational manifold. The analysis presented here gives us a rare opportunity to compare several different thermodynamic properties of local unfolding within the native state of AK with what would be expected for global unfolding of that same region.

Published values are available for the expected entropic consequences of a Val/Ala to Gly mutation for an unfolding reaction (D'Aquino *et al.*, 1996). In addition to the backbone contributions emphasized in Section 2, unfolding of Val increases the rotational freedom of the side chain (Lee *et al.*, 1994). We have used published values to simulate the expected destabilizing effect of Val

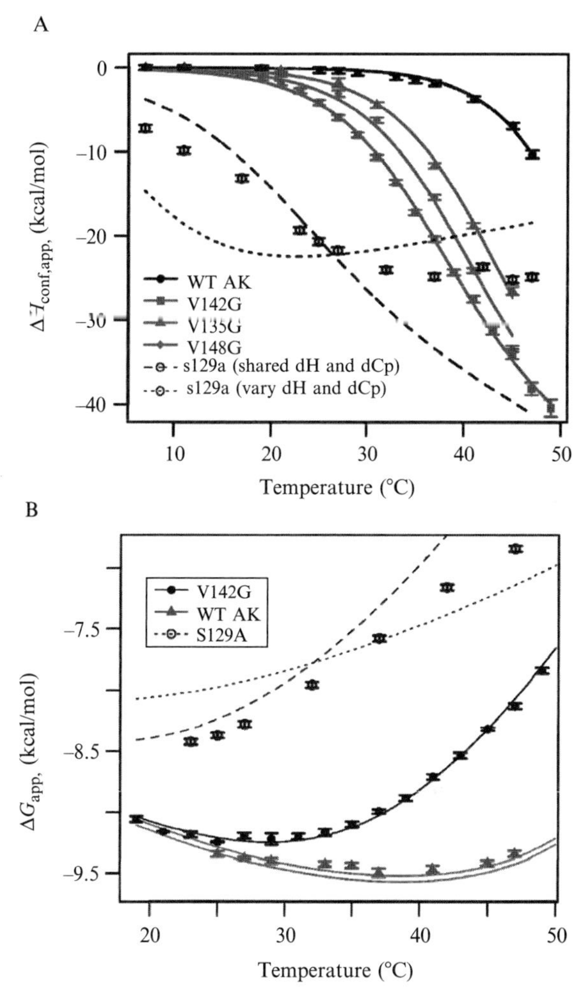

Figure 9.9 ITC measurement of the s129a/Ap5A Binding Reaction. (A) Corrected data (see text) and fitting functions representing $\Delta H_{conf,app}$. Error bars are two times average standard error for all similar experiments. *Solid lines*: Two-state fitting functions. *Long dashed lines*: Failed fitting (best) of Eq. (9.16) to the s129a data using the shared ΔH and ΔC_p used to fit WT and Gly variants of AK. *Short dashed lines*: Failed fitting (best) of Eq. (9.16) to the s129a data varying ΔH and ΔC_p. (B) The apparent free energy of binding for s129a, v142g and WT AK. *Solid lines*: Prediction of $\Delta\Delta G_{bind}$ for WT and v142g, as in Fig. 9.7. *Dashed lines*: Failure of either of the corresponding fits ($\Delta H_{conf,app}$) to reproduce the temperature dependence of the s129a binding affinity.

to Gly mutation in AK, considering the background of the measured thermodynamics of the local unfolding in WT AK. Because of the expected variability in side chain contributions, we have simulated both the case of an average surface-exposed Val to Gly as well as the backbone contribution of

Val to Gly mutation alone (see Fig. 9.5). In both cases, the effects of all mutations were found to be greater than or equal to the expected effect. We reiterate that the destabilization induced by the mutations appears to be strictly entropic. As demonstrated above, mutation effects should be decreased relative to the prediction if residues in the locally unfolded regions are partially constrained. Although the origin of the additional entropy gain is unknown, our data are in agreement with a model where the mutated positions, upon local unfolding, gain the full amount of conformational freedom available to the same residue in the fully unfolded state.

This assertion is also buttressed by success of the v142g local unfolding data to describe the difference between the v142g and WT AK in terms of global stability. If additional conformational freedom were gained at the mutated site upon global unfolding, as might be expected, the three-state unfolding model presented would underestimate the effect of the mutation on global unfolding. In this case, the wild-type protein should be more stable than predicted. Thus, the ability of the local unfolding parameters to describe the global unfolding data also supports the equivalence of the entropy (and conformational diversity) of local and global unfolding for the mutated positions in question.

Similarly, it is interesting to investigate the solvent associated enthalpy and heat capacity change of local unfolding to determine if these values are also compatible with an unfolding model. Fortunately, excellent parametric equations exist that relate changes in solvent exposed (polar and apolar) surface area (ASA) upon unfolding to the heat and heat capacity of the unfolding reaction (Murphy *et al.*, 1992). These relationships are based on calorimetric unfolding experiments. Estimation of the ΔH and ΔC_p expected for a given local unfolding reaction can be made, based on the crystallographic structure, by calculating the expected ΔASA of transferring a segment of the protein from the context of the fold into the solvent (i.e., the surface area of the molecular interface) as well as the ΔASA of unfolding the isolated segment. Recent reviews have described the full computational procedure for making such calculations (COREX algorithm), so the details of these methods will not be discussed further here (Hilser and Freire, 1997; Hilser *et al.*, 2006).

In the case of the local unfolding in AK, the amplification of the locally unfolded state has allowed us to map the residues involved in the local unfolding reaction using NMR, simply assigning resonances with (greatly) enhanced chemical exchange broadening in the HSQC spectrum of the v142g mutant (Schrank *et al.*, 2009). This well-described effect is due to the local unfolding reaction, which happens to be on the chemical shift time scale. Therefore, using the surface area based approximation described above, predictions of the local and global unfolding of AK were made, and are summarized in Table 9.3. As is evident, the calculation reproduces the measured heats of local unfolding as well as unfolding the remainder of

Table 9.3 Experimental and predicted thermodynamics of local and global unfolding

		Local[a]		Global[b]		Sum[c]	
	T (°C)	ITC[d]	COREX[‡]	CD (v142g)[e]	COREX[‡]	Experimental[f]	COREX[‡]
ΔH	35.2	33.3 ± 0.5	32.8	34 ± 9[†]	40.9	67 ± 10[†]	73.7
	54.7	46.2 ± 1.1[†]	46.5	89 ± 4	88.2	135.2 ± 5[†]	134.6
ΔC_p		0.66 ± 0.03	0.7	2.8 ± 0.2	2.4	3.5 ± 0.2[†]	3.1

[‡]Values calculated from the expected change in exposed polar and apolar surface area upon unfolding, using the COREX energy function and PDB ID 4AKE (Müller *et al.*, 1996), confidence is generally ±10% for estimation of ΔC_p and ΔH for a differential scanning calorimetric unfolding experiment. [±]Experimental 95% confidence interval (CD), or standard error (ITC). [†]Propagated error. Adapted from previously published results (Schrank *et al.*, 2009).

[a] Experimental and computational measures of a possible local unfolding reaction (residues 110–164).

[b] Experimental and computational measures of unfolding the remainder of the protein (residues 1–109, 165–214).

[c] Sum of local and global reactions. Represents unfolding all residues in the protein (residues 1–214).

[d] Parameters estimated from the ITC data fitting, or calculated from these estimates.

[e] Parameters estimated from the CD data fitting, or calculated from these estimates.

[f] Sum of heat or heat capacity estimated for the BI/BC transition (ITC) and the native/denatured transition (CD).

the protein, as approximated by the global unfolding experiment for v142g. Therefore, in terms of solvent accessibility and the related thermodynamic parameters ΔH and ΔC_p, the locally unfolded state appears to be indistinguishable from a true protein unfolding event.

These data demonstrate that local unfolding within the context of a folded protein at times can be highly similar in thermodynamic and conformational character to much studied global unfolding reactions. This result, of course, does not imply that the protein is completely extended. Instead, it simply means that the thermodynamic character of local and global unfolding is similar.

9. Summary

The exploration of the structural, energetic, and kinetic details of the native state ensemble is one of the key goals of modern biochemistry. However, the experimental tool box available to the investigator is still rather limited, especially when searching for techniques that will yield physical insight into the conformational nature of fluctuations and/or allow the investigation of minor populations typically hidden in ensemble (average property) based experimental approaches.

In this work, we have demonstrated an experimental strategy that allows one to selectively probe the native state ensemble for highly disordered states. The advantages of the surface-exposed, binding site associated Gly mutation strategy used are threefold. First, the structure of the ground state and the associated biochemical and functional properties of the protein (i.e., the intrinsic binding affinity) need not be altered, thus allowing unambiguous interpretation of the functional effects of high-energy states. Second, studying binding site associated regions of the protein allows depopulation of high-energy states by the addition of ligand. This external tool for population modulation greatly facilitates the calorimetric/thermodynamic investigation of partially populated states. Third, Gly is a somewhat selective probe for highly disordered states, as large regions of φ/ψ space must be explored for a strong effect. Thus, the effectiveness of the mutation strategy itself provides qualitative conformational information.

The example mutations discussed here have been successful for the study of conformational fluctuations within the native state of AK (Schrank *et al.*, 2009). We have found that in this case, our strategy preserves the cooperative substructure of local unfolding present within the WT protein. Furthermore, the ΔH and ΔC_p of native state unfolding of the LID are preserved between WT and mutant proteins. This greatly controlled set of mutations has allowed the otherwise impossible estimation of the population of LID unfolding in WT AK ($\sim$5% at 37 °C; Schrank *et al.*, 2009). In ongoing research in our group, these mutant proteins have allowed us to enhance the population of local unfolding above the detection limit of relaxation dispersion NMR experiments (unpublished results), thus revealing detailed kinetic and conformational information about these states.

This experimental framework has also been successfully applied to conformational fluctuations of the SEM-5 SH3 domain, as well as experiments designed to probe the polyproline-II content of peptides free in solution (Ferreon and Hilser, 2003b; Ferreon *et al.*, 2004; Manson *et al.*, 2009). Therefore, this approach is likely to be widely applicable to many other systems and compatible with many other experimental techniques such as NMR, FRET, and single molecule techniques.

REFERENCES

Babu, C. R., Hilser, V. J., and Wand, A. J. (2004). Direct access to the cooperative substructure of proteins and the protein ensemble via cold denaturation. *Nat. Struct. Mol. Biol.* **11,** 352–357.

Bae, E., and Phillips, G. N. (2004). Structures and analysis of highly homologous psychrophilic, mesophilic, and thermophilic adenylate kinases. *J. Biol. Chem.* **279,** 28202–28208.

Bai, Y. (2006). Protein folding pathways studied by pulsed- and native-state hydrogen exchange. *Chem. Rev.* **106,** 1757–1768.

Baldwin, A. J., and Kay, L. E. (2009). NMR spectroscopy brings invisible protein states into focus. *Nat. Chem. Biol.* **5,** 808–814.
Boehr, D. D., McElheny, D., Dyson, H. J., and Wright, P. E. (2010). Millisecond timescale fluctuations in dihydrofolate reductase are exquisitely sensitive to the bound ligands. *Proc. Natl. Acad. Sci. USA* **107,** 1373–1378.
Cliff, M. J., Williams, M. A., Brooke-Smith, J., Barford, D., and Ladbury, J. E. (2005). Molecular recognition via coupled folding and binding in a TPR domain. *J. Mol. Biol.* **346,** 717–732.
D'Aquino, J. A., Gómez, J., Hilser, V. J., Lee, K. H., Amzel, L. M., and Freire, E. (1996). The magnitude of the backbone conformational entropy change in protein folding. *Proteins* **25,** 143–156.
Eftink, M. R., Anusiem, A. C., and Biltonen, R. L. (1983). Enthalpy–entropy compensation and heat capacity changes for protein–ligand interactions: General thermodynamic models and data for the binding of nucleotides to ribonuclease A. *Biochemistry* **22,** 3884–3896.
Englander, S. W. (1975). Measurement of structural and free energy changes in hemoglobin by hydrogen exchange methods. *Ann. NY Acad. Sci.* **244,** 10–27.
Ferreon, J. C., and Hilser, V. J. (2003a). Ligand-induced changes in dynamics in the RT loop of the C-terminal SH3 domain of Sem-5 indicate cooperative conformational coupling. *Protein Sci.* **12,** 982–996.
Ferreon, J. C., and Hilser, V. J. (2003b). The effect of the polyproline II (PPII) conformation on the denatured state entropy. *Protein Sci.* **12,** 447–457.
Ferreon, J. C., Hamburger, J. B., and Hilser, V. J. (2004). An experimental strategy to evaluate the thermodynamic stability of highly dynamic binding sites in proteins using hydrogen exchange. *J. Am. Chem. Soc* **126,** 12774–12775.
Henzler-Wildman, K. A., Lei, M., Thai, V., Kerns, S. J., Karplus, M., and Kern, D. (2007). A hierarchy of timescales in protein dynamics is linked to enzyme catalysis. *Nature* **450,** 913–916.
Hilser, V. J., and Freire, E. (1997). Predicting the equilibrium protein folding pathway: Structure-based analysis of staphylococcal nuclease. *Proteins* **27,** 171–183.
Hilser, V. J., García-Moreno, E. B., Oas, T. G., Kapp, G., and Whitten, S. T. (2006). A statistical thermodynamic model of the protein ensemble. *Chem. Rev.* **106,** 1545–1558.
Huyghues-Despointes, B. M., Langhorst, U., Steyaert, J., Pace, C. N., and Scholtz, J. M. (1999). Hydrogen-exchange stabilities of RNase T1 and variants with buried and solvent-exposed Ala -> Gly mutations in the helix. *Biochemistry* **38,** 16481–16490.
Korzhnev, D. M., Neudecker, P., Zarrine-Afsar, A., Davidson, A. R., and Kay, L. E. (2006). Abp1p and Fyn SH3 domains fold through similar low-populated intermediate states. *Biochemistry* **45,** 10175–10183.
Krishna, M. M. G., Hoang, L., Lin, Y., and Englander, S. W. (2004). Hydrogen exchange methods to study protein folding. *Methods* **34,** 51–64.
Leach, S. J., Némethy, G., and Scheraga, H. A. (1966). Computation of the sterically allowed conformations of peptides. *Biopolymers* **4,** 369–407.
Lee, K. H., Xie, D., Freire, E., and Amzel, L. M. (1994). Estimation of changes in side chain configurational entropy in binding and folding: General methods and application to helix formation. *Proteins* **20,** 68–84.
LeMaster, D. M., Anderson, J. S., and Hernández, G. (2009). Peptide conformer acidity analysis of protein flexibility monitored by hydrogen exchange. *Biochemistry* **48,** 9256–9265.
Liu, T., Whitten, S. T., and Hilser, V. J. (2007). Functional residues serve a dominant role in mediating the cooperativity of the protein ensemble. *Proc. Natl. Acad. Sci. USA* **104,** 4347–4352.
Maity, H., Lim, W. K., Rumbley, J. N., and Englander, S. W. (2003). Protein hydrogen exchange mechanism: Local fluctuations. *Protein Sci.* **12,** 153–160.

Manson, A., Whitten, S. T., Ferreon, J. C., Fox, R. O., and Hilser, V. J. (2009). Characterizing the role of ensemble modulation in mutation-induced changes in binding affinity. *J. Am. Chem. Soc.* **131,** 6785–6793.
Miller, D. W., and Dill, K. A. (1995). A statistical mechanical model for hydrogen exchange in globular proteins. *Protein Sci.* **4,** 1860–1873.
Müller, C. W., and Schulz, G. E. (1992). Structure of the complex between adenylate kinase from *Escherichia coli* and the inhibitor Ap5A refined at 1.9 Å resolution. A model for a catalytic transition state. *J. Mol. Biol.* **224,** 159–177.
Müller, C. W., Schlauderer, G. J., Reinstein, J., and Schulz, G. E. (1996). Adenylate kinase motions during catalysis: An energetic counterweight balancing substrate binding. *Structure* **4,** 147–156.
Murphy, K. P., Bhakuni, V., Xie, D., and Freire, E. (1992). Molecular basis of co-operativity in protein folding III. Structural identification of cooperative folding units and folding intermediates. *J. Mol. Biol.* **227,** 293–306.
Pan, H., Lee, J. C., and Hilser, V. J. (2000). Binding sites in *Escherichia coli* dihydrofolate reductase communicate by modulating the conformational ensemble. *Proc. Natl. Acad. Sci. USA* **97,** 12020–12025.
Ponder, J. W., and Richards, F. M. (1987). Tertiary templates for proteins. Use of packing criteria in the enumeration of allowed sequences for different structural classes. *J. Mol. Biol.* **193,** 775–791.
Prabhu, N. V., and Sharp, K. A. (2005). Heat capacity in proteins. *Annu. Rev. Phys. Chem.* **56,** 521–548.
Ramachandran, G. N., and Sasisekharan, V. (1968). Conformation of polypeptides and proteins. *Adv. Protein Chem.* **23,** 283–438.
Ramakrishnan, C., and Ramachandran, G. N. (1965). Stereochemical criteria for polypeptide and protein chain conformations. II. Allowed conformations for a pair of peptide units. *Biophys. J.* **5,** 909–933.
Rogero, J. R., Englander, J. J., and Englander, S. W. (1986). Individual breathing reactions measured by functional labeling and hydrogen exchange methods. *Methods Enzymol.* **131,** 508–517.
Rundqvist, L., Adén, J., Sparrman, T., Wallgren, M., Olsson, U., and Wolf-Watz, M. (2009). Noncooperative folding of subdomains in adenylate kinase. *Biochemistry* **48,** 1911–1927.
Schrank, T. P., Bolen, D. W., and Hilser, V. J. (2009). Rational modulation of conformational fluctuations in adenylate kinase reveals a local unfolding mechanism for allostery and functional adaptation in proteins. *Proc. Natl. Acad. Sci. USA* **106,** 16984–16989.
Shapiro, Y. E., Sinev, M. A., Sineva, E. V., Tugarinov, V., and Meirovitch, E. (2000). Backbone dynamics of *Escherichia coli* adenylate kinase at the extreme stages of the catalytic cycle studied by (15)N NMR relaxation. *Biochemistry* **39,** 6634–6644.
Vallurupalli, P., Hansen, D. F., Stollar, E., Meirovitch, E., and Kay, L. E. (2007). Measurement of bond vector orientations in invisible excited states of proteins. *Proc. Natl. Acad. Sci. USA* **104,** 18473–18477.
Whitten, S. T., Wooll, J. O., Razeghifard, R., García-Moreno, E. B., and Hilser, V. J. (2001). The origin of pH-dependent changes in m-values for the denaturant-induced unfolding of proteins. *J. Mol. Biol.* **309,** 1165–1175.
Whitten, S. T., Yang, H., Fox, R. O., and Hilser, V. J. (2008). Exploring the impact of polyproline II (PII) conformational bias on the binding of peptides to the SEM-5 SH3 domain. *Protein Sci.* **17,** 1200–1211.
Wolf-Watz, M., Thai, V., Henzler-Wildman, K., Hadjipavlou, G., Eisenmesser, E. Z., and Kern, D. (2004). Linkage between dynamics and catalysis in a thermophilic–mesophilic enzyme pair. *Nat. Struct. Mol. Biol.* **11,** 945–949.
Wright, P. E., and Dyson, H. J. (2009). Linking folding and binding. *Curr. Opin. Struct. Biol.* **19,** 31–38.

CHAPTER TEN

Fluorescence-Detected Sedimentation in Dilute and Highly Concentrated Solutions

Jonathan S. Kingsbury* *and* Thomas M. Laue†

Contents

Abstract

Analytical ultracentrifugation (AUC) is a powerful, first-principles method for characterizing macromolecules in solution. The recent development of fluorescence-detected sedimentation for the AUC (AU-FDS) has extended the sensitivity and selectivity of the instrument which, in turn, has enabled the study of both higher affinity interactions and the sedimentation of one component in

* Therapeutic Protein Research, Genzyme Corporation, Framingham, Massachusetts, USA
† Department of Biochemistry and Molecular Biology, University of New Hampshire, Durham, New Hampshire, USA

Methods in Enzymology, Volume 492
ISSN 0076-6879, DOI: 10.1016/B978-0-12-381268-1.00021-5

© 2011 Elsevier Inc.
All rights reserved.

complex, concentrated solutions. While still in its infancy, AU-FDS is becoming more widespread as shown by the increasing number of literature reports citing its use. While AU-FDS enables the analysis of systems not amenable to absorbance or interferometric detection, its use is not without limitations. In most cases, preparing samples for AU-FDS analyses requires chemical conjugation with fluorescent dyes, a step that may influence the size or shape of a molecule sufficiently to alter its transport during sedimentation. Careful preparation and characterization of the amount of free dye and the degree and site specificity of labeling is required for robust interpretation of AU-FDS data. In some cases, studies of the effect of labeling on the structure, activity, or association properties of the macromolecule may be warranted. However, these complications are of minor consequence compared to the unique information that can be obtained by AU-FDS. In particular, its ability to provide direct, physical characterization of the thermodynamic behavior of molecules in complex and concentrated solutions makes AU-FDS a powerful technology for understanding the physical underpinnings of living systems.

1. Overview of AUC

Analytical ultracentrifugation (AUC) is a powerful method for characterizing solutions of macromolecules and is an indispensable tool for the quantitative analysis of macromolecular interactions (Cole and Hansen, 1999; Hansen *et al.*, 1994; Hensley, 1996; Howlett *et al.*, 2006). Because it relies on the principle property of mass and the fundamental laws of gravitation, AUC has broad applicability and can be used to analyze the solution behavior of a variety of molecules in a wide range of solvents and over a wide range of solute concentrations. In contrast to many commonly used methods, during analytical ultracentrifugation, samples are characterized in their native state under biologically relevant solution conditions. The experiments are performed in free solution, so there are no complications due to interactions with matrices or surfaces. Because it is nondestructive, samples may be recovered for further tests following AUC. For many questions, there is no satisfactory substitute method of analysis.

Two complementary views of solution behavior are available from AUC. Sedimentation velocity provides first-principles, hydrodynamic information about the size and shape of molecules (Howlett *et al.*, 2006; Laue and Stafford, 1999; Lebowitz *et al.*, 2002), while sedimentation equilibrium provides first-principles, thermodynamic information about the solution molar masses, stoichiometries, association constants, and solution nonideality (Howlett *et al.*, 2006; Laue, 1995). Different experimental protocols are used to conduct these two types of analyses (Cole *et al.*, 2008).

Analytical ultracentrifugation has very few restrictions on the type of sample or the nature of the solvent. The fundamental requirements for the sample are (1) that it has an optical property that distinguishes it from other solution components, (2) that it sediments or floats at a reasonable rate at an experimentally achievable gravitational field, and (3) that it is chemically compatible with the sample cell. The fundamental solvent requirements are its chemical compatibility with the sample cell and its compatibility with the optical systems. The range of molecular weights suitable for AUC exceeds that of any other solution technique, from a few hundred daltons (e.g., peptides, dyes, oligosaccharides, etc.), to several hundred-million daltons (e.g., viruses, organelles, etc.).

The fundamental measurements in AUC are radial concentration distributions. These concentration distributions, called 'scans,' are acquired at intervals ranging from minutes (for velocity sedimentation) to hours (for equilibrium sedimentation). As the rotor spins, each cell passes through the optical paths of detectors capable of measuring the concentration of molecules at closely spaced radial intervals in the cell. There are three commercially available optical detectors for the XLI to measure the concentration distributions: an absorbance spectrophotometer and Rayleigh interferometer from Beckman Coulter (Fullerton, California, USA), and the fluorescence detector retrofit from Aviv Biomedical (the AU-FDS, Lakewood, New Jersey, USA). All subsequent analysis of sedimentation data relies on the quantity and quality of data available from these detectors. A detailed comparison of the virtues and limitations of the three optical systems is available (Cole *et al.*, 2008). This chapter focuses on the unique capabilities available with fluorescence detection.

2. Fluorescence Optics for the Ultracentrifuge

A prototype fluorescence detector was first reported for the Beckman Model E analytical ultracentrifuge in 1976 (Crepeau *et al.*, 1976). In the 1980s, another prototype was developed and used to study the sedimentation of ethidium-intercalated nucleic acids (Goodman *et al.*, 1984; Kapahnke *et al.*, 1986) as well as viruses labeled with fluorescent antibodies (Schmidt *et al.*, 1990). In the late 1990s, a prototype system was developed for the Beckman XLA/XLI (MacGregor *et al.*, 2004). A reengineered version of this optical system, the AU-FDS, is available commercially (Laue *et al.*, 2006) and may be retrofitted to existing XLA/XLIs.

The AU-FDS is a scanning confocal fluorimeter (Fig. 10.1) whose design derives from a prototype instrument that used a stationary argon ion gas laser coupled into a narrow-bore single mode fiber optic, which carried the light through the centrifuge vacuum chamber wall and connected to the radially translated optical track (MacGregor *et al.*, 2004). This design was functional, but was limited by variations in the excitation

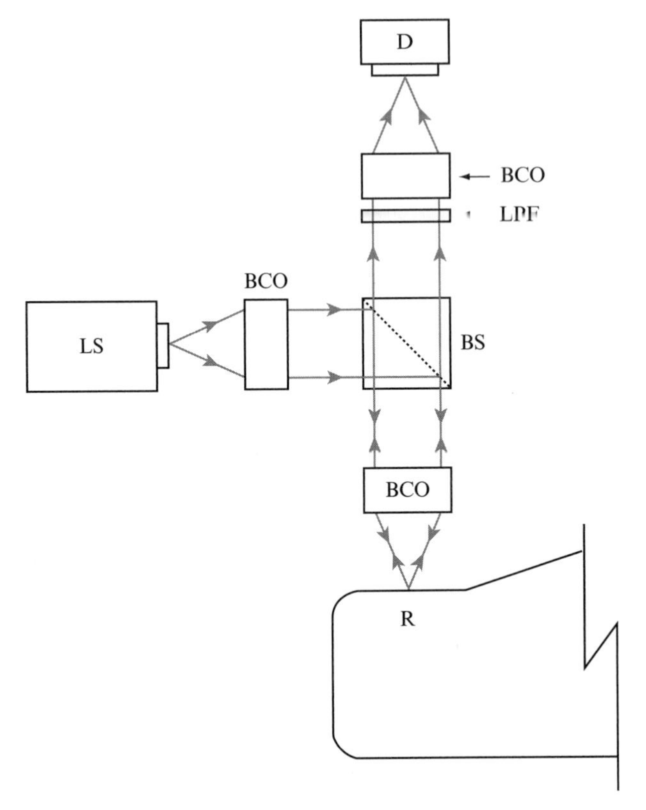

Figure 10.1 Schematic of the AU-FDS optical system. Divergent light from the light source (LS) consisting of a continuous wave solid-state laser and beam-spreading optics is collimated by the first beam collimating assembly (BCO). The beam is reflected from a beam splitting unit (BS), consisting of a dichroic mirror, and is focused by the second BCO assembly (serving as the objective lens) onto the fluorescent sample. Emission radiation that falls within the second BCO focal cone is collected and collimated before entering the BS, where it passes through the dichroic mirror, then through a long pass filter (LPF), and into a third BCO assembly. The third BCO focuses the light onto a photomultiplier within the detector assembly (D). The entire optical assembly is translated along the radial axis of the centrifuge by a stepping motor as the sample revolves below it.

intensity caused by the flexing of the optical fiber. The commercial instrument incorporates an intense cyan (15 mW, 488 nm) continuous wave solid-state laser as the excitation source fixed directly to the optical track within the vacuum chamber. The laser beam is focused to a point-source that is imaged through the confocal optical path into the sedimenting sample, thereby illuminating a narrow radial slice. A portion of the emission radiation is collected back through the optics and redirected onto a photomultiplier assembly, which converts the emission intensity into a voltage that is digitized and stored in the computer (Laue *et al.*, 2006). The entire optical system is seated in the centrifuge vacuum chamber and the illuminated spot is scanned radially over the rotor as it revolves at speeds of up to 60,000 rpm. The data acquisition system monitors the photomultiplier signal continually during the experiment, and a software platform, the Analytical Ultracentrifuge Advanced Operating System (AU-AOS), parses the sample data from this total signal (Laue *et al.*, 2006). This process is rapid, and a complete radial scan of all of the samples can be acquired simultaneously in less than 2 min. The AU-AOS provides output files in a format that is amenable to analysis with all standard sedimentation data analysis platforms.

3. Advantages of AU-FDS

Each of the three different optical systems for the AUC provides advantages for certain types of analyses, and each has properties that limit its use in other situations. Determining when to use each system is critical in designing successful sedimentation experiments (Laue, 1996). Prior to the commercial availability of fluorescence detection, the absorbance and interference systems were used extensively to produce complementary and orthogonal data. The absorbance system is easier to use and provides greater sensitivity and selectivity, while the interferometer provides much higher precision, eliminates the need for having a usable absorbance signal, and has greater acquisition speed. AU-FDS provides a different set of complementary properties, namely, exquisite sensitivity and the ability to discriminate one component type among many. Although the limitations of fluorescence detection are not trivial, AU-FDS can provide significant advantages over both absorbance and interferometric detection in the experimental investigation of certain macromolecular systems.

3.1. Sensitivity

Fluorescence is an intrinsically sensitive phenomenon due in part to its efficiency, as defined by the quantum yield (Φ), which is given as the ratio of the number of photons emitted to the number of photons absorbed, and due to the fact that it only requires the detection of photons against a dark background (rather than determining the number of photons against a bright background as is the case for the absorbance system). Many fluorophores demonstrate Φ approaching unity, indicating that nearly all energy absorbed in the excitation band is reemitted in the emission band. In addition, due to the confocal optical path of the AU-FDS, excitation photons do not reach the detector. Therefore, the resulting signal is a function only of emission radiation and system noise, which may include dark current from the photomultiplier as well as stray light (Cole *et al.*, 2008).

3.2. Selectivity

The ability to detect solute components independently of others is not required for all types of sedimentation analyses. However, in some cases, selective detection can provide significant advantages. The interferometric system provides virtually no selectivity, since the contribution to the refractive increment is largely independent of solute composition. However, the absorbance optics can provide selective detection since absorptivity is chromophore-specific. Thus, intrinsic chromophores such as heme groups or aromatic amino acid side chains can be used to differentiate the target biomolecule from other solutes or buffer components which do not contain these. AU-FDS extends this discrimination further since it detects only fluorescent molecules that have the appropriate excitation and emission bands. Therefore, in a complex mixture of macromolecules, the transport of fluorescent molecules may be isolated from that of the background cosolutes.

4. Sample Requirements for Fluorescence Detection

4.1. Labeling

Several notable proteins, including the green fluorescent protein and related mutants, are intrinsically fluorescent (Tsien, 1998). These are ideal molecules for experimentation with AU-FDS since they can be expressed, purified, and studied as is without further manipulation. The emerald GFP mutant (EGFP) has optimal fluorescence properties for the AU-FDS and its use has been reported in several methodological studies (Kroe and Laue, 2009; MacGregor *et al.*, 2004). Since EGFP-containing plasmids are widely available, fusion constructs with nonfluorescent proteins can be genetically engineered and studied by AU-FDS. In such studies, it is important to consider the effect that the large fluorescent molecule (~30 kDa) will have on the sedimentation and diffusion of the target protein.

In cases where minimal change in macromolecular transport is required, where the molecule of interest is not a protein, or where it is a protein that cannot be labeled with GFP, fluorescence may be imparted through chemical conjugation. Small-molecule fluorophores with a wide variety of chemically active groups are available to extrinsically label macromolecules (Waggoner, 2006). Due to the cyan excitation source of the AU-FDS, fluorescein or other dyes with similar properties, including Rhodamine Green™, Oregon Green® 488, BODIPY® FL, and Alexa Fluor® 488 (all available from Invitrogen, Carlsbad, California, USA), are optimal ($\lambda_{EX} \approx 490$ nm, $\lambda_{EM} \approx 515$ nm). The conjugation of a fluorophore to a target biomolecule allows a highly sensitive and selective detection, but may introduce an additional source of molecular heterogeneity that must be accounted for and controlled. The structural specificity of the conjugation site may be increased by using certain sequence-specific labels (e.g., the FLAsH system, Invitrogen). However, in most applications, the extrinsic dye is covalently conjugated to the native molecule through relatively nonspecific chemistry (e.g., amine coupling, thiol coupling, etc.). In addition to introducing heterogeneity, conjugation may induce structural rearrangement of the macromolecule that can perturb self- and hetero-interactions. Preceding AU-FDS measurements, these possibilities must be explored. Fortunately, a variety of methods are available for evaluating and controlling these effects. Ultimately, it is the context of the AU-FDS experiment that will determine the extent of characterization that is required.

4.2. Characterization of products

Biological macromolecules are intrinsically heterogeneous. Proteins, in particular, may vary in composition within a single preparation due to many factors, including glycosylation pattern and variable posttranslational modification of specific amino acid functional groups. While it is generally good practice to characterize and report this microheterogeneity whenever possible, isolation of chemically pure biomolecules is seldom a viable option. Thus, measured properties of biological molecules are to be interpreted within the context that the measurement represents an average value of the bulk material prepared as described by the experimenters. Fluorescent labeling of biomolecules creates subpopulations that simply reflect one of the many other sources of microheterogeneity. However, chemical variation in this case is directly induced by the experimenter and therefore suggests a more important role for characterization of the product. As such, details of the amount of free dye remaining in the preparation as well as the degree of labeling and the site(s) of conjugation may be important for reliable interpretation of data. In addition, such details are beneficial when reporting data in the literature so that samples can be more closely replicated when studies are repeated by other investigators. Each of these complications will be discussed separately.

4.2.1. Free dye

The AU-FDS is sensitive to all fluorescence of the appropriate excitation and emission bands regardless of molecular size. Thus, any unconjugated dye that remains in the sample during analyses may complicate data interpretation. Although the presence of free dye may not directly interfere with macromolecular transport or analysis of the resulting data, the background signal may limit the sensitivity of the measurement. In addition, accurate data modeling would require the inclusion of the dye as an additional solute component. This added complexity is unwarranted since the removal of unconjugated dye to undetectable levels is straightforward and can be achieved with simple fractionation techniques available in most laboratories. The preferred method of free dye removal is with gel filtration chromatography where the column retention characteristics are selected such that the labeled molecule is baseline-resolved from the column total volume, where the dye will likely be retained. HPLC is the preferred strategy due to the capability of high resolution. However, standard pressure systems are sufficient in most cases. Where chromatographic instrumentation is not available, disposable gravity-fed or centrifugal desalting columns suffice (Bailey *et al.*, 2009). These may be differentiated from the gel filtration products as they incorporate a stationary phase designed for molecular weight cutoff (usually 5000–6000 Da) rather than selective permeation.

4.2.2. Degree of labeling

Most commercially available fluorophores have chemical specificities (e.g., amines, sulfhydryls, etc.) that may result in conjugation at more than one location on the target biomolecule. This not only results in mixed populations where relative concentrations may vary from batch to batch, but also may result in loss of structure or function if the degree of labeling is high or if conjugation occurs at key sites. Determination of the degree of labeling is straightforward and is recommended in support of AU-FDS analyses.

In one approach specific to fluorescent labeling of proteins, the absorbance of the conjugate is measured both in the aromatic amino acid band (A_{280}) and in the band corresponding to the maximum absorbance of the fluorophore (A_{max}). The degree of labeling (DOL) in moles of dye per mole protein is given as

$$\mathrm{DOL} = \frac{A_{\mathrm{max}}}{c_{\mathrm{protein}}\varepsilon_{\mathrm{dye}}} \tag{10.1}$$

where ε_{dye} is the molar extinction coefficient of the dye (generally provided by the manufacturer) and $c_{protein}$ is the protein concentration in molar units. The concentration of the protein is determined by the Beer–Lambert law after first correcting for the overlapping absorbance contribution of the dye:

$$A_{\text{protein}} = A_{280} - A_{\max} \times \text{CF} \tag{10.2}$$

where CF is a correction factor that is equal to $A_{280}/A_{\max}$ of the free dye. This quantity can be measured, but is generally available from the dye manufacturer.

An orthogonal and more general approach is to measure the intact mass of the conjugate ($M_{\text{r,conj}}$) using mass spectrometry and compare to the unlabeled molecule (M_{r}). The difference is equivalent to the molecular weight of the dye ($M_{\text{r,dye}}$) multiplied by the molar stoichiometry (n):

$$M_{\text{r,conj}} = M_{\text{r}} + \left(M_{\text{r,dye}} \times n\right) \tag{10.3}$$

This approach is not specific to the chemical composition of the biomolecule and is therefore applicable to conjugates derived from proteins, lipids, nucleic acids, and carbohydrates in so much as the target molecule is amenable to intact mass spectrometric analysis. Although complications may arise with very large or highly heterogeneous molecules, complex accurate-mass instruments are not generally required, since highly accurate values of n are not necessary to establish a reasonable estimate for DOL. In most cases, simple and widely available configurations such as MALDI-TOF, ESI-TOF, or ESI-q will suffice.

4.2.3. Site specificity of conjugation

In some cases, the above studies may be sufficient to characterize the conjugate prior to AU-FDS. However, in cases where the specific site of fluorophore conjugation may affect interpretation of the AU-FDS data, characterization in greater detail must be performed. Studies based on the interaction of labeled macromolecules or relying on the correlation of AU-FDS results with functional assays will certainly require greater detail in the characterization of the product. Generally, in studies that are interpreted under the assumption that the conjugate reasonably replicates the structure and function of the native protein, it may be important not only to establish the site(s) of conjugations, but also to limit the extent of conjugation so that subpopulation variability is minimal.

The evaluation of conjugation site specificity is best understood for proteins of known amino acid sequence. Here, the multiplicity of functional groups used to conjugate fluorophores (primary amines on lysine and at the N-terminus, sulfhydryls on cysteines, etc.) can lead to site heterogeneity. The best approach to establishing the conjugation site(s) is by comparative peptide mapping of the conjugated and unconjugated proteins using LC-MS or LC-MSMS. For these methods, an endoprotease such as trypsin is used to generate specific peptides, which will have known masses due to prior knowledge of the amino acid sequence and endoprotease specificity. The proteolytic fragments are monitored by their elution position on a reversed-phase column (typically, C18) and by their mass determined in the coupled mass spectrometer. Proteolytic fragments within the peptide map that contain one or more fluorophore-conjugated amino acids will generally be shifted in their column retention compared to the corresponding position in the map of the unmodified protein. In addition, the mass of the shifted peak should reflect an increase of a multiple of the mass of the fluorophore. At this level of detail, the site will have been deduced if there is only one compatible chemical group within the given fragment. If there are multiple potential sites, more details will be required. However, it may suffice to pinpoint conjugation to within the fragment without elucidation of the specific amino acid(s), particularly if the fragment can be located on an available crystal structure. Alternatively, if LC-MSMS is available, the specific amino acid can be determined by evaluation of the distribution of *b* and *y* ions resulting from collisionally induced or electron transfer dissociation of the parent ion of the conjugated fragment. It should be noted that some fluorophore–amino acid linkages such as the thiourea bond formed from isothiocyanate dyes may be labile and not well suited to this analysis (Schnaible and Przybylski, 1999).

4.3. Dye–protein interaction

Some studies using AU-FDS are interpreted under the assumption that the conjugate reasonably replicates the structure and function of the native protein. In these cases, it is important not only to establish the site(s) of conjugations as discussed above, but also to demonstrate experimentally that the assumption is valid. This can be accomplished in several ways. First, the conjugation chemistry should be carefully selected to balance DOL and sensitivity of detection. Clearly, with higher DOL, greater signal will be imparted per mole of target molecule. However, higher DOL means there is also a greater likelihood that the macromolecular structure will be perturbed. In some cases, it may be required to target fluorophore conjugation to a single site. This might be accomplished when labeling proteins with a single free cysteine. More generally, primary amine-reactive chemistries can be targeted to favor the α-amine of the N-terminus over the ε-amine of lysine residues because of the difference in pKa of these groups (~7 vs. ~10, respectively and depending on local conformation).

Once the conjugation chemistry is selected and the derivative has been prepared and purified, the effect of the fluorophore on the global protein conformation can be evaluated with solution-based structural methods such as circular dichroism (CD) spectroscopy. CD spectra are acquired for the derivative and for the unconjugated protein and then qualitatively compared to assess relative changes. The optical activity of the dye itself should not interfere with acquisition of far-UV spectra (~190–250 nm).

Unfortunately, the corresponding experiment is not possible in the near-UV range (~250–350) due to an overlapping signal from the fluorophore, which convolutes interpretation of protein conformational changes. Thus, the comparative analysis is limited primarily to secondary structure.

For enzymes, it may be informative to conduct side-by-side functional assays with labeled and unlabeled material to determine whether the conjugation perturbs the conformation around the active site. This assessment is critical if the AU-FDS studies are to be correlated with activity measurements.

In the determination of thermodynamic parameters for associating proteins with AU-FDS, it is important to investigate the effect that the fluorophore has on the interaction. This can be accomplished by following up analyses of the labeled sample with identical analyses using an equilibrated mixture of labeled and unlabeled proteins. While limited in sensitivity due to the population of unlabeled species, these experiments may be sufficient to demonstrate comparability with respect to the trend in dissociation at higher concentrations.

5. Applications of AU-FDS

The development of the AU-FDS is a recent advance in AUC technology, and the majority of studies using the technique have emerged only within the last 3 years. The first hypothesis-driven reports using the detector (second-generation prototype and commercial AU-FDS) were published in 2008 and followed with several other reports over the next 2 years. To our knowledge, presently, there have been 12 research papers published in which AU-FDS was used (Bailey *et al.*, 2009; Burgess *et al.*, 2008; Demeule *et al.*, 2009; Hainzl *et al.*, 2009; Kingsbury *et al.*, 2008; Kroe and Laue, 2009; MacGregor *et al.*, 2004; Rajagopalan *et al.*, 2008; Ryan *et al.*, 2008; van Dieck *et al.*, 2009, 2010; Zhu *et al.*, 2010). While this represents a small fraction of the total AUC research published during that time (over 300 published reports), use of AU-FDS will certainly grow as the method becomes more established in the AUC community.

In 2009, Kroe and Laue published a comprehensive report in which the experimental capabilities of AU-FDS were investigated (Kroe and Laue, 2009). In this chapter, the terms, NUTS (normal use tracer sedimentation) and BOLTS (biological online tracer sedimentation), were proposed as a means of categorizing AU-FDS-based studies within the context of the two main advantages of the detector: sensitivity and selectivity. Here, the experimental basis for these methodologies will be outlined and pertinent literature reports will be discussed.

5.1. NUTS

Characterization of self- or hetero-associating systems by sedimentation velocity or sedimentation equilibrium requires optical detection at loading concentrations where appreciable dissociation has occurred. Of the two, sedimentation equilibrium is more forgiving since the exponential concentration gradient provides concentrations much lower than the loading concentration near the air–liquid meniscus. However, the greater signal-to-noise ratio available with sedimentation velocity analysis means that it may be better suited for low concentration work. For both approaches, the sensitivity of absorbance and interference optical systems makes them of limited utility in the analysis of very high-affinity interactions. Since AU-FDS provides greater sensitivity, lower concentrations of macromolecule can be detected and therefore, stronger association constants can be determined. Thus, AU-FDS can be used to extend the lower limit of detection for standard analyses in dilute solution. This methodology has become known as normal use tracer sedimentation or NUTS (see Fig. 10.2).

AU-FDS was employed by Burgess *et al.* (2008) to evaluate the quaternary structure of Alexa Fluor® 488-labeled dihydrodipicolinate synthase from methicillin-resistant *Staphylococcus aureus* (MRSA-DHDPS) at low concentrations. Detection of this protein with absorbance optics was reported to be limited to low micromolar concentrations, whereas AU-FDS extended the data to subnanomolar concentrations. The authors determined that at low concentrations, MRSA-DHDPS exists in a monomer–dimer equilibrium. Furthermore, incubation with one of the substrates of the enzyme, pyruvate, stabilized the dimeric form. Sedimentation equilibrium AU-FDS was used to quantitatively determine K_D values and indicated a 20-fold higher affinity in the presence of pyruvate.

Similarly, Zhu *et al.* (2010) investigated the quaternary structure of Alexa Fluor® 488-labeled pyruvate kinate type 1 (PK) from *Escherichia coli* in the presence and absence of its allosteric regulator fructose-1,6-bisphosphate (FBP) and substrate phosphoenol pyruvate (PEP). Here, the authors used

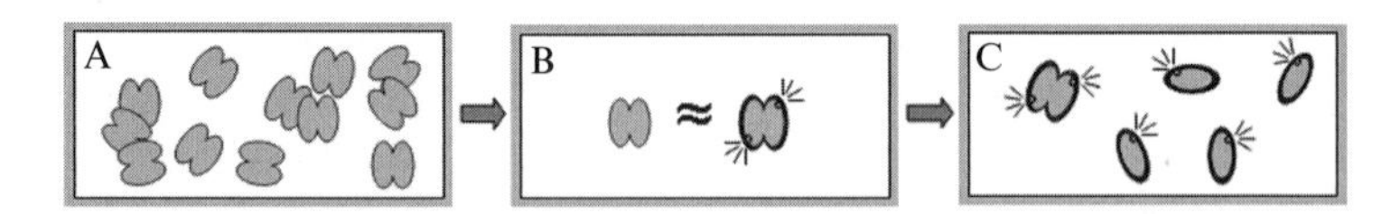

Figure 10.2 Principle of NUTS. (A) At the low detection limit of conventional optical systems, high-affinity self-associated complexes may not be significantly dissociated. Here, such a system is depicted as a dimer. (B) Labeling the protein with a fluorophore allows greater signal per mole of sample due to the sensitivity of AU-FDS. For NUTS, it is assumed that the labeled protein reasonably approximates the native protein at least with respect to the self-association properties. (C) At the low detection limit of the AU-FDS (up to ~1000 fold lower than conventional optics), the dissociated state may be sufficiently populated to allow characterization of the affinity.

AU-FDS to permit detection at PK concentrations within the range employed for enzymatic assays (low nanomolar). With this capability, the authors were able to test the hypothesis that the mechanism of allostery was through modulation of the quaternary structure. The results of the study indicated that PK from *E. coli* was a tightly associated tetramer and that neither FBP nor PEP modulated the quaternary structure at the concentrations tested.

Ryan *et al.* (2008) identified a tetrameric intermediate in the amyloid fibrillization pathway of Alexa Fluor® 488-labeled apolipoprotein C-II (apoC-II) in a study that was enabled partly by AU-FDS. At low concentrations, under conditions where fibrillogenesis was negligible, apoCII was observed to sediment as a monomer. In the presence of submicellar concentrations of short-chain phospholipids, a discrete tetrameric form was observed. The formation of the tetramer correlated with increased β-structure compared to the monomeric form. Furthermore, conditions favoring tetramer formation increased the rate of fibril formation, suggesting that the tetramer is an intermediate in the fibrillogenic pathway.

In the investigation of another amyloidogenic protein, transthyretin (TTR), Kingsbury *et al.* (2008) used AU-FDS to generate weight-average *s* isotherms of fluorescein-labeled TTR at low concentrations not amenable to absorbance or interference detection. Dissociation of the TTR tetramer was observed below $\sim$7 n*M* and the isotherm could be modeled as a monomer–dimer–tetramer equilibrium. S-sulfonated and S-cysteinylated forms of the protein were assessed in the same manner, and the data suggested that S-sulfonation stabilized the tetramer, while S-cysteinylation destabilized the tetramer.

5.2. BOLTS

As a first-principles method, AUC provides an extraordinary amount of physicochemical information about the target biomolecule based on the seemingly simple phenomenon of transport in the centrifugal field. Whether sedimentation velocity or equilibrium is employed, the mathematical framework for the analysis of data generally limits interpretation to dilute solutions. Although extremely informative and rich in biophysical information, in some cases, dilute-solution constraints limit the quantitative analysis of some important and interesting systems. For example, high concentration protein solutions are frequently encountered in the formulation of biopharmaceutics, yet our current understanding of hydrodynamics limits us to qualitative and semiquantitative interpretations (Shire *et al.*, 2004). If we could overcome the current limitations, it will be possible to consider the quantitative study of molecular interactions in the presence of high concentration of cosolutes, thus providing a more physiologically relevant context for understanding the interaction energies (Rivas and

Minton, 2003, 2004). Furthermore, models that incorporate a more complete hydrodynamic theory will be needed to allow a complete and general understanding of molecular thermodynamics (Chebotareva *et al.*, 2004; Ellis, 2001; Minton, 2000, 2001, 2005; van den Berg *et al.*, 1999; Wills and Winzor, 2001). As evidenced by the elegant work of Minton and coworkers, sedimentation provides a unique and vital experimental tool in this area (Fodeke and Minton, 2010; Rivas and Minton, 2003, 2004; Rivas *et al.*, 1999). Due to the limited selectivity of the absorbance and interference optical systems, the technique of tracer sedimentation has been developed to facilitate the study of molecular transport in complex solutions. Prior to the availability of AU-FDS, these studies were conducted by fractionation of sedimentation–diffusion equilibrium concentration gradients, followed by offline analysis with highly selective methods such as radiolabel, fluorescence, or enzymatic activity assays (Fodeke and Minton, 2010). The AU-FDS provides a higher precision alternative since a separate fractionation step is not required. The real-time detection available with AU-FDS also allows sedimentation velocity analyses, which may be the preferred mode in some cases. To emphasize the potential physiological implications and the real-time nature of detection, this approach using AU-FDS has been termed biological online tracer sedimentation or BOLTS (see Fig. 10.3).

The use of AU-FDS to study transport in complex solutions represents a promising, but as yet underdeveloped methodology. To our knowledge, only three reports with experiments fitting within the context of BOLTS have been published at this time (Demeule *et al.*, 2009; Kingsbury *et al.*, 2008; Kroe and Laue, 2009). However, this area is expected to become much more active as AU-FDS becomes more widely known.

While experimental application is limited, one may classify BOLTS analyses in at least two broad theoretical groups. In the first, information about the effect of the complex solution on the macromolecule is desired.

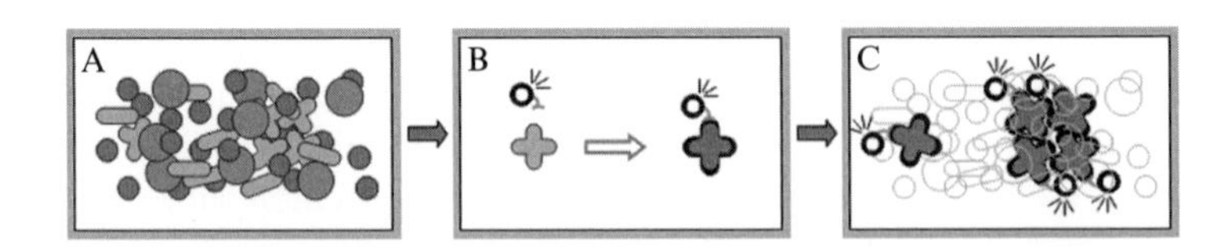

Figure 10.3 Principle of BOLTS. (A) With conventional optics, a target molecule is not readily distinguished from a high background of nontarget molecules such as what might compose a biological fluid. (B) The target molecule can be labeled either directly through chemical conjugation followed by introduction into the matrix or indirectly through a fluorescent probe which binds specifically and noncovalently to the target molecule while within the matrix. (C) The AU-FDS selectively detects the fluorescently labeled target molecule including all association states or multicomponent complexes with which the target may interact.

These experiments are biophysical in nature and may be utilized in order to address the theoretical implications of nonideality or to investigate the effect that crowded solutions have on molecular structure or interactions. In the second category, one may seek to determine the state of an endogenous protein in a physiologic solution such as blood serum, urine, or cerebrospinal fluid. Here, the experiment is used to test hypotheses involving the association state or cotransport of target molecules with respect to pathology.

5.2.1. BOLTS for biophysical studies

Traditionally, sedimentation analyses are conducted in dilute solution. However, in some cases, it is desirable to deviate experimentally from the ideal case and to investigate transport in highly concentrated or heterogeneous solutions. For instance, therapeutic proteins such as monoclonal antibodies are often formulated at concentrations of 100 mg/mL or greater (Shire *et al.*, 2004). One might employ sedimentation at such concentrations in order to characterize the size distribution of aggregates formed under those conditions. Further, biophysical characterization after dilution into a biological fluid may provide some insight into what may occur structurally to the therapeutic after it has been infused into a patient. AU-FDS can be used to address these issues due to the ability to track the sedimentation of the fluorescently labeled target molecule within a high background of unlabeled like molecules or within a biological fluid. An example is provided by Demeule *et al.* (2009) who investigated the complexes formed between a fluorescently labeled therapeutic IgG (omalizumab) and human IgE in dilute solution as well as in human serum and plasma samples. The authors report that the biological fluid does indeed have a measureable effect both on the affinity, which is increased in serum, and on the size distribution of the complexes. To date, this is a unique study and highlights the potential that AU-FDS, and in particular, BOLTS analyses, have for the biophysical characterization of therapeutic proteins.

5.2.2. BOLTS for pathophysiologic studies

The above category of BOLTS represents the study of exogenous molecules diluted into the complex solution. An alternative strategy is to impart fluorescence by way of a probe to endogenous molecules already present in the solution and determine the transport properties directly within that native biological fluid. This type of BOLTS has obvious clinical significance when considering a target molecule that undergoes complex formation or aggregation as a result of some pathologic mechanism. In this mode, BOLTS has the potential to differentiate clinical samples on the basis of detecting the disease-associated complexes. Currently, there are no literature reports involving this type of research, so the actual utility of the approach remains to be determined.

5.3. Other experimental approaches

While efforts have been made here to group AU-FDS studies in accordance with the application of NUTS or BOLTS analyses, clearly not all studies are amenable to such categorization. Nor should AU-FDS studies be limited to within the constraints of nomenclature. In general, AU-FDS is used to achieve high sensitivity or selectivity, and uses other than those described above have been reported. For example, AU-FDS has been used to qualitatively assess protein–protein interactions in dilute solution (Hainzl *et al.*, 2009; Rajagopalan *et al.*, 2008; van Dieck *et al.*, 2009, 2010; van Dieck *et al.*, 2010). Here, the sedimentation coefficient of the fluorescently labeled protein is determined in the absence of and in the presence of various concentrations of cosolute. Binding is detected as an increase in the sedimentation coefficient when the mixture is analyzed. Corresponding experiments are possible with the other optical systems. However, AU-FDS simplifies the analysis since the cosolute itself does not contribute to signal intensity.

6. Current Challenges for AU-FDS

The distinct advantages that AU-FDS offers sedimentation analyses have been outlined and discussed. It is equally important, however, to consider the properties of the optical system that limit its broader applicability. Furthermore, several considerations arise when interpreting data derived from fluorophore-labeled, as opposed to native, biomolecules both in dilute and complex solution. Presently, these factors may be viewed as necessary experimental trade-offs in the light of enhanced sensitivity and selectivity. It is anticipated that as the use of AU-FDS increases, so too will the understanding of these factors so that control measures can be improved and incorporated into common practice. Here, we discuss several key challenges that must be addressed as AU-FDS advances to more routine usage.

6.1. Signal nonlinearity and accuracy

The fundamental measurement in AUC is the concentration as a function of radial position, and it is the concentration which is required to interpret molecular associations and nonideality. Each of the three optical systems provides a signal that is proportional to concentration and must be converted to concentration through a defined mathematical relationship. In the absence of any other information, it is assumed that the signal is zero at zero concentration and that the signal changes in direct proportion to the concentration.

In AU-FDS experiments, the fluorescence signal is the number of counts that is proportional to the fluorescence intensity. The electronics are designed so that there will be a signal of $\sim$300 counts in the absence of any fluorescence. The proportionality between the fluorescence intensity and the signal depends on both instrumental factors (the gain settings, the photomultiplier sensitivity, the optical focus and alignment, etc.) and the properties of the fluorophore (its absorbance characteristics at the excitation wavelength and its quantum yield). Furthermore, the fluorescence signal is relative (not referenced to an optical blank), and it should be expected that the same sample may produce different signal amplitudes on different machines. Finally, fluorescence intensity signals may be prone to nonlinearity with fluorophore concentration.

Factors leading to nonlinearity may include collisional quenching, photobleaching, binding of the sample to surfaces, and the inner filter effect (absorbance of the excitation or emission signals by the nonilluminated portions of the sample that lie along the optical path). All of these processes will decrease the total signal intensity relative to what it would be in their absence. The result of these processes is the suppression of the signal and consequent distortion of its radial disposition compared to the actual radial concentration distribution. Such effects may lead to poor accuracy if they are not detected and minimized, and in the case of the inner filter effect may lead to a decreasing signal with increasing concentration.

The best way to detect nonlinearity is to construct a graph of the signal as a function of concentration for a dilution series of the sample. Our experience is that for a given gain setting, the AU-FDS is linear over at least one order of magnitude in concentration (MacGregor *et al.*, 2004). The range of useable concentrations varies, but successful experiments have been conducted on 700 f*M* samples (protein concentration, 3 n*M* dye concentration) and on 10 μ*M* samples. More typically, the useable concentration range is from 100 p*M* to 1 μ*M*. For high concentrations, it is recommended that the AU-FDS objective lens be focused as close to the bottom surface of the window as possible in order to minimize the inner filter effect. In addition, for very dilute samples (<1 n*M*), binding of sample to the windows and centerpiece is often observed, and for these cases, it is recommended that 0.1 mg/mL of a carrier protein be included in the solution (Kroe and Laue, 2009).

In AU-FDS and all other fluorescence-based methods, photobleaching and collisional quenching are dependent on properties of the fluorophore, its conjugation to the macromolecule, and solvent conditions. Modern high-quality fluorophores are resistant to both photobleaching and quenching (at least autoquenching), and so their effects are not likely to be detectable if DOL is low and compatible solvents are used. High DOL may improve the lower limit of detection by providing more signal per biomolecule. However, high DOL is more likely to perturb molecular structure and may result in *decreased* linear range due to increased quenching and inner-filter effect. Thus, DOL may need to be optimized depending on the analytical requirements. In general, careful development of fluorescent constructs beginning with fluorophore selection and continuing with molecular characterization and system testing will insure accurate interpretation of AU-FDS experiments. While time-consuming, these preparatory efforts will help prevent spurious interpretations and lead to better quality data.

While it may seem that concentration measurements are problematic for the AU-FDS, it is important to note that corresponding absorbance and refractive measurements are not free from equivocal effects (Laue, 1996). The absorbance system is subject to nonlinearity in the extinction coefficient at high concentrations. In addition, stray light resulting from scattering and fluorescence in addition to polychromicity from the optical bandpass can erode accuracy. The interferometer is less subject to nonlinear optical effects, but accuracy is limited by numerical aperture and lens quality and is subject to Weiner-skewing errors at high concentrations (Laue, 1996).

Interacting systems may provide a different concern if the quantum yield changes with the labeled protein's association state. In this case, one must determine the change in the quantum yield and incorporate the different values in the fitting model. Work has been conducted on this matter in J. J. Correia's laboratory (personal communications), and the most recent version of Sedanal (Stafford and Sherwood, 2004) (available at http://www.rasmb.bbri.org/software/windows/sedanal-stafford/) allows analyses of associating systems with variable quantum yields.

6.2. Dye–protein interaction in NUTS

Of critical importance in the interpretation of NUTS data for interacting systems is the assumption that fluorophore conjugation does not measurably alter the interaction affinities. This assumption can be tested directly with solute-mixing experiments, as described previously (Kingsbury *et al.*, 2008). To limit the effect, in some cases, it may be possible to fine-tune the specificity of the conjugation site. Alternatively, fluorophores with similar optical properties but different chemical structures can be tested. Despite these efforts, in some cases, extrinsic labeling of the molecule may not be possible without altering the conformation such that the interacting surfaces are perturbed. There is no data to suggest how often these instances may occur or which biomolecules may be more susceptible, but it is best if there are estimates of the association strength made by other means. In the absence of other estimates for comparison, association strengths determined by NUTS analyses must be interpreted with caution. On the other hand, stoichiometries determined using NUTS are less susceptible to the effects of signal nonlinearity. In some cases, only relative changes in affinities may be sought regardless of whether the native affinity is perturbed by dye conjugation. For instance, the relative effects of various posttranslational modification, amino acid substitution, excipient concentration, or ionic strength on affinity may all be studied in conjugates that deviate slightly from native conformation.

6.3. Lamm equation modeling in BOLTS

The evolution of sedimenting boundaries in dilute solution is a well-defined process described accurately by the Lamm equation, and there are several data analysis programs that fit sedimentation data based on solutions or approximations of the Lamm equation (Cole *et al.*, 2008). Conversely, sedimentation in highly concentrated or heterogeneous fluids cannot be adequately described using existing Lamm equation solutions. Nonfluorescent cosolutes do not contribute to the signal intensity, but their redistribution in the centrifugal field will affect the sedimentation of the labeled molecules. At the very least, high concentrations of macromolecular cosolute will form density and viscosity gradients that will cause the sedimentation coefficient of the fluorescent species to be a function of both the radial position and time. One hydrodynamic artifact caused by high concentrations of cosolutes is the Johnston–Ogston effect in which the concentration of a slower sedimenting material builds up behind the boundary of a faster sedimenting material, leading to a hypersharp boundary of the slower material, or even to a peak in its concentration at the faster material's boundary position (Johnston and Ogston, 1946). Several AU-FDS analyses conducted in blood serum have demonstrated the Johnston–Ogston effect (Kingsbury *et al.*, 2008; Kroe and Laue, 2009). For the purposes of these studies, serum may be considered a concentrated and heterogeneous cosolute mixture with a single predominant component, albumin, which may be present at up to 50 mg/mL or greater. In serum, the sedimentation of GFP (Kroe and Laue, 2009) and fluorescently labeled transthyretin (Kingsbury *et al.*, 2008) (both slightly smaller than albumin) were reportedly not well described by the superposition of ideal Lamm equation solutions in c (s) analyses (Schuck, 2000). However, sedimentation of IgG, which is more than twice the size of albumin, is less perturbed in serum since its boundary is in the plateau concentration of the slower sedimenting albumin (Demeule *et al.*, 2009). While disconcerting to experimenters unfamiliar with it, the Johnston–Ogston effect can be described completely by appropriate solutions of the Lamm equation. Analytical programs that could model transport in highly nonideal fluids accurately would be a tremendous advance for BOLTS studies.

7. Conclusion

The recent development of fluorescence optics for the AUC has extended the applicability of the technique, enabling the direct analysis of much tighter macromolecular interactions and permitting the selective detection of target molecules in a high background of nonfluorescent solution components. The former application represents the bulk of the present AU-FDS literature reports. However, it is the latter application that will prove most useful in biochemistry and molecular biology. The use of the instrument in the investigation of complex macromolecular solutions is promising, but has not yet been widely reported. While enabling unique studies of interesting and important molecular systems, the challenges of AU-FDS analyses are not to be overlooked. With careful sample preparation and characterization, complications that may lead to data misinterpretation can be mitigated or controlled. It is expected that use of the AU-FDS will continue to grow as more researchers become aware of the unique experimental strategies provided by the instrument.

REFERENCES

Bailey, M. F., *et al.* (2009). Methods for sample labeling and meniscus determination in the fluorescence-detected analytical ultracentrifuge. *Anal. Biochem.* **390,** 218–220.

Burgess, B. R., *et al.* (2008). Structure and evolution of a novel dimeric enzyme from a clinically important bacterial pathogen. *J. Biol. Chem.* **283,** 27598–27603.

Chebotareva, N. A., *et al.* (2004). Biochemical effects of molecular crowding. *Biochemistry Mosc.* **69,** 1239–1251.

Cole, J. L., and Hansen, J. C. (1999). Analytical ultracentrifugation as a contemporary biomolecular research tool. *J. Biomol. Tech.* **10,** 163–176.

Cole, J. L., *et al.* (2008). Analytical ultracentrifugation: Sedimentation velocity and sedimentation equilibrium. *In* "Methods in Cell Biology," Vol. 84, (J. J. Correia, *et al.*, eds.) pp. 143–179. Chapter 6, Elsevier, Holland.

Crepeau, R. H., *et al.* (1976). UV laser scanning and fluorescence monitoring of analytical ultracentrifugation with an on-line computer system. *Biophys. Chem.* **5,** 27–39.

Demeule, B., *et al.* (2009). A therapeutic antibody and its antigen form different complexes in serum than in phosphate-buffered saline: A study by analytical ultracentrifugation. *Anal. Biochem.* **388,** 279–287.

Ellis, R. J. (2001). Macromolecular crowding: Obvious but underappreciated. *Trends Biochem. Sci.* **26,** 597–604.

Fodeke, A. A., and Minton, A. P. (2010). Quantitative characterization of polymer-polymer, protein-protein, and polymer-protein interaction via tracer sedimentation equilibrium. *J. Phys. Chem. B* **114,** 10876–10880.

Goodman, T. C., *et al.* (1984). Viroid replication: Equilibrium association constant and comparative activity measurements for the viroid-polymerase interaction. *Nucleic Acids Res.* **12,** 6231–6246.

Hainzl, O., *et al.* (2009). The charged linker region is an important regulator of Hsp90 function. *J. Biol. Chem.* **284,** 22559–22567.

Hansen, J. C., *et al.* (1994). Analytical ultracentrifugation of complex macromolecular systems. *Biochemistry* **33,** 13155–13163.

Hensley, P. (1996). Defining the structure and stability of macromolecular assemblies in solution: The re-emergence of analytical ultracentrifugation as a practical tool. *Structure* **4,** 367–373.

Howlett, G. J., *et al.* (2006). Analytical ultracentrifugation for the study of protein association and assembly. *Curr. Opin. Chem. Biol.* **10,** 430–436.

Johnston, J. P., and Ogston, A. G. (1946). A boundary anomaly found in the ultracentrifugal sedimentation of mixtures. *Trans. Faraday Soc.* **42,** 789–799.

Kapahnke, R., *et al.* (1986). The stiffness of dsRNA: Hydrodynamic studies on fluorescence-labelled RNA segments of bovine rotavirus. *Nucleic Acids Res.* **14,** 3215–3228.

Kingsbury, J. S., *et al.* (2008). The modulation of transthyretin tetramer stability by cysteine 10 adducts and the drug diflunisal: Direct analysis by fluorescence-detected analytical ultracentrifugation. *J. Biol. Chem.* **283,** 11887–11896.

Kroe, R. R., and Laue, T. M. (2009). NUTS and BOLTS: Applications of fluorescence-detected sedimentation. *Anal. Biochem.* **390,** 1–13.

Laue, T. M. (1995). Sedimentation equilibrium as a thermodynamic tool. *Methods Enzymol.* **259,** 427–452.

Laue, T. M. (1996). Choosing which optical system of the Optima™ XL-I analytical ultracentrifuge to use, Application Note A-1821A. Beckman Instruments.

Laue, T. M., and Stafford, W. F., III. (1999). Modern applications of analytical ultracentrifuge. *Annu. Rev. Biophys. Biomol. Struct.* **28,** 75–100.

Laue, T. M., *et al.* (2006). A light intensity measurement system for the analytical ultracentrifuge. *Prog. Colloid Polym. Sci.* **131,** 1–8.

Lebowitz, J. L., *et al.* (2002). Modern analytical ultracentrifugation in protein science: A tutorial review. *Protein Sci.* **11,** 2067–2079.

MacGregor, I. K., *et al.* (2004). Fluorescence detection for the XLI analytical ultracentrifuge. *Biophys. Chem.* **108,** 165–185.

Minton, A. P. (2000). Implications of macromolecular crowding for protein assembly. *Curr. Opin. Struct. Biol.* **10,** 34–39.

Minton, A. P. (2001). The influence of macromolecular crowding and macromolecular confinement on biochemical reactions in physiological media. *J. Biol. Chem.* **276,** 10577–10580.

Minton, A. P. (2005). Influence of macromolecular crowding upon the stability and state of association of proteins: Predictions and observations. *J. Pharm. Sci.* **94,** 1668–1675.

Rajagopalan, S., *et al.* (2008). 14-3-3 activation of DNA binding of p53 by enhancing its association into tetramers. *Nucleic Acids Res.* **36,** 5983–5991.

Rivas, G., and Minton, A. P. (2003). Tracer sedimentation equilibrium: A powerful tool for the quantitative characterization of macromolecular self- and hetero-associations in solution. *Biochem. Soc. Trans.* **31,** 1015–1019.

Rivas, G., and Minton, A. P. (2004). Non-ideal tracer sedimentation equilibrium: A powerful tool for the characterization of macromolecular interactions in crowded solutions. *J. Mol. Recognit.* **17,** 362–367.

Rivas, G., *et al.* (1999). Direct observation of the self-association of dilute proteins in the presence of inert macromolecules at high concentration via tracer sedimentation equilibrium: Theory, experiment, and biological significance. *Biochemistry* **38,** 9379–9388.

Ryan, T. M., *et al.* (2008). Fluorescence detection of a lipid-induced tetrameric intermediate in amyloid fibril formation by apolipoprotein C-II. *J. Biol. Chem.* **283,** 35118–35128.

Schmidt, B., *et al.* (1990). A fluorescence detection system for the analytical ultracentrifuge and its application to proteins, nucleic acids, and viruses. *Colloid Polym. Sci.* **268,** 45–54.

Schnaible, V., and Przybylski, M. (1999). Identification of fluorescein-5'-isothiocyannate-modification sites in proteins by electrospray-ionization mass spectrometry. *Bioconj. Chem.* **10,** 861–866.

Schuck, P. (2000). Size-distribution analysis of macromolecules by sedimentation velocity ultracentrifugation and Lamm equation modeling. *Biophys. J.* **78,** 1606–1619.

Shire, S. J., *et al.* (2004). Challenges in the development of high protein concentration formulations. *J. Pharm. Sci.* **93,** 1390–1402.

Stafford, W. F., and Sherwood, P. J. (2004). Analysis of heterologous interacting systems by sedimentation velocity: Curve fitting algorithms for estimation of sedimentation coefficients, equilibrium and kinetic constants. *Biophys. Chem.* **108,** 231–243.

Tsien, R. Y. (1998). The green fluorescent protein. *Annu. Rev. Biochem.* **67,** 509–544.

van den Berg, B., *et al.* (1999). Effects of macromolecular crowding on protein folding and aggregation. *EMBO J.* **18,** 6927–6933.

van Dieck, J., *et al.* (2009). Modulation of the oligomerization state of p53 by differential binding of proteins of the S100 family to p53 monomers and tetramers. *J. Biol. Chem.* **284,** 13804–13811.

van Dieck, J., *et al.* (2010). Molecular basis of S100 proteins interacting with the p53 homologs p63 and p73. *Oncogene* **29,** 2024–2035.

Waggoner, A. (2006). Fluorescent labels for proteomics and genomics. *Curr. Opin. Chem. Biol.* **10,** 62–66.

Wills, P. R., and Winzor, D. J. (2001). Studies of solute self-association by sedimentation equilibrium: Allowance for effects of thermodynamic non-ideality beyond the consequences of nearest-neighbor interactions. *Biophys. Chem.* **91,** 253–262.

Zhu, T., *et al.* (2010). The quaternary structure of pyruvate kinase type 1 from *Escherichia coli* at low nanomolar concentrations. *Biochimie* **92,** 116–120.

Author Index

H

M

N

O

P

T

U

V

W

X

Y

Z

Subject Index

N

T

X

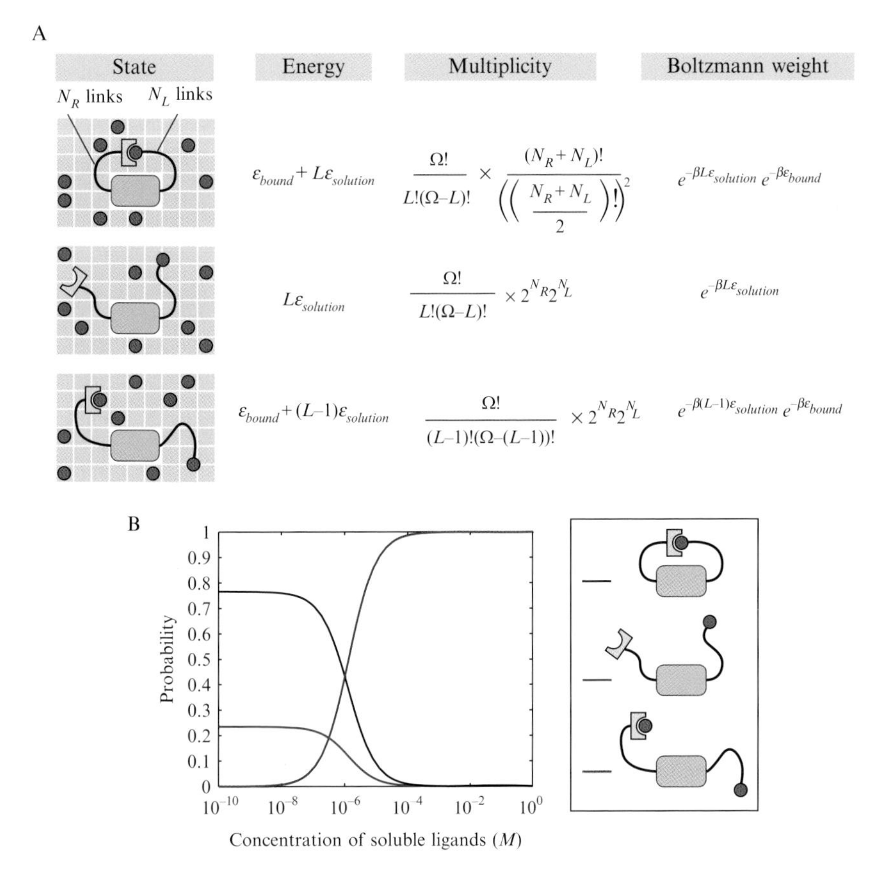

Hernan G. Garcia *et al.*, Figure 2.13 Random-walk models of tethered ligand–receptor pairs. (A) Schematic of the tethered ligand–receptor pair and the associated statistical weights. The multiplicities for the polymer are computed using a one-dimensional toy model of the tether. N_R and N_L represent the number of right- and left-pointing segments out of the total $N = N_R + N_L$. The rest of the parameters are defined as in Fig. 2.3. For more realistic calculations, see Van Valen *et al.* (2009). (B) Concentration dependence of the probabilities of the different states that can be realized by the tethered ligand–receptor pair. The parameters used in the plot are $\Delta\varepsilon = \varepsilon_{bound} - \varepsilon_{solution} = -15\ K_BT$ and a probability of looping given by the one-dimensional random walk of 10^{-6}. The volume of the elementary box v has been chosen to be approximately 1.67 nm^3.

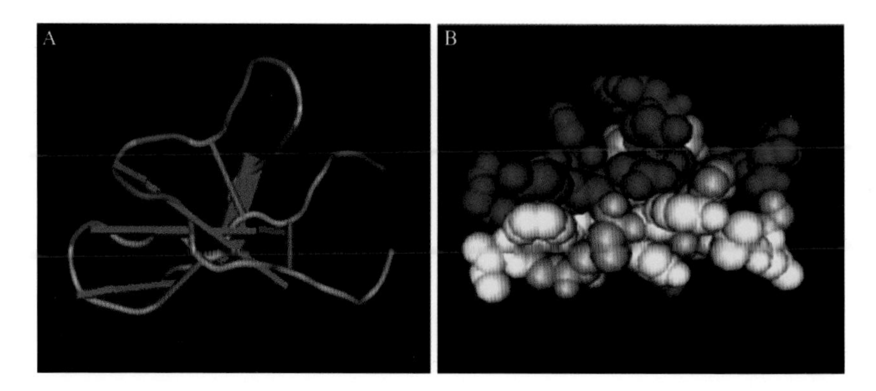

Abhijit Mishra *et al.*, Figure 4.1 Solution structure of mouse Paneth cell alpha-defensin Crp-4 (protein data bank ID 2GW9) obtained by NMR. (A) Structure shown in worm rendering. The three disulfide bonds from six cysteines are displayed in orange and Beta-sheets are represented by purple arrows. Blue regions denote cationic amino acids (arginine, lysine, and histidine), while the anionic glutamic acid is red. (B) Space-filled structure illustrates the cationic (blue) and hydrophobic (yellow) patches of amphipathic Crp-4. Here, hydrophobic amino acids include leucine, isoleucine, valine, phenylalanine, and tyrosine. Neutral residues are colored gray.

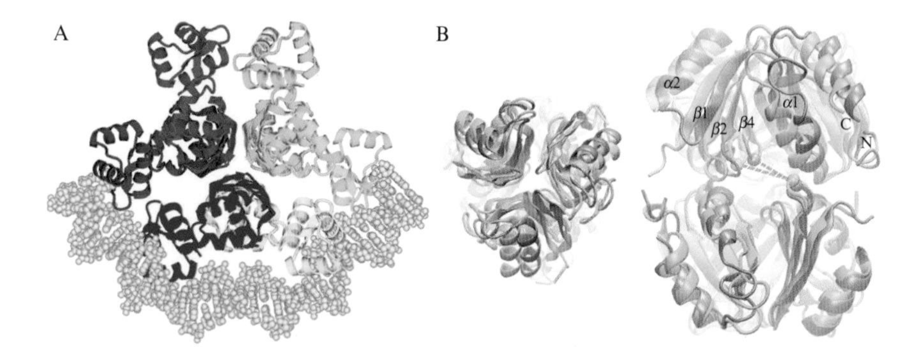

Rebecca Strawn *et al.*, Figure 5.1 ArgR structure. A. Model of DNA complex. Intact ArgR viewed down the three-fold axis with central ArgRC domains and peripheral ArgRN domains docked with bent B-form DNA. Subunits A, yellow; B, green; C, magenta; D, cyan; E, blue; and F, red correspond to those in Fig. 5.2. Protein structure prepared from PDB ID 1B4A (apoBstArgR (Ni *et al.*, 1999)), DNA from PDB ID 1J59 (cAMP receptor) as described in (Sunnerhagen *et al.*, 1997). B. ArgRC rotation. Overlay of average hexamer structures from the equilibrated part of the simulations showing the conformational shift from the holoArgRC (orange) structure that occurs uniquely in apoArgRC (blue). The bottom trimer (ABC of panel A) was used for C_α RMSD minimization. Left, top view. The size and viewpoint are chosen to match panel A. Right, side view. Selected secondary structure elements and N- and C-termini are labeled for orientation. CPK spheres mark C_α atoms of two Gly103-Asp128 residue pairs whose interatomic distances (dashed) are measured to quantify rotation.

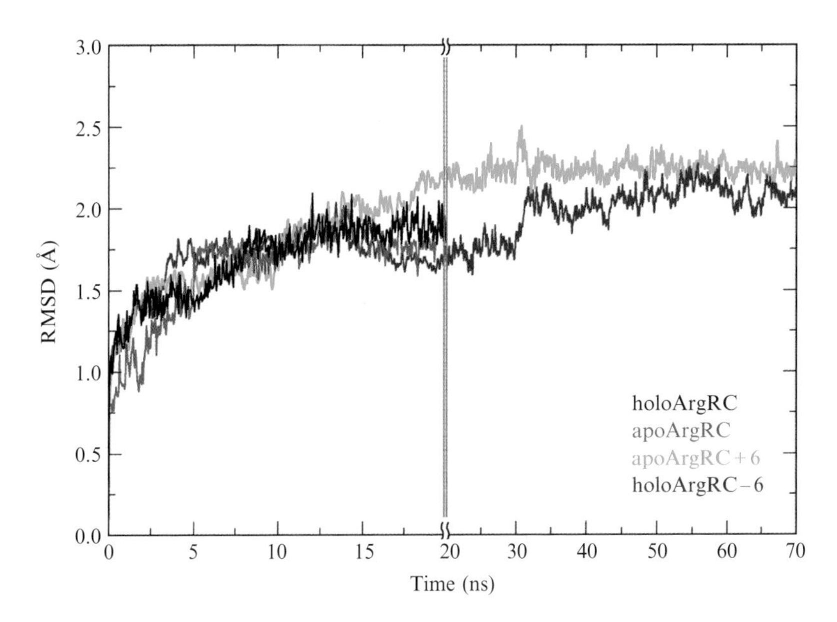

Rebecca Strawn *et al.*, Figure 5.2 Equilibration. Every 50 ps during the trajectory, each simulation structure is compared to the corresponding initial simulation structure after overlaying by C_α superposition to derive root-mean-square deviation (RMSD, Å) of C_α positions. Each color represents the individual simulation indicated. Note the compressed time scale after 20 ns (vertical line).

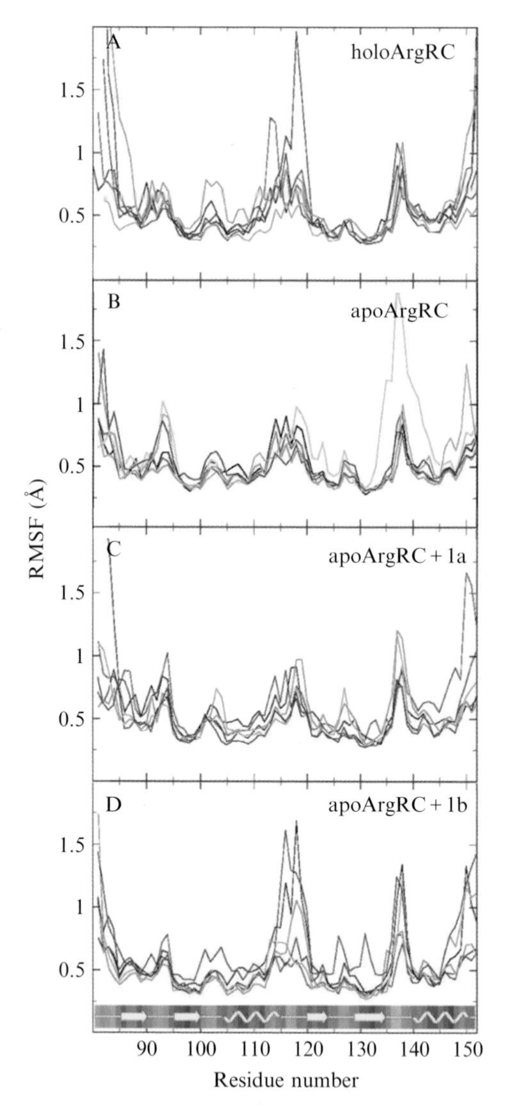

Rebecca Strawn *et al.*, Figure 5.3 C_α fluctuations. Each indicated simulation structure A-D was compared every 50 ps with the corresponding average structure calculated during the last 10 ns of the trajectory. The maximum displacement of each C_α during the last 10 ns is shown (RMSF, Å). Colors indicate individual monomers as in Fig. 5.1 except monomer A is black for clarity; the outlier referred to in the text for apoArgRC is monomer E, shown in grey. The secondary structure of ArgRC is indicated below: arrows, strands; folded tape, helices.

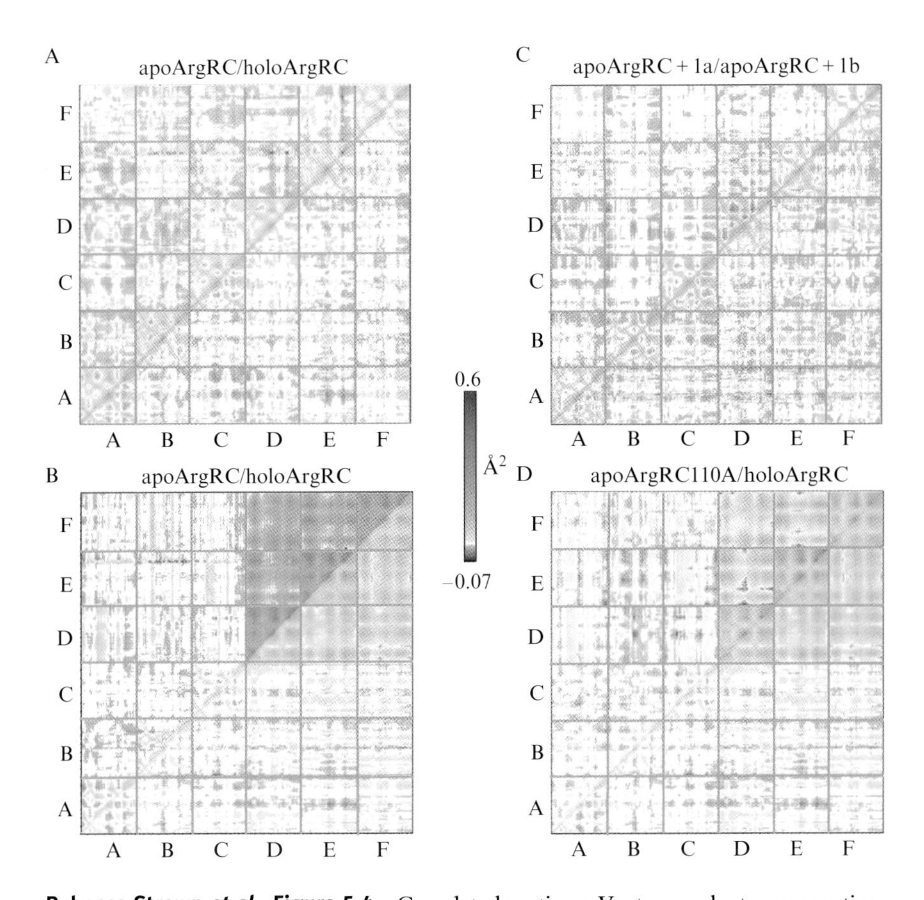

Rebecca Strawn *et al.*, Figure 5.4 Correlated motions. Vector products representing the maximum extent of correlated motion (Å^2) for each C_α pair are plotted. The color scale indicates the degree of correlation: red, positively correlated; blue, negatively correlated; white, uncorrelated. Each panel (A)-(D) displays the six subunits of each hexamer, A-F, arrayed along the axes. In each panel two covariance matrices are fused along the diagonal to facilitate comparison of the simulations indicated above the panel; the simulation presented on the upper left half is indicated first, followed by a slash (/) representing the diagonal, followed by the simulation on the lower right half. The reference state for calculating the extent of C_α motion is the average hexamer for panels A and C; for B and D the reference state is the trimer composed of ABC monomers.

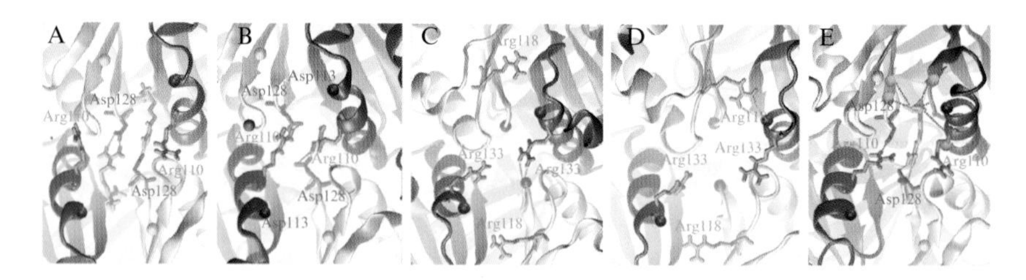

Rebecca Strawn *et al.*, Figure 5.5 L-Arg binding sites. The view is approximately that of panel B but zoomed in on the trimer interface, showing four subunits in front with the two in back faded for clarity. The viewpoint is fixed in all panels by minimizing RMSD of the dark grey helix shown at lower left. Each panel is one snapshot from the indicated simulation resembling the mean state. A. HoloEcArgRC. The guanidino group of each L-arg (cyan) forms a doubly-H-bonded salt-bridge with Asp128 (dashed lines). Approximate locations of His99 (yellow) and Asp113 (purple) are marked by unlabeled dots for clarity. B. ApoEcArgRC. Each Arg110 forms a doubly-H-bonded salt-bridge with Asp128. Gly103 used for distance measurements to Asp128 is marked by unlabeled blue dot. C. ApoMtArgRC clockwise. Rotation is promoted by Arg132-Asp146 salt bridges, the equivalent of EcArgRC Arg110-Asp128. Asp132 and Asp146 positions marked by purple CPK spheres. D. ApoMtArgRC counterclockwise. Rotation is promoted by Arg118-Asp132 salt bridges; Asp132 is equivalent to EcArgRC Asp113 but Arg118 corresponds to EcArgRC His99. E. EcArgRC, singly-bound L-arg. Some residues and their interactions with L-arg C_{α} substituents are marked by unlabeled cyan dots and dashes; others are omitted for clarity.

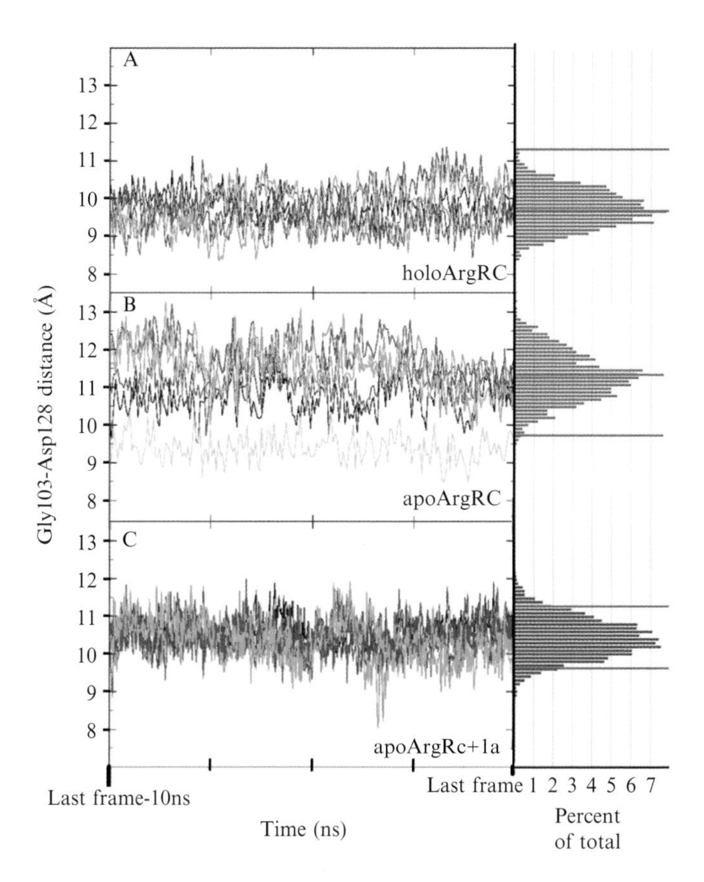

Rebecca Strawn *et al.*, Figure 5.6 Species distribution. Left, distances between Gly103 and Asp128 residues. Distances are measured for each of the six residue pairs in the hexamer every 50 ps during the final 10 ns of each indicated simulation. Monomer colors, defined by the subunit housing Gly103, correspond to those of Fig. 5.3. Right, frequency histograms of Gly–Asp distances shown on left panels. Distances are binned by size in groups of 0.1 Å. On each histogram, lower red line marks the mean crystal distance; upper line marks mean distance in the rotated ensemble of apoArgRC.

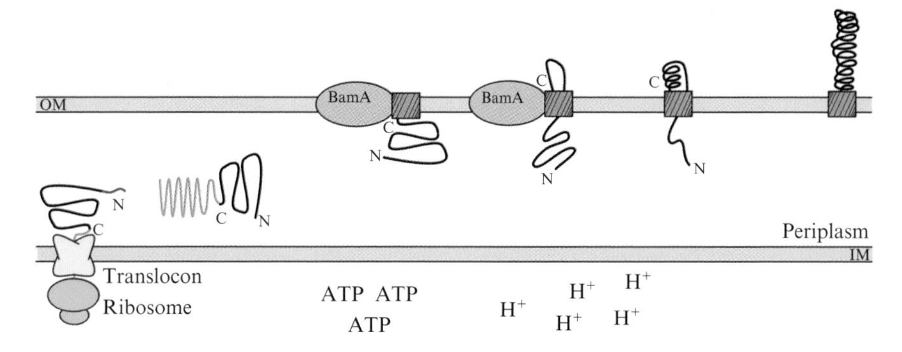

Jonathan P. Renn and Patricia L. Clark, Figure 8.1 The autotransporter secretion pathway. The N-terminal signal sequence (pink) directs transport across the OM in an ATP dependent manner. The passenger domain (black) represents the mature folded virulence protein. The C-terminal porin domain (blue) is required for OM secretion.

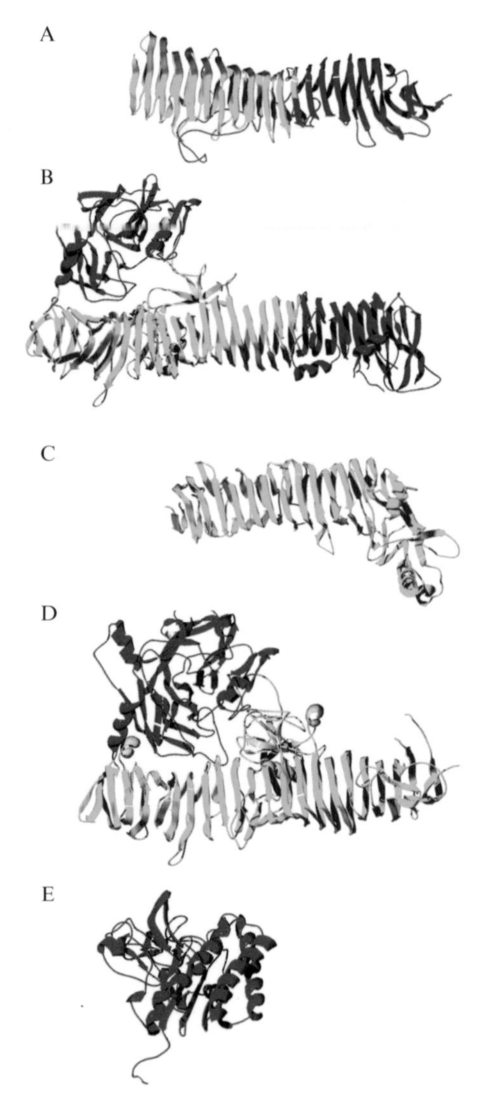

Jonathan P. Renn and Patricia L. Clark, Figure 8.3 Crystal structure of AT passenger domains. There are five crystal structures available for AT passenger domains: (A) pertactin (PDB ID:1dab; Emsley *et al.*, 1996), (B) hemoglobin protease (1wxr; Otto *et al.*, 2005), (C) VacA (2qv3; Gangwer *et al.*, 2007), (D) IgA protease (3h09; Johnson *et al.*, 2009), and (E) EstA (3kvn; van den Berg, 2010) are shown. β-Helical structure is shown in green, although the C-terminal stable core structures of pertactin (Junker *et al.*, 2006) and the Hbp homolog Pet (Renn and Clark, 2008) are shown in blue. Non-β-helical domains containing AT virulence function are shown in red, while other non-β-helical structure is shown in orange.